Resilient Optical Network Design:

Advances in Fault-Tolerant Methodologies

Yousef S. Kavian
Shahid Chamran University of Ahvaz, Iran

Mark Stephen Leeson
University of Warwick, UK

Information Science
REFERENCE

Managing Director:	Lindsay Johnston
Senior Editorial Director:	Heather Probst
Book Production Manager:	Sean Woznicki
Development Manager:	Joel Gamon
Development Editor:	Michael Killian
Acquisitions Editor:	Erika Gallagher
Typesetters:	Adrienne Freeland
Print Coordinator:	Jamie Snavely
Cover Design:	Nick Newcomer, Greg Snader

Published in the United States of America by
Information Science Reference (an imprint of IGI Global)
701 E. Chocolate Avenue
Hershey PA 17033
Tel: 717-533-8845
Fax: 717-533-8661
E-mail: cust@igi-global.com
Web site: http://www.igi-global.com

Library of Congress Cataloging-in-Publication Data

Resilient optical network design: advances in fault-tolerant methodologies /
Yousef S. Kavian and Mark Stephen Leeson, editors.
      p. cm.
  Includes bibliographical references and index.
  ISBN 978-1-61350-426-0 (hardcover) -- ISBN 978-1-61350-427-7 (ebook) -- ISBN
978-1-61350-428-4 (print & perpetual access) 1. Optical communications--
Reliability. 2. Fault tolerance (Engineering) I. Kavian, Yousef S., 1978-
II. Leeson, Mark Stephen, 1963-
  TK5103.59.R46 2012
  621.382'7--dc23
                            2011038349

British Cataloguing in Publication Data
A Cataloguing in Publication record for this book is available from the British Library.

All work contributed to this book is new, previously-unpublished material. The views expressed in this book are those of the authors, but not necessarily of the publisher.

Table of Contents

Detailed Table of Contents

WDM optical networks are widely viewed as the most appropriate choice for the future Internet backbone with the potential to fulfill the ever-growing demands for bandwidth. A failure in a network such as a cable cut may result in a tremendous loss of data. Therefore, network survivability, the ability for a network to continue to provide services in the event of failures, is a very important issue in WDM optical networks. This chapter introduces the principles and state-of-the-art of survivability provisioning in optical networks, in particular, in optical networks that employ wavelength division multiplexing (WDM). Concepts of survivability provisioning in optical networks such as protection and restoration, dedicated versus shared survivability, path-based, link-based, segment-based, cycle-based survivability, and so on, are covered to provide multiple classes of quality of protection against single failure, dual-failure, multiple simultaneous failures, or shared risk link group failures, in WDM mesh networks. Recent developments in survivable service provisioning are summarized, such as survivability provisioning that takes into account the connection holding-time, survivability in WDM light-trail networks and optical burst switched networks. Finally, the chapter briefly examines future research directions.

Survivable routing serves as one of the most important issues in optical backbone design. Due to the high data rates enabled by the wavelength division multiplexing technology, any interruption in the service results in the loss of a large amount of application data. Thus, making efforts to calculate and signal the protection resources promptly after the failure occurred would lead to an unacceptable high delay. As the main purpose of this chapter, the principles of pre-planned protection approaches in mesh optical backbone networks are discussed. The Shared Risk Link Group (SRLG) concept is introduced modeling physical and geographical dependency among seemingly unrelated link failures. Finally, methods are presented for calculating the exact end-to-end availability of a connection.

Chapter 3

Anusha Sivakumar, Indian Institute of Technology Madras, India & India UK Advanced Technology
Center of Excellence in Next Generation Networks, Systems and Services, India
Ganesh C. Sankaran, Indian Institute of Technology Madras, India & India UK Advanced
Technology Center of Excellence in Next Generation Networks, Systems and Services, India
Krishna M. Sivalingam, Indian Institute of Technology Madras, India & India UK Advanced
Technology Center of Excellence in Next Generation Networks, Systems and Services, India
Gerard Parr, University of Ulster, UK & India UK Advanced Technology Center of Excellence
in Next Generation Networks, Systems and Services, UK

Passive Optical Networks (PON) support subscribers with bandwidth requirements more than 10 Mbps. Fiber and node failures in a PON network can lead to large amounts of data loss, while isolating the central office from the subscribers. Hence, high network availability is desired when a PON is used for business enterprises and for providing mobile backhaul services. To maximize network availability, several protection architectures have been proposed in literature. In this chapter, we critically analyze and compare novel WDM PON protection architectures amongst those proposed in the literature. The comparison is done from topology, resource utilization and power budget perspectives. We also discuss protection mechanisms that are typically used in the architectures and their impact on restoration.

Chapter 4

Carmen Mas Machuca, Technische Universität München, Germany

The advantages of transparent optical networks such as high capacity and low cost can be outweighed by their complex fault management and the high impact of the faults occurring within them. Indeed, transparent optical networks reduce unnecessary, complex, and expensive opto-electronic conversion, to the cost of having faults more deleterious and affecting longer distances than in opaque networks. Moreover, transparent optical networks have limited monitoring capabilities, which could hinder efficient and accurate fault detection and localization. Different approaches have been proposed in the literature to perform fault localization, targeting different fault scenarios (e.g. single/multiple faults or looking at the optical/higher layers), and considering different assumptions (e.g. ideal/existence of false or lost alarms). Furthermore, fault management depends on the placement of monitoring equipment, whose optimization has been studied and also presented in this chapter.

Chapter 5

Z. Ghassemlooy, Northumbria University, UK
W. P. Ng, Northumbria University, UK
H. Le Minh, Northumbria University, UK

In traditional optical networks, configured as static physical pipes, the carrier-grade network resilience is provided by means of protection and restoration capabilities. However, there is a need to develop a new generation of dynamic reconfigurable all optical networks with built in network resilience capabilities. In the next generation, high-speed photonic packet switching networks, ultrafast packet header processing, and packet switching are the vital building blocks. In this chapter, a review of different routing schemes for high-speed photonic packet switching networks and the concept of reducing the size of the look-up routing table are presented. A novel PPM signal format has been introduced in order to reduce the size of the routing table in order reduce packet switching and processing time compared to the conventional routing tables. A failure self detection and a routing table reconfiguration in the optical domain are introduced, and a number of factors such as system performance, reliability, and complexity are also discussed.

Chapter 6

Abdelhamid Eshoul, University of Ottawa, Canada
Hussein T. Mouftah, University of Ottawa, Canada

The chapter outlines the different survivability approaches for mesh networks under static and dynamic traffic environments. It describes the different solution options and their implementations. Also included are detailed performance analyses and evaluations for the difference survivability approaches under both traffic environments. Finally, we present a performance comparison between the different survivability approaches and end the chapter with some concluding remarks.

Chapter 7

Arun K. Somani, Iowa State University, USA
David W. Lastine, Iowa State University, USA

Achieving low blocking probability and connection restorability in the presence of a link failure is a major goal of network designers. Typically fault tolerant schemes try to maintain low blocking probability by maximizing the amount of primary capacity in the network. In this chapter, we assume the total capacity on each link is fixed, and then it is allocated into primary or backup capacity. The distribution of primary capacity affects blocking probability for dynamic traffic. This can be seen by simulating dynamic traffic with different ways to distribute capacities in a network. A Hamiltonian p-cycle is a capacity optimal way of allocating primary and backup capacity. However, different Hamiltonian p-cycle may deliver different blocking probability for dynamic traffic. In general, more evenly distributing the backup and primary capacity lowers the blocking probability. This chapter provides upper bounds on how much primary capacity a network can provide if it uses a link based protection strategy to guarantee survivability for one or more link failures. Using integer linear programs we show that requiring preconfiguring carries a cost in terms of capacity if the solution is structured as a set of cycles.

Chapter 8

Taisir E.H. El-Gorashi, University of Leeds, UK
Jaafar M. H. Elmirghani, University of Leeds, UK

Due to its huge bandwidth, optical fibre is currently widely deployed to provide a variety of telecommunications services and applications. Wavelength-division multiplexing (WDM) has emerged as the technology of choice to harness the huge bandwidth available in an optical fibre. Traffic grooming supports efficient utilization of network resources by allowing sub-wavelength granularity connections to be groomed onto a single lightpath. Fault-tolerance for WDM networks is a major architectural and design issue as a single link failure can cause loss of an enormous amount of information. However, providing 100% guaranteed resilience to all types of traffic supported by existing and future networks may be unnecessary and wasteful in terms of resource utilization and cost efficiency. This chapter investigates the problem of dynamic traffic grooming for WDM networks under a differentiated resilience scheme. We propose two differentiated resilience schemes at different grooming levels— Differentiated Resilience at Lightpath (DRAL) level scheme, and Differentiated Resilience at Connection (DRAC) level scheme. These schemes explore different ways of provisioning backup paths and tradeoff between bandwidth efficiency and the number of required grooming ports. Both schemes support three resilience classes: dedicated protection, shared protection, and restoration. Simulation is carried out to evaluate and compare the two differentiated resilience schemes. Simulation results show that the DRAL scheme is not very sensitive to the changes in the number of grooming ports, while the DRAC scheme utilizes grooming ports more aggressively as it trades grooming ports for bandwidth efficiency in routing and grooming.

Chapter 9

Hussein T. Mouftah, University of Ottawa, Canada
Burak Kantarci, University of Ottawa, Canada

High capacity advantage of optical networks also introduces the risk of huge data loss in case of a service interruption, even if the outage lasts a short time. Therefore, survivable and reliable design and management of optical networks is urgent. However, deployment of efficient survivability policies does not always guarantee the continuity of the service. Long failure restoration delays, multiple failures, and lack of protection resources may lead to service unavailability. Hence, connection availability arises as a design constraint, and it is defined as the probability of a connection being in the operating state at any time. Availability-constrained optical network design and availability-constrained connection provisioning are two important problems to guarantee robustness of connections in a survivable network.

Chapter 10

Paolo Monti, Royal Institute of Technology, Sweden
Cicek Cavdar, Royal Institute of Technology, Sweden
Jiajia Chen, Royal Institute of Technology, Sweden
Lena Wosinska, Royal Institute of Technology, Sweden
Andrea Fumagalli, The University of Texas at Dallas, USA

Originally, networks were engineered to provide only one type of service, i.e. either voice or data, so only one level of resiliency was requested. This trend has changed, and today's approach in service provisioning is quite different. A Service Level Agreement (SLA) stipulated between users and service providers (or network operators) regulates a series of specific requirements, e.g., connection set-up times and connection availability that has to be met in order to avoid monetary fines. In recent years this has caused a paradigm shift on how to provision these services. From a "one-solution-fits-all" scenario, we witness now a more diversified set of approaches where trade-offs among different network parameters (e.g., level of protection vs. cost and/or level of protection vs. blocking probability) play an important role. This chapter aims at presenting a series of network resilient methods that are specifically tailored for a dynamic provisioning with such differentiated requirements. Both optical backbone and access networks are considered. In the chapter a number of provisioning scenarios - each one focusing on a specific Quality of Service (QoS) parameter - are considered. First the effect of delay tolerance, defined as the amount of time a connection request can wait before being set up, on blocking probability is investigated when Shared Path Protection is required. Then the problem of how to assign "just-enough" resources to meet each connection availability requirement is described, and a possible solution via a Shared Path Protection Scheme with Differentiated Reliability is presented. Finally a possible trade off between deployment cost and level of reliability performance in Passive Optical Networks (PONs) is investigated. The presented results highlight the importance of carefully considering each connection's QoS parameters while devising a resilient provisioning strategy. By doing so the benefits in terms of cost saving and blocking probability improvement becomes relevant, allowing network operators and service providers to maintain satisfied customers at reasonable capital and operational expenditure levels.

Chapter 11

    Emad M. Al Sukhni, University of Ottawa, Canada
    Hussein T. Mouftah, University of Ottawa, Canada

This chapter provides new distributed frameworks to support Quality of Service (QoS) differentiation. These frameworks provide differentiated protection services to meet customers' availability requirements effectively. We describe the availability-analysis for connections with different protection schemes. Through this analysis, we show how connection availability is affected by resource sharing. Based on the availability analysis, the proposed framework provisions each connection in which an appropriate level of protection is provided according to its predefined availability requirement. We consider the networks without wavelength conversion capability as well as dynamic traffic environment. In these distributed frameworks we propose several distributed schemes to provision and manage connections cost-effectively while satisfying the existing and new connections availability requirements.

Chapter 12

    Iván S. Razo-Zapata, ITESM, Mexico
    Gerardo Castañón, ITESM, Mexico
    Carlos Mex-Perera, ITESM, Mexico

This work presents a novel approach for dealing with failures and attacks on Transparent Optical Packet Switching (TOPS) mesh networks. The approach is composed of two phases, whereas the first one dynamically dimensions the resources in the network, the second one applies an incremental learning algorithm that generates an intelligent policy. At each node, such a policy allows a self-healing behavior when there are failures or attacks in the network. Finally, the performance of this approach is presented as well as future research lines.

The idea of this chapter is to give an overview on optical communication systems. The most important devices for fiber-optic transmission systems are presented, and their properties discussed. In particular, we consider such systems working with those basic components which are necessary to explain the principle of operation. Among them is the optical transmitter, consisting of a light source, typically a low speed LED or a high speed driven laser diode. Furthermore, the optical receiver has to be mentioned; it consists of a photodiode and a low noise, high bit rate, front-end amplifier. Yet, in the focus of the considerations, you will find the optical fiber as the dominant element in optical communication systems. Different fiber types are presented, and their properties explained. The joint action of these three basic components can lead to fiber-optic systems, mainly applied to data communication. The systems can operate as transmission links with bit rates up to 40 Gbit/s. But communication systems are also used for recent application areas in the MBit/s region, e.g. in aviation, automobile, and maritime industry. Therefore—besides pure glass fibers—polymer optical fibers (POF) and polymer-cladded silica (PCS) fibers have to be taken into account. Moreover, even different physical layers like optical wireless and visible light communication can be a solution

Preface

In the last two decades, optical networks have been potentially considered as the most appropriate solutions for developing high speed backbones of past, current, and future client networks, including asynchronous transfer mode (ATM), synchronous optical networking (SONET) or synchronous digital hierarchy (SDH, SONET/SDH) networks, and internet protocol (IP) networks.

Dense Wavelength Division Multiplexing (DWDM) optical transport networks are the bulk carriage mechanism to convey data, voice, and video over the Global Internet with the potential to fulfill the ever-growing demands for bandwidth. DWDM technology provides an excellent platform to exploit the huge capacity of optical fibre by multiplexing non-overlapping wavelength channels offering multiple terabits per second (Tb/s) transmission rates. The DWDM technique is emerging as a promising technology for the next generation of high-speed communication networks.

Optical networks are prone to failures or may face attacks. When these come unexpectedly in network components, such as link and node failures at the optical layer, they can potentially lead to a catastrophic loss of data and revenue, thus producing an unacceptable deterioration in the delivered quality of service. Therefore, one of the most important optical network design issues is survivability, also called resilience or fault tolerance. This is the ability of a network to provide continuous service at an acceptable level in the presence of different failure scenarios.

Design of optical networks subject to service restoration and survivability requirements has become a crucial issue in network planning, which is known as an NP-hard problem. The integration of resilience into optical core networks is a complicated problem, which requires some redundant resources such as additional bandwidths. Typically in addition to the working lightpaths established between the original and destination node pairs, some spare lightpaths are also established in static or dynamic manners to restore traffic demands in the event of failures at working ligthpaths.

There are several approaches for designing resilient optical networks, which are mainly based on protection and restoration architectures. In protection architectures, both working and spare ligthpaths are established during configuration of the network for arrival requests, while in the restoration methods, the path planner uses network state variables such as link state variables and wavelength state variables to assign spare lightpaths after the occurrence of failures. Generally, the protection architecture is static, while the restoration architecture is dynamic; consequently the latter is efficient for bandwidth trading, and the former is efficient for restoration time. Network resilience approaches are assigned to different classes according to various criteria such as network cost planning, design complexity, bandwidth trading, traffic recovery time, robustness, quality of protection, scalability, type and number of failures, et cetera.

In this book we offer a collection of the latest contributions to the area of survivability in optical networks. These have been written by a number of well-established researchers in optical networks. There is also a special section which deals with the latest issues in survivability in each chapter. The book contains several chapters and is preceded by a preview and state-of-the-art introductory chapter. Each of the chapters focuses on some theoretical and practical aspects of network survivability methodologies applied to real world problems.

Chapter 1 introduces the principles and state-of-the-art of survivability provisioning in optical networks, and in particular, in optical networks that employ wavelength division multiplexing (WDM). Concepts of survivability provisioning in optical networks such as protection and restoration, dedicated versus shared survivability, path-based, link-based, segment-based, cycle-based survivability, and so on, are covered to provide multiple classes of quality of protection against single failure, dual-failure, multiple simultaneous failures, or shared risk link group failures, in WDM mesh networks. Recent developments in survivable service provisioning are summarized, such as survivability provisioning that takes into account the connection holding-time, survivability in WDM light-trail networks, and optical burst switched networks. Finally, the chapter briefly examines future research directions.

Survivable routing serves as one of the most important issues in optical backbone design. Due to the high data rates enabled by the wavelength division multiplexing technology, any interruption in the service results the loss of a large amount of application data. Thus, making efforts to calculate and signal the protection resources promptly after the failure occurred would lead to an unacceptable high delay. As the main purpose of **Chapter 2**, the principles of pre-planned protection approaches in mesh optical backbone networks are discussed. The Shared Risk Link Group (SRLG) concept is introduced for modeling physical and geographical dependency among seemingly unrelated link failures. Finally, methods are presented for calculating the exact end-to-end availability of a connection.

Passive Optical Networks (PON) support access network subscribers with bandwidth requirements more than 10 Mbps. Fiber and node failures in a PON network can lead to large amounts of data loss, while isolating the central office from the subscribers. Hence, high network availability is desired when a PON is used for business enterprises and for providing mobile backhaul services. To maximize network availability, several protection architectures have been proposed in literature. In **Chapter 3**, the novel WDM PON protection architectures amongst those proposed in the literature are critically analyzed and compared. The comparison is performed from topology, resource utilization, and power budget perspectives. The protection mechanisms that are typically used in the architectures and their impact on restoration are also discussed.

The advantages of transparent optical networks, such as high capacity and low cost, can be outweighed by their complex fault management and the high impact of the faults occurring within them. Indeed, transparent optical networks reduce unnecessary, complex, and expensive opto-electronic conversion, at the cost of having more damaging faults that affect longer distances than in opaque networks. Moreover, transparent optical networks have limited monitoring capabilities which could hinder efficient and accurate fault detection and localization. Different approaches have been proposed in the literature to perform fault localization, targeting different fault scenarios (e.g. single/multiple faults or looking at the optical/higher layers), and considering different assumptions (e.g. ideal/existence of false or lost alarms). Furthermore, fault management depends on the placement of monitoring equipment, whose optimization is studied and also presented in **Chapter 4.**

In traditional optical networks, configured as static physical pipes, the carrier-grade network resilience is provided by means of protection and restoration capabilities. However, there is a need to develop a new generation of dynamic reconfigurable all optical networks with built in network resilience capabilities. In the next generation high-speed photonic packet switching networks, ultrafast packet header processing and packet switching are the vital building blocks. In **Chapter 5**, a review of different routing schemes for high-speed photonic packet switching networks, as well as the concept of reducing the size of the look-up routing table, is presented. A novel PPM signal format is presented in order to reduce the size of the routing table so as to reduce packet switching and processing time compared to conventional routing tables. A failure self detection and a routing table reconfiguration in the optical domain is introduced, and a number of factors such as system performance, reliability, and complexity are also discussed.

Chapter 6 outlines the different survivability approaches for mesh networks under static and dynamic traffic environments. It describes the different solution options and their implementations. Also included are detailed performance analyses and evaluations for the difference survivability approaches under both traffic environments. Finally, a performance comparison between the different survivability approaches is presented, and the chapter ends with some concluding remarks.

Achieving low blocking probability and connection restorability in the presence of a link failure is a major goal of network designers. Typically, fault tolerant schemes try to maintain low blocking probability by maximizing the amount of primary capacity in the network. In **Chapter 7**, maximizing primary capacities in survivable networks is proposed. It is assumed that the total capacity on each link is fixed, and then it is allocated into primary or backup capacity. The distribution of primary capacity affects blocking probability for dynamic traffic. This is seen by simulating dynamic traffic with different ways to distribute capacities in a network. A Hamiltonian p-cycle is a capacity optimal way of allocating primary and backup capacity. However, different Hamiltonian p-cycles may deliver different blocking probabilities for dynamic traffic. In general, more evenly distributing the backup and primary capacity lowers the blocking probability. This chapter provides upper bounds on how much primary capacity a network can provide if it uses a link based protection strategy to guarantee survivability for one or more link failures. Using integer linear programs, it is shown that requiring pre-configuring carries a cost in terms of capacity if the solution is structured as a set of cycles.

Traffic grooming supports efficient utilization of network resources by allowing sub-wavelength granularity connections to be groomed onto a single lightpath. **Chapter 8** investigates the problem of dynamic traffic grooming for WDM networks under a differentiated resilience scheme. Two differentiated resilience schemes at different grooming levels – the Differentiated Resilience at Lightpath (DRAL) level scheme and the Differentiated Resilience at Connection (DRAC) level scheme - are presented. These two explore different ways of provisioning backup paths and tradeoff between bandwidth efficiency and the number of required grooming ports. Both schemes support three resilience classes: dedicated protection, shared protection, and restoration. Simulation is carried out to evaluate and compare the two differentiated resilience schemes. Simulation results show that the DRAL scheme is relatively insensitive to the changes in the number of grooming ports, while the DRAC scheme utilizes grooming ports more aggressively as it trades grooming ports for bandwidth efficiency in routing and grooming.

The high capacity advantage of optical networks also introduces the risk of substantial data loss in case of a service interruption even if the outage lasts only a short time. Therefore, survivable and reliable design and management of optical networks is urgent. However, deployment of efficient survivability policies does not always guarantee the continuity of the service. Long failure restoration delays, multiple failures, and lack of protection resources may lead to service unavailability. Hence, connection availability arises as a design constraint, and it is defined as the probability of a connection being in the operating state at any time. Availability-constrained optical network design and availability-constrained connection provisioning are two important problems to guarantee robustness of connections in a survivable network which are discussed in **Chapter 9.**

Originally, networks were engineered to provide only one type of service, i.e. either voice or data, so only one level of resiliency was requested. This trend has changed, and today's approach in service provisioning is quite different. A Service Level Agreement (SLA) stipulated between users and service providers (or network operators) regulates a series of specific requirements, e.g., connection set-up times and connection availability that has to be met in order to avoid monetary fines. In recent years this has caused a paradigm shift on how to provision these services. From a "one-solution-fits-all" scenario, we witness now a more diversified set of approaches where trade-offs among different network parameters (level of protection vs. cost and/or level of protection vs. blocking probability) play an important role. **Chapter 10** aims at presenting a series of network resilient methods that are specifically tailored for a dynamic provisioning with such differentiated requirements. Both optical backbone and access networks are considered. In the chapter, a number of provisioning scenarios - each one focusing on a specific Quality of Service (QoS) parameter - are considered. First, the effect of delay tolerance, defined as the amount of time a connection request can wait before being set up, on blocking probability is investigated when Shared Path Protection (SPP) is required. Then, the problem of how to assign "just-enough" resources to meet each connection availability requirement is described, and a possible solution via a SPP scheme with Differentiated Reliability is presented. Finally, a possible trade off between deployment cost and level of reliability performance in Passive Optical Networks (PONs) is investigated. The results presented highlight the importance of carefully considering each connection's QoS parameters whilst devising a resilient provisioning strategy. By doing so, the benefits in terms of cost saving and blocking probability improvement become relevant, allowing network operators and service providers to maintain satisfied customers at reasonable capital and operational expenditure levels.

Chapter 11 provides new distributed frameworks to support Quality of Service (QoS) differentiation. These frameworks provide differentiated protection services to meet the availability requirements of customers in an effective manner. The availability-analysis for connections with different protection schemes is described. Through this analysis, it is shown how connection availability is affected by resource sharing. Based on the availability analysis, the proposed framework provisions each connection, in which an appropriate level of protection is provided according to its predefined availability requirement. Networks without wavelength conversion capability are considered as well as dynamic traffic environment. In these distributed frameworks several distributed schemes to provision and manage connections cost-effectively while satisfying the existing and new connections availability requirements are proposed.

Chapter 12 presents a novel approach for dealing with failures in and attacks on Transparent Optical Packet Switching (TOPS) mesh networks. The approach is composed of two phases, wherein the first one dynamically dimensions the resources in the network and the second one applies an incremental learning algorithm, which generates an intelligent policy. At each node, such a policy allows a self-healing behavior when there are failures or attacks in the network. Finally, the performance of this approach is presented as well as future research lines.

The idea of **Chapter 13** is to give an overview of fiber-optic communication systems. The most important devices for fiber-optic transmission systems are presented and their properties discussed. In particular, there is consideration of such systems working with those basic components necessary to explain the principle of operation. Among them is the optical transmitter, consisting of a light source, typically a high speed driven laser diode. Furthermore, the optical receiver has to be mentioned; it consists of a photodiode and a low noise high bit rate front-end amplifier. Nevertheless, in the focus of the considerations the optical fiber is found as the dominant element in optical communication systems. Different fiber types are presented and their properties explained. The joint action of these three basic components leads to a fiber- optic systems, mainly applied for data communication. The systems operate as transmission links with bit rates up to 40 Gbit/s. Furthermore, optical communication systems have also been used for recent application areas in the MBit/s region, e.g. in aviation, automobile, and maritime industry. Therefore, besides pure glass fibers, polymer optical fibers (POF) and polymer-cladded silica (PCS) fibers have to be taken into account.

Thus, this collection provides a wide ranging overview of many of the salient issues in modern optical networks from the point of view of resilience. It is intended to appeal to a wide range of readers, and for this reason, it includes both material of a tutorial nature and reports on state of the art methods. It is essential that the key differences and similarities between optical and other networks are appreciated, particularly with regard to all optical networks. Moreover, given the importance of WDM in the provision of multiple channels, it receives substantial coverage. Furthermore, optical networks make it necessary to take a broad view of resilience, and this collection has done so, including several layers of the OSI model. For readers new to many aspects of the topic, there is coverage of many of the essentials fundamentals, particularly in the first few chapters. Needless to say, these can be skipped over quickly by those more familiar with the subject area.

When considering transparent optical networks with their advantages such as high capacity, it was also part of the aim to bring to the attention of the reader their complex fault management and the high impact of the faults occurring within them. Thus, the collection here considers cutting-edge aspects of this problem such as incremental learning and the optimum placement of monitoring equipment.

Resilience is not just of significance within the core network, and in this collection, high speed access networks are not forgotten, with a particular emphasis on Passive Optical Networks (PON) supporting relatively demanding end users. Given the tree structure implicit in PONs, failures can lead to large data losses and subscriber isolation, so protection is of paramount importance. Moreover, the differentiated services offered throughout modern optical networks (including the access part) mean that differentiated resilience must also be offered. For the future, it will be necessary to provide an appropriate level of protection across the same network for users from impoverished students tackling homework assignments to major corporations handling voluminous and confidential data processing tasks. Whereas in the past, networks were often designed for just one type of service, modern service level provision is via a totally different model. In the networks of today and tomorrow, each user may have a different level of agreed service, leading to inevitable tradeoffs that may include price as well as purely technical factors. In this respect, distributed frameworks are germane to the delivery of the differentiated QoS required. Moreover, traffic grooming offers a route to the efficient of network resource utilization, and thus, features significantly in the book in the context of differentiated resilience.

Achieving low blocking probability and connection restorability when failures occur also figure prominently in the work. Within these topics, methods for bounding the capacities available and for offering rapid restoration are covered. In addition, the traffic changes with time in modern high speed networks, so the increasing prevalence of dynamic traffic environments figures within the collection

Research and development engineers, graduate students studying optical networks, and senior undergraduate students with a background in algorithms and networking will find this book interesting and useful. This work may also be used as supplemental readings for graduate courses on internetworking, routing, survivability, and network planning algorithms.

Yousef S. Kavian
Shahid Chamran University of Ahvaz, Iran

Mark Stephen Leeson
University of Warwick, UK

Acknowledgment

We would like to thank the chapter authors for providing us with the fundamental constituents of this volume and the reviewers who provided detailed and constructive feedback. Also, thanks are due to members of staff at the publisher who have guided us through the publishing process.

Finally, Yousef would like to thank his wife, Mandana, for her patience, understanding and support during the preparation of this manuscript.

Yousef S. Kavian
Shahid Chamran University of Ahvaz, Iran

Mark Stephen Leeson
University of Warwick, UK

Chapter 1
Survivability in Optical Networks:
Principles and State-of-the-Art

Bin Wang
Wright State University, USA

ABSTRACT

WDM optical networks are widely viewed as the most appropriate choice for the future Internet backbone with the potential to fulfill the ever-growing demands for bandwidth. A failure in a network such as a cable cut may result in a tremendous loss of data. Therefore, network survivability, the ability for a network to continue to provide services in the event of failures, is a very important issue in WDM optical networks. This chapter introduces the principles and state-of-the-art of survivability provisioning in optical networks, in particular, in optical networks that employ wavelength division multiplexing (WDM). Concepts of survivability provisioning in optical networks such as protection and restoration, dedicated versus shared survivability, path-based, link-based, segment-based, cycle-based survivability, and so on, are covered to provide multiple classes of quality of protection against single failure, dual-failure, multiple simultaneous failures, or shared risk link group failures, in WDM mesh networks. Recent developments in survivable service provisioning are summarized, such as survivability provisioning that takes into account the connection holding-time, survivability in WDM light-trail networks and optical burst switched networks. Finally, the chapter briefly examines future research directions.

INTRODUCTION

Optical fiber offers much higher bandwidths (nearly 50 terabits per second (Tb/s)) than copper cables and is less susceptible to various kinds of electromagnetic interferences and other undesirable effects. Optical fiber transmission has played a key role in increasing the bandwidth of telecommunication networks through wavelength division multiplexing (WDM), particularly in the last two decades as the Internet has increasingly penetrated daily life. WDM divides the enormous bandwidth of an optical fiber into many

DOI: 10.4018/978-1-61350-426-0.ch001

non-overlapping wavelength channels, each of which may operate at the rate of 10 Gigabit per second or higher. WDM optical networks are widely viewed as the most appropriate choice for the future Internet backbone with the potential to fulfill the ever-growing demands for bandwidth. A failure in a network such as a cable cut may result in a tremendous loss of data. For example, on December 26, 2006, a serious undersea earthquake off the coast of Taiwan that measured 7.1 on the Richter scale caused significant damage to submarine optical cable systems (Kitamura, Lee, Sakiyama, & Okamura, 2007). The resulting fiber cable failures shut down communications in several countries in the Asia Pacific networks. Several Internet service providers (ISPs) were affected because each cable system is shared by multiple ISPs. Therefore, *network survivability*, the ability for a network to continue to provide services in the event of failures, is a very important issue in WDM optical networks.

Network survivability provisioning requires the commitment of additional resources from the viewpoint of service providers, sometimes a significant amount depending on the desirable levels of tolerance to failures and end user requirements. Practically, there is a strong need for providing a spectrum of quality of protection in the event of failure to suite the requirements of different applications/end users and, at the same time, to find a balance point between provisioning cost, speed of recovery, and scalability. Quality of protection can range from best-effort only to assurance of complete restorability of single, dual-failure, or multiple simultaneous failures on a per-demand basis. This allows a network operator to tailor the investment in capacity to provide ultra-high availability on a selective basis while avoiding the very high investment required for complete failure restorability for all. At the same time, the ability to provide different grades of protection to services provided offers the ISP a competitive advantage.

Many factors impact efficient network survivability provisioning. A WDM optical network needs to keep and exchange network state information for network operations and restorations in case of failure, e.g., topological information, primary paths, backup paths, resources used by each path, available wavelengths over a fiber link, residual bandwidth on each wavelength, tuning ranges of wavelength converters, shared risk link groups and so on. Network state information needs to be exchanged or updated when the state of the network changes, e.g., the addition or the removal of a primary path and its corresponding backup path, the occurrence of a fault resulting in the activation of the backup path and the change of network topology and so forth. The amount of information that needs to be kept by the network becomes overwhelming as the network size, the number of wavelengths, and the number of service requests for paths gets larger. In addition, information update takes a non-negligible amount of time in a real network environment. As a result, network state information is inherently imprecise. Since speed of protection/restoration is paramount, key component functions, such as the routing and restoration algorithm, connection setup strategy, and wavelength assignment strategy and so on, need to achieve scalability and fast performance. The need to provide survivability to services with different quality of protection requirements further complicates the scalable survivability provisioning. Scalability is a real issue that hinders effective network operations and is thus critically needed to be addressed. There appears to be a tradeoff between the scalable provisioning of survivable services and the cost in terms of resource use, network state information maintenance, and network control complexity. A balance point may be achieved through intelligent integrated information sharing and designing effective component mechanisms.

This chapter introduces the principles and state-of-the-art of survivability provisioning in WDM optical networks. Concepts of survivability provisioning in optical networks such as protection and restoration, dedicated versus shared survivability, path-based, link-based, segment-based, cycle-based survivability, and so on, are covered. Recent developments in survivable service provisioning are summarized, such as survivability provisioning that takes into account the connection holding-time, survivability in WDM light-trail networks and optical burst switched networks. The chapter also briefly examines future research directions in this field.

BACKGROUND

Optical fiber transmission has played a key role in increasing the bandwidth of telecommunications networks, especially in recent two decades as the Internet penetrates our daily lives. The evolution of optical communication systems has gone through several generations. In the first-generation of optical networks, optical fibers were used purely as a transmission medium, serving as a replacement for copper wires with all the switching and processing of the bits being handled by electronics. Examples of first-generation optical networks are SONET (synchronous optical network) and SDH (synchronous digital hierarchy) networks (Ramswami & Sivarajan, 2002). Incorporating some of the switching and routing functions that were performed by electronics into the optical part of the network, the second-generation and future generation optical networks were capable of providing more functions than simple point-to-point transmission, for example, lightpath service (a circuit-switched end-to-end all-optical channel), dynamic service provisioning, traffic grooming, and so on, in WDM wavelength-routed networks (Ramswami & Sivarajan, 2002). Driven by the increasing demands on communication bandwidth, WDM technology has been widely deployed in the Internet infrastructure. When the bandwidth demand exceeds the capacity in existing fibers, WDM can be more cost-effective than

laying more fibers, especially over long distances because more wavelength channels can be lit up as necessary. The tradeoff is between the cost of installation/burial of additional fibers and the cost of additional line terminating equipment.

Metro Optical Ring Networks

Much of today's optical ring networks are built around SONET rings. A pair of fibers is used in unidirectional path-switched ring (UPSR) where one fiber is used as the working fiber and the other as the protection fiber. Traffic from node A to node B is sent simultaneously on the working fiber in the clockwise direction and on the protection fiber in the counterclockwise direction. As a result, if a link fails on one fiber, node B will be able to receive from the other fiber. The bi-directional line-switched ring (BLSR) connects adjacent nodes through one or two pairs of optical fibers, corresponding to BLSR/2 and BLSR/4, respectively. BLSRs are much more sophisticated than UPSRs by incorporating additional protection mechanisms. Unlike a UPSR, working traffic in a BLSR can be carried on different fibers in both directions and is routed along the shortest path in the ring. Half of the capacity of each fiber is reserved for carrying the protection traffic in BLSR/2. In the event of a link failure, the traffic on the failed link is rerouted along the other part of the ring using the protection capacity available in the two fibers. A BLSR with 4 fibers (i.e., BLSR/4) uses a pair of fibers for protection and employs a span switching protection mechanism first. If a transmitter or receiver on a working fiber fails, the traffic is routed on the protection fibers between the two nodes on the same span. BLSRs provide spatial reuse capabilities by allowing protection capacity to be shared between spatially separated connections. BLSRs are significantly more complex to implement than UPSRs due to the extensive signaling required between the nodes. WDM technology has provided the ability to support multiple SONET rings on a single fiber pair by using wavelength add/drop multiplixers (WADMs) to separate the multiple SONET rings. This tremendously increases the capacity as well as the flexibility of the optical ring networks. However, additional electronic multiplexing equipment is needed which dominates the cost component and needs to be minimized via traffic grooming.

Wavelength-Routed Optical Networks

The massive increase in network bandwidth due to WDM has heightened the need for faster switching at the core of the network (i.e. long-haul networks) to move from point-to-point WDM transmission systems to an all-optical backbone network that eliminates the need for per-hop packet forwarding. Wavelength-routed networks have become a major focus area since the early 1990s. Wavelength routed networks are considered to be an ideal candidate for wide area backbone networks.

A wavelength-routed network physically consists of a number of optical cross-connects (OXCs) or wavelength routers, taking an arbitrary topology. Each wavelength router takes in a signal at each of the wavelengths at an input port, and routes it to a particular output port, independent of the other wavelengths. The wavelength routers may also be equipped with wavelength converters that allow the optical signal on an incoming wavelength of an input fiber to be switched to some other wavelength on an output fiber link. The basic mechanism of communication in a wavelength-routed network is a *lightpath*. This is an all-optical communication channel that may span more than one fiber link between two nodes in the network. The intermediate nodes in the physical fiber path route the lightpath in the optical domain using the wavelength routers. If no wavelength converters are used, a lightpath must use the same wavelength on each hop of its physical fiber link. This is known as the *wavelength continuity constraint*. However, if converters are available, a different wavelength on each fiber link

may be used to create a lightpath. A fundamental requirement of a wavelength-routed optical network is that two or more lightpaths traversing the same fiber link must use different wavelengths so that they do not interfere with each other. The end-nodes of the lightpath access the lightpath with transmitters and receivers that are tuned to the wavelength used by the lightpath.

Because of limitations on the number of wavelengths that can be used, and hardware constraints at the network nodes, it is not possible to set up a lightpath between every pair of source and destination nodes. The particular set of lightpaths that are established on a physical network constitutes the virtual topology or logical topology. Careful design of virtual topologies over a WDM network is to combine the best features of optics and electronics. The tradeoff is between bandwidth flexibility and electronic processing overhead. The traffic on the lightpath does not have to undergo optoelectronic conversion at intermediate nodes. Traffic delay can be reduced through the use of virtual topologies and appropriate routing. However, because lightpaths are circuit-switched, forming lightpaths locks up bandwidth in the corresponding links on the assigned wavelength. A good virtual topology trades some of the ample bandwidth inherent in the fiber to obtain a solution that is the best of both worlds. Different virtual topologies can be set up on the same physical topology, which allows operators to choose or reconfigure a virtual topology that achieves the best network performance given network conditions such as average traffic between network nodes.

Optical networks will essentially serve as optical transport networks (OTNs) that enable "everything over optics" integration, e.g., IP over WDM integration, leading to the building of the next generation optical Internet. In addition, new optical network architectures, such as optical burst switching (Chen, Qiao, & Yu, 2004), optical packet switching (Ramaswami & Sivarajan, 2002), light-trail (Gumaste & Chlamtac, 2004), optical split-and-direct switching, and so on, have been proposed, and may potentially change the landscape of future optical networking. Interested readers are referred to the respective references for more details.

As WDM optical networks provide substantial capacity, providing resilience against failures is an especially important requirement for optical networks since the amount of disruption caused by failures is considerable. Therefore, network survivability is a very important issue in WDM optical networks. Network survivability is the ability of a network to recover traffic and services affected by failures. There is a large body of literature on survivability in traditional as well as optical networks. Much recent work focuses on survivability in optical WDM networks. Standard 1+1, 1:1, 1:N protection, and automatic protection switching (APS) have been well studied (Ramswami & Sivarajan, 2002). These techniques have also been applied successfully in SONET rings (Ramswami & Sivarajan, 2002). However, the techniques used for SONET are not immediately applicable to WDM systems. Survivability in transport WDM mesh networks is more complicated than that in point-to-point links or ring networks, and therefore is the focus of this chapter.

FAILURE AND QUALITY OF PROTECTION

Failure

Network protection and survivability are of paramount importance to today's optical transport networks. To address the survivability issues, we need to first understand the types of network failure. Several types of network failures are often considered, including (1) link failure, (2) node failure, (3) shared risk link group (SRLG) failure, and (4) network control system failure.

A link failure means that the fiber link between a pair of nodes is interrupted which causes the service disruption to all the traffic between the node pair. Fiber links between node pairs are often vulnerable due to some unintentional digs or natural disasters (Kitamura, Lee, Sakiyama, & Okamura, 2007). Thus, compared to other network failures, the link failure is often more common. A simple failure scenario often assumed in many research studies is a single link/span failure. A dual-link/span failure scenario has also been considered.

A node failure is less frequent than a link failure, which means the failure of a switch node. The failure can be caused by reasons such as power turnoff or a fire. Although less frequent, a node failure is generally more serious than a link failure. Most modern node devices have built-in redundancy which greatly improves their reliability. Therefore link failure is more of a concern than node failure.

In a network, a single factor can cause more than one link failure. In practice, for instance, multiple fibers are bundled into the same underground duct, or span. A fiber cut usually occurs due to a duct cut during construction or destructive natural events such as earthquakes (Kitamura et al., 2007). When a duct is cut, normally all the fibers in the duct fail at the same time. Hence, a network survivable under a single fiber failure is not necessarily survivable in duct failure scenarios. More generally, transport network carriers use the notation of *shared risk link group* (SRLG) which associates all the links with a failure, to describe this type of network failure phenomenon. Obviously the fiber links in the same duct belong to the same SRLG because they all share the same risk of a duct cut. Shared risk link group failure is considered a general network failure concept. As special cases, it can model single, dual, or multiple link and node failures. It can also model more complicated network failure scenarios such as a subnetwork on some region that may incur high risk of disaster such as a region that has a high frequency of earthquake.

The last type of network failure is related to network control system. It is often more disastrous as the whole network will be out of control.

Specifically, a network control system failure can disable the function of establishing a new service connection or releasing an expired connection even if existing network services are not affected.

Quality of Protection

While survivable service provisioning is a must in WDM optical networks, not all demand flows in a transport network necessarily need the same level of protection or restoration as in SONET rings, e.g., fully restorable (or protected) against any single failure scenario. In fact, that being able to support a whole range of survivability requirements and service classes in a flexible and efficient manner would provide a competitive edge for service providers and allow different charge schemes to be implemented to suit the needs of different service demands.

Various quality of protection (QoP) classes have been proposed by numerous researchers (Yoo & Qiao, 1999; Gerstel & Sasaki, 2001; Doucette & Grover, 2002). While there is no standardized definition of quality of protection classes, a good example of QoP classes in which a number of QoP classes are defined (Doucette & Grover, 2002), is described as follows in order of decreasing importance (also see Figure 1):

- **Platinum service (dual span-failure restoration):** this service provides assured restoration to any dual span-failure scenario.
- **Gold service (assured restoration):** this service ensures that working capacity units are fully restored under any single span-failure scenario.
- **Silver service (best-effort restoration):** in this service, working capacity units should be restored if possible within existing spare capacity, after the restoration of any higher-class service capacity, i.e., platinum-class and gold-class services.
- **Bronze service (non-protected service):** in this service, working capacity units are not restored at all. However, they are also not interrupted if they do not themselves experience a failure.
- **Economy service (preemptible services):** this class of service uses working capacity units that are not protected. Furthermore, the capacity units of this class may be preempted and logically converted to spare capacity on spans that are not affected by a current failure, if they are needed to satisfy the restoration requirements of other higher classes of service. Therefore, this class of service may be disrupted.

Figure 1. An example of multiple classes of quality of protection

Platinum	Assured restoration to any dual span-failure scenario	Additional QoP class
Gold	Assured restoration in any single span-failure scenario	
Silver	Best effort restoration after restoration of class Gold	Basic multi-QoP paradigm
Bronze	No restoration attempt if affected by a failure	
Economy	No restoration attempt and possible preemption to satisfy Gold-class (and possibly also Silver-class) restorability requirements	

Note that other variations of the above sample classes of QoP are certainly plausible, depending on the need, benefit, and complexity of such service classification.

SURVIVABILITY PROVISIONING IN WDM OPTICAL NETWORKS

Taxonomy and Principles

Various approaches to provisioning survivability exist for WDM optical networks. These approaches can generally be divided into *protection* (a.k.a. pre-planned approaches, or proactive approaches) and *restoration* (a.k.a. reactive approaches) (Mohan & Murthy, 2000) (Figure 2). Specifically, protection refers to techniques that use pre-assigned capacity to ensure survivability while techniques that re-route affected traffic after a failure occurrence using available capacity are referred to as restoration (Mohan & Murthy, 2000). In general, protection based approaches offer faster recovery times while restoration based approaches may be more resource efficient. Various fault-resilient schemes can be designed at IP and/or WDM layers to protect users' traffic from disruptions due to failures. WDM layer survivability is desirable due to its many advantages: speed, simplicity, effectiveness, and transparency (Zhou & Subramaniam, 2000).

Protection based approaches can be further divided into path-based, link-based schemes and a few other variations such as segment-based schemes, protection cycles based schemes, and so on. In path-based protection schemes, upon a failure, the source and the destination of a connection affected by the failure switch to its corresponding protection path that is routed on a preferably fiber-disjoint path from the affected working path, as shown in Figure 3(a). Link-based protection schemes re-route (or loop back) traffic around the failed link, and involve only the nodes adjacent to the link failure (i.e., the end nodes of the failed link are responsible for recovery), as shown in Figure 3(b). On the other hand, path-based protection schemes need a mechanism to notify the source and the destination of the affected connection of the failure. However, path-based protection schemes may be more resource efficient and usually offer a shorter end-to-end recovery route than link-based protection techniques (Gerstel & Ramaswami, 2000; Zhang, Taira, Takagi, & Das, 2002).

Based on whether or not backup network resource sharing is allowed, path-based protection schemes can be classified into dedicated path-based protection approaches and shared path-based protection approaches. In the former case, the resources on the links of a protection path are reserved for protecting the working capacity of a given connection. Two protection paths sharing

Figure 2. A taxonomy of survivability provisioning schemes in WDM optical networks

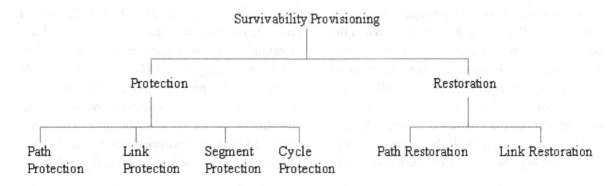

Figure 3. Protection schemes

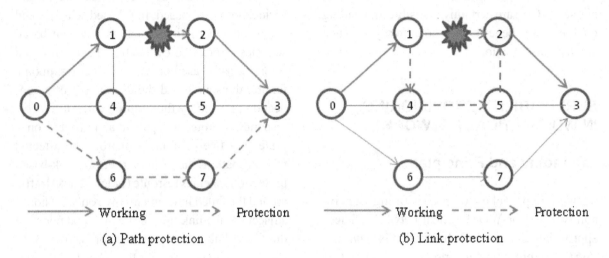

(a) Path protection   (b) Link protection

some common fiber links must use different wavelengths even if their corresponding working paths are link-disjoint. On the other hand, in the latter case resource sharing among protection paths is allowed, and multiple protection paths can go through common fiber links and share common resources as long as their working paths satisfy certain constraints, e.g., their working paths are link-disjoint. For example, suppose that three connections are to be established in a network shown in Figure 3 with the objective of minimizing the total wavelength-links used in the network by employing a shared path protection scheme. Assume that the connection requests are processed sequentially and the wavelength continuity constraint is required in the network. In Figure 4(a), the first connection (0, 1) has been established in the network with a working path $(0 \rightarrow 1)$ and a protection path $(0 \rightarrow 4 \rightarrow 1)$. Both paths use wavelength 0 on all the links they traverse. When the second connection request (0, 2) arrives, paths $(0 \rightarrow 1 \rightarrow 2)$ and $(0 \rightarrow 4 \rightarrow 5 \rightarrow 2)$ are chosen to be the working path and the protection path, respectively. The two working paths and the two protection paths (Figure 4(b)) use a common link (0, 1) and link (0, 4), respectively. Since any two link-joint working paths must use different wave-

lengths, wavelength 1 on link (0, 1) and link (1, 2) are assigned to working path WP_2. We use $(i, j: k)$ to refer to the k-th wavelength-link on link (i, j). Since these two working paths are link-joint, protection path PP_2 and protection path PP_1 cannot share the same wavelength-link (0, 4: 0). Instead PP_2 has to use wavelength 1 in order to guarantee 100% restorability in case that link (0, 1) fails. Suppose a third connection request (0, 3) arrives after these two connections have been established in the network. If we jointly solve the routing and wavelength assignment (RWA) problem given this request, the optimal solution is to choose path $(0 \rightarrow 6 \rightarrow 7 \rightarrow 3)$ as the working path WP_3 and $(0 \rightarrow 4 \rightarrow 5 \rightarrow 3)$ as the protection path PP_3. The total number of additional wavelength-links to accommodate this connection request with shared path protection is 4 which is the minimum among all feasible solutions. Figure 4(c) shows the network state after the third request has been established. In the figure, protection path PP_3 shares wavelength-links (0, 4: 1) and (4, 5: 1) with protection path PP_2 and wavelength-link (5, 3: 1) is the only additional wavelength-link needed by the new protection path. Therefore, overall 4 additional wavelength-links are needed to accommodate the third request. By employing

Figure 4. Wavelength assignment for working and protection paths in shared path protection

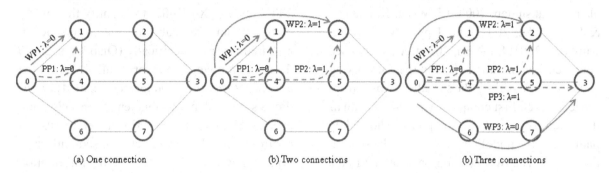

(a) One connection   (b) Two connections   (b) Three connections

shared protection, therefore, it is possible to utilize network resources more efficiently while still achieving 100% restorability against single failure. The recovery time for shared path protection schemes may be longer. However the overall resource utilization is much better compared with that of dedicated protection schemes (Zang & Mukherjee, 2001; De Patre, Maier, Pattavina, & Martinelli, 2002).

Path- and Link-Based Survivability Provisioning

Path- and link-based survivability schemes have been extensively researched. Integer linear program (ILP) formulations of path- and link-protection schemes to protect against all possible single-link failures were presented in (Ramamurthy & Mukherjee, 1999). The network topology and a demand matrix that consists of the number of connections to be established between each node-pair are given. The set of alternate routes that are used to satisfy any demand between each node-pair is pre-computed and given. The objective is to minimize the total number of wavelengths used on all the links in the network for both the primary/working paths and backup/protection paths. The ILP solution also determines the routing and wavelength assignment of the primary and backup paths. Since the routing and wavelength assignment (RWA) problem with no protection for any demands has been shown to be *NP*-complete,

the ILP problem formulations that consider protection are also *NP*-complete. Shared-path protection provides significant savings in capacity utilization over dedicated-path and shared-link protection schemes, and dedicated-path protection provides marginal savings in capacity utilization over shared-link protection. Other protection approaches to optimizing resource utilization for a given traffic matrix have also been studied by numerous researchers.

As the number of variables and the number of equations for the ILPs grow rapidly with the size of the network, the ILP formulations are practical only for small networks. For large networks, heuristic methods are needed. Therefore, on-line control and management schemes for setting up lightpaths with protection paths were proposed in (Zang & Mukherjee, 2001) under a dynamic traffic pattern in which a lightpath is taken down after the connection-holding time. The results show that, with shared-path protection, lower call-blocking probability with fast fault-recovery can be achieved.

Indeed, dynamic survivable lightpath provisioning using shared-path protection has been the focus of much research due to the fact that backup paths can share wavelength-links when their corresponding working paths are mutually diverse. Consequently, how to increase backup resource sharing based on different cost models has generated a lot of interest and results (Bouillet, Labourdette, Ramamurthy, & Chaudhuri, 2002;

Elie-Dit-Cosaque Ali, & Tancevski, 2002; Kodialam & Lakshman, 2000; Li, Wang, Kalmanek, & Doverspike, 2002; Liu, Tipper, & Siripongwutikorn, 2001; Su & Su, 2001). A widely used approach for computing a feasible pair of working and protection paths is the so-called two-step approach, which first computes a least-cost path as the working path and then computes as the backup path a link (or node) disjoint path of least additional cost. The two-step approach cannot find a solution in a trap topology even though a solution exists. An improvement is to compute K working/backup path pairs by applying a K-shortest-path algorithm to compute K candidate working paths and computing a backup path for each candidate working paths, and then computing a backup path for each candidate working path and selecting the pair of minimal cost. Since backup resource sharing depends on the routes of working paths, sequential computation of a backup path after the working path is determined may lead to using more resources than necessary. It has been shown that the problem of finding an eligible pair of working and backup paths under shared-path-protection constraints for a lightpath request with respect to existing lightpaths is NP-complete (Ou, Zhang, & Zang, 2004). This calls for efficient heuristics. The work of (Ou et al., 2004) designs a backtracking-based heuristic, called CAFES (compute a feasible solution), to compute an eligible pair of working and backup paths for a lightpath request. A general optimization procedure is developed to iteratively optimize the resource consumption of the working and backup paths for a given solution (i.e., two link-disjoint paths).

While it is desirable to have complete information about the routing and wavelength assignment of the existing lightpaths in a network to decide backup resource sharing, complete network state information may not always be available due to control and management concerns. Three network state availability scenarios – complete, partial, and no information – have been examined in (Kodialam & Lakshman, 2000; Kodialam & Lakshman, 2001; Qiao & Xu, 2002) to quantify the impact of the amount of available information on backup resource sharing. Specifically, (Qiao & Xu, 2002) describes a novel framework, called *distributed partial information management* (DPIM). It addresses several major challenges in achieving efficient shared path protection under distributed control with only *partial* network state information, including (1) how much partial information about existing working and backup paths (or active paths (APs) and backup paths (BPs), respectively) is maintained and exchanged; (2) how to obtain a good estimate of the bandwidth needed by a candidate BP, called backup bandwidth (BBW), and subsequently select a pair of AP and BP for a connection establishment request so as to minimize total bandwidth consumption and/or maximize revenues; (3) how to distributively allocate minimal BBW (and deallocate maximal BBW) via distributed signaling; and (4) how to update and subsequently exchange the partial network state information. This work proposes two new distributed control schemes under the DPIM framework, with one using an ILP formulation and the other using the AP first two-step heuristic. As a result, the $O(E^2)$ complete and aggregated information maintained by each node in survivable routing is now partitioned among all the nodes in a network, and thus each node only maintains and use $O(E)$ partial information where E is the number of links in the network. The proposed schemes are able to allocate (and deallocate) *minimal* (and *maximal*) BBW on a chosen BP even though only partial information is available at each node. As a result, they can achieve remarkable improvement over shared protection with partial information.

The work of (Zang, Ou, & Mukherjee, 2003) considers path-based protection with so called duct-layer constraints. Specifically, in path based protection, the primary path and the backup path of a connection must be fiber-disjoint so that the network is survivable under single-fiber failures.

In practice, fibers are put into cables, which are buried into *ducts* under the ground. A fiber cut usually occurs due to a duct cut during construction or destructive natural events, such as earthquakes. When a duct is cut, normally all of the fibers in the duct fail at the same time. Hence, a network survivable to a single-fiber failure is not necessarily survivable in duct-failure scenarios. A desired backup path of a given connection should not share any duct with the primary path of the same connection. Zang (Zang et al., 2003) develops ILP formulations for the RWA problem under both dedicated-path protection and shared-path protection. Then a three-stage heuristic is presented. In the first stage, it computes two duct-disjoint (and link-disjoint) paths for each connection demand; in the second stage, the heuristic assigns a wavelength to each path computed in the first stage; and in the final stage, the heuristic optimizes the resource utilization by iteratively rearranging, if necessary, the primary and backup lightpaths. This paper focuses on static traffic only. With dynamic traffic, a pair of duct-disjoint routes and a pair of wavelengths have to be determined for a given connection demand. However, with existing lightpaths in the network, some wavelengths on some links might be occupied by other connections. It is therefore necessary to decide which ducts should be considered for routing dynamically.

In general, the consideration of duct-layer constraints can be extended to consider shared risk link group (SRLG) constraints. Therefore, for path-based protection, it amounts to determining a pair of SRLG-disjoint paths for the working path and protection path, respectively. This problem is also known as diverse routing problem. For the link/node disjoint diverse routing problem with no SRLG failure constraint, the work of (Suurballe, 1974; Suurballe & Tarjan, 1984; Bhandari, 1998) proposed polynomial time complexity solutions. The diverse routing problem in networks with generally defined SRLG failures has been proved to be *NP*-complete (Hu, 2003) where an SRLG may include an arbitrary group of links. The work

in (Datta & Somani, 2004) considers the diverse routing problem subject to the SRLG constraint that a link cannot belong to more than one SRLG except that a link incident to the two endpoints of a common link can belong to two SRLGs. The work in (Doucette & Grover, 2002) also considers a similar problem in which an SRLG is defined as an incident-SRLG that has exactly two links incident to a common node. A link, however, is allowed to belong to more than one incident-SRLG. This type of SRLG failures covers the dual-link failure scenario only. In (Luo & Wang, 2005), we study the diverse routing problem in WDM optical networks with SRLG failures where all the optical links in an SRLG share a common endpoint. In addition, a link can belong to arbitrary number of SRLGs and an SRLG may include more than two links. This problem can be reduced to some commonly known diverse routing problems. For example, if each SRLG includes all the links incident to a common node, the problem is transformed to the node-disjoint diverse routing problem. On the other hand, if each SRLG contains only a single link, this problem becomes the link-disjoint diverse routing problem. We have developed a polynomial time optimal algorithm to find a pair of least cost SRLG-disjoint paths between a source and a destination. We also prove the correctness of the algorithm which is also shown to be more time efficient than earlier algorithms.

Segment-Based Survivability Provisioning

Earlier discussions indicate that protection schemes can in general recover from a single failure faster, but are less bandwidth efficient than restoration schemes. On the other hand, restoration schemes can survive one or multiple failures (as long as the destination is still reachable, with sufficient connectivity and bandwidth), but they cannot guarantee the recovery time, and/or the amount of information loss for real-time applications, making them unsuitable for mission-critical applications.

Link protection uses "local" rerouting and can recover from a failure faster than path protection. However, path-based protection schemes may be more resource efficient compared with link protection. To combine the best of link and path protection schemes (e.g., bandwidth efficiency and fast recovery), a new approach to survivability provisioning, termed as, *segment protection*, has been proposed. The basic mechanism of shared segment protection is illustrated in Figure 5. The network contains a working path WP_1 (A→B→C→D→E→F→G) and its protection path PP_1 along the physical route (A→H→C→J→E→K→G). Let another working path WP_2 (C→D→E) be allocated with its protection path. The protection capacity taken by the protection path of WP_1 can never be shared by the protection path of WP_2 if path based protection is employed because both working paths share the same links. For the shared segment protection shown in Figure 5, on the other hand, WP_1 is divided into multiple *segments*, each of which is assigned with switching/merging nodes and a protection path segment. The protection path of WP_2 can share the protection capacity taken by the protection path segments in the first and the third protection domain, resulting in more efficient network resource utilization.

In some sense, segment-based protection schemes are similar to failure-dependent path protection (or link protection). However, rerouted traffic only needs to go through the node that starts a backup segment (*BS*) which protects the failed link, as opposed to the source as in path based protection (or the immediate upstream node of the failed link as in link based protection). In general, both link and path based protection can, thus, be considered as a special case of segment-based protection. A segment based protection approach in general can recover faster than shared-path protection schemes because, just as in link protection, its *BS* is usually shorter than the backup path used in path protection.

Segment based protection has received a lot of attention. The approach developed in (Saradhi & Murthy, 2002) is characterized by separating the tasks of routing and spare capacity allocation into two subsequent processes, in which the spare capacity sharing for each link is not considered until the physical routes of the backup segments are defined. In (Ho & Mouftah, 2002; Ho, Tapolcai, & Cinkler, 2004) a framework called *short leap shared protection* (SLSP) along with a dynamic cascaded diverse routing (CDR) is proposed to realize segmented shared protection, in which the enumeration of *K*-shortest paths in each segment is performed. In (Ou et al., 2004), a subpath protection schemes that generalizes shared-path

Figure 5. Shared segment protection

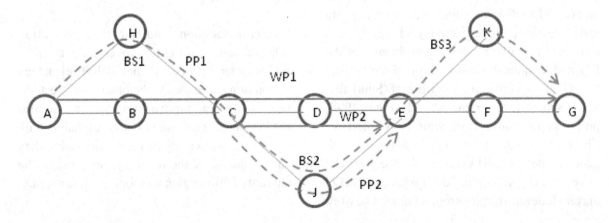

protection is proposed. The subpath protection scheme partitions a large optical network into smaller domains and applies shared-path protection to the optical network such that an intradomain lightpath does not use resources of other domains and the primary/backup paths of an interdomain lightpath exit a domain (and enter another domain) through a common domain-border node, which results in high scalability and fast recovery time for a modest sacrifice in resource utilization.

The work of (Xu, Xiong, & Qiao, 2003) proposes a scheme termed as *protection using multiple segments* (PROMISE) that can significantly reduce recovery time while achieving the bandwidth efficiency of shared-path protection schemes. In this approach, an active path (*AP*) or working path is divided into several active segments or *AS*s and each *AS* is then protected with a backup segment (*BS*) instead of protecting the *AP* as a whole as in path protection schemes. The approach allows for overlapping *AS*s (and *BS*s), and exploits bandwidth sharing not only among the *BS*s for different *AP*s, but also among those for the same *AP*. An ILP model is developed that can determine the optimal partition of a given *AP* into *AS*s and find the corresponding *BS*s. This ILP model is then used to obtain an optimal solution for a medium-to-large network. In addition, a fast dynamic programming based heuristic is presented to obtain a near-optimal set of *AS*s and their *BS*s for a given *AP*. Comprehensive performance evaluation and comparison of various link, path, and segment protection schemes conducted by the authors found that the proposed heuristic approaches can achieve similar or higher bandwidth efficiency, and at the same time, be able to recover from a failure much faster than the best-performing shared-path protection schemes.

Cycle-Based Survivability Provisioning

Before *p*-Cycles (pre-planned, preconfigured-cycles) were first reported by Grover and Stamatela-

kis (1998), ring or protection cycle based schemes have been proposed (Ellinas, Hailemariam, & Stern, 2000; Stamatelakis & Grover, 2000). Path based protection schemes are capacity efficient whereas the ring based schemes are fast. *p*-Cycle based protection schemes strive to obtain the ring like speed with path-like capacity efficiency based on the property of a *p*-Cycle being able to provide shared protection not only to its on-cycle spans, but also to any possible straddling spans.

Specifically, the spare capacity of the network is pre-configured to form the ring-like structures called *p*-cycles. Consequently only two switching actions (as in rings), at the end nodes of the failed span, are needed in the event of failure, to switch the traffic to the protection path provided by the pre-configured cycle. Figure 6 shows a simple network with an established *p*-Cycle (dashed line). In the example, the *p*- Cycle protects five on-cycle spans, and two straddling spans. When one of the on-cycle spans (*B−C*) fails, as shown in Figure 7(a), the span is protected by redirecting all traffic on it around the cycle (*B−E−F−D−C*), similar to a line switched self-healing ring. For a straddling span failure, the *p*-Cycle can provide protection capacity for two working paths on the failed span. In Figure 7(b), for instance, paths (*B−E−F*) and (*B−C−D−F*) can be used to protect two working paths on failed span (*B−F*). A span refers to all links between two adjacent nodes. A straddling span is a span connecting two non-adjacent nodes

Figure 6. A network with one p-Cycle

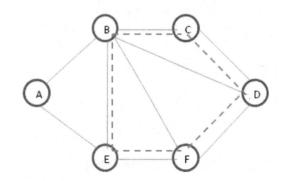

on the ring. From the example, *p*-Cycle based protection schemes can be considered as a form of shared span protection.

The main difference between any ring-based scheme and *p*-Cycle based schemes is the protection of straddling spans by the *p*-Cycles as shown in Figure 7. Another important benefit of *p*-Cycles over rings is that working paths may be freely routed over the network topology, and are not required to follow ring-constrained routings in order to be protected.

In recent years, many techniques have been developed to construct different types of *p*-Cycles such as Hamiltonian *p*-Cycle, simple *p*-Cycle, span *p*-Cycle, node encircling *p*-Cycle, path protecting *p*-Cycle, flow *p*-Cycle, and so on, to enable different types of protection. In general, three optimization objectives are considered when constructing *p*-Cycles (Asthana, Singh, & Grover, 2010):

1. **Spare capacity optimization.** In this approach, working capacities of demands are routed first. The objective is to minimize the capacity required for the formation of *p*-Cycles while protecting all the working capacities on the span of the network.

2. **Optimization of total capacity.** In this approach, demands are not routed in advance

to find the working paths. Instead the set of eligible candidate working paths are found for each source-destination pair. From the set, the working paths for each source-destination pair are selected along with the placement of spare capacity to optimize the total capacity, spare plus working together.

3. **Protected working capacity envelope (PWCE) optimization.** In this approach, *p*-Cycles are determined to maximize the total volume of working capacity protected. The PWCE concept is proposed for accommodating dynamic traffic. The incoming demands are routed freely in the available capacity which is protected by the PWCE (Zhang, Zhong, & Bose, 2005). To further provide optimization to the dynamic traffic scenario, the adaptive PWCE (APWCE) has been proposed in (Shen & Grover, 2005). The APWCE slowly adapts to the changing traffic scenario with re-optimization process using ILP based on the previous protected working capacity envelope. The objective is to provide protection to as many demands as possible within the available capacity constraints

p-Cycle based protection has been demonstrated to be such a versatile concept that various

Figure 7. p-Cycle span protection

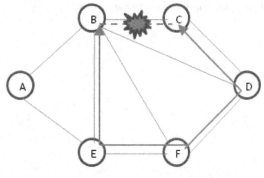

(a) Failure of an on-cycle span

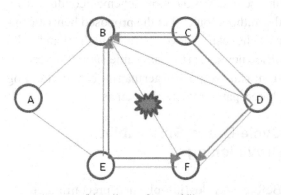

(b) Failure of a straddling span

p-Cycle based schemes provide the ability to implement span, node, path, and path-segment protection. The *p*-Cycle concept is general enough to be applicable to provide protection at any connection-oriented transport layer, including SONET, ATM, MPLS, or IP layers. Many strategies have been proposed to provide dual or multi failure (such as SRLG failures) network survivability using *p*-Cycles (Liu & Ruan, 2005). Recently, *p*-cycles have also been extended to protect multicast traffic.

RECENT DEVELOPMENTS

Maximum Reliable Path Routing

As discussed earlier, a widely used approach to achieving survivability is through path protection in which a link-disjoint protection path is pre-computed and reserved for each working path. Such schemes provide 100% survivability against single-link failures. However, various risk factors such as natural and man-caused catastrophes introduce the possibility that when a network failure occurs, multiple links that belong to the same shared risk link group (SRLG) fail simultaneously (Kaminow & Koch, 1997; Ramanurthy & Mukherjee, 2003; Strand, Chiu, & Tkach, 2001; Grover, 2003). In this case, only a working and a protection path that are SRLG-disjoint can survive the failure (Sebos, Yates, Hjalmtysson, & Greenberg, 2001. However, the study in (Yuan & Jue, 2004) proved it *NP*-hard to find SRLG-disjoint working and protection paths. Consequently, attempts were made to minimize the probability of simultaneous failures of the pair. Yuan et al. proved this problem also *NP*-hard and proposed heuristic algorithms for the special cases in which all SRLGs have equal failure probabilities (Yuan, Varma, & Jue, 2008; Yuan, Varma, & Jue, 2005). Various heuristic solutions were also proposed for the general case in which the SRLGs have different failure probabilities

(Huang, Li, & Srinivasan, 2007; She, Huang, & Jue, 2006; Guo, 2007).

In addition to the difficulties of finding the maximum reliable working and protection paths, another limitation of the protection schemes is that they require the reservation of a significant amount of network resources. Shared path protection may be used to improve resource utilization at the expense of signaling and network management (Guo, 2007; Yuan & Jue, 2002). Therefore for certain types of traffic and customers, it may be more resource-efficient and cost-effective to use a single path without protection, if the reliability of the path can be maximized.

For networks with only single failures, finding the maximum reliable path without protection is equivalent to a minimum-cost-path problem and can be easily solved. However, simultaneous multiple failures complicate the matter and make the problem *NP*-hard (Yuan et al., 2008; Yuan et al., 2005). The work in (Yuan et al., 2008; Yuan et al., 2005) proposed heuristic solutions for the special cases in which all SRLGs have equal failure probabilities. We have studied the maximum reliable path problem for the general case in which the SRLGs have arbitrary failure probabilities. The significance this problem lies in the difficulties of efficiently finding the maximum survivable working and protection paths.

Maximum Reliable Path Problem

The problem is defined as follows. Given network $G = (N, L, S, P_s)$ where N is the set of nodes, L is the set of links (assume they are bidirectional), and S is the set of SRLGs in the network, $P_s = \{p_1, p_2, p_3, ..., p_i ...p_{|S|}\}$ is the set of SRLGs in the network, $P_s = \{p_1, p_2, p_3, ..., p_i ...p_{|S|}\}$ is the set of non-failure probabilities of each SRLG $s_i \in S$ and $0 < p_i < 1$. Also given $S^l \; \forall \; l \in L$, and $S^l \subseteq S$ is the set of SRLGs to which link l belongs, find one path P from source node s to destination node t such that P has the maximum reliability. The reliability of a path is defined as follows. Let S^p be the set of all SRLGs to which the links of a path P belong. Then the reliability of P is,

$$r^P = \prod_{s_i^P \in S^P} p_i, p_i \in P_s. \tag{1}$$

We have developed both heuristic and optimal solutions for this problem. In the special case in which every SRLG contains only one link, the maximum reliable path can be readily obtained using Dijkstra's algorithm. This problem becomes *NP*-hard if each SRLG contains multiple links that may fail simultaneously (Yuan et al., 2008; Yuan et al., 2005). However, for networks in which the number of SRLGs is small, we proposed optimal solutions whose running times are polynomial in the number of network nodes and are exponential only in the number of SRLGs (Yuan, Wang, Waller, & Delavina, 2010). For networks with arbitrarily large number of SRLGs, we proposed heuristic solutions whose running times are polynomial in both the number of network nodes and SRLGs (Yuan et al., 2010).

Simulation studies have shown that the amount of time for the heuristic algorithms to generate the reliable paths was small compared to that for the optimal solution. The simulations demonstrate that the performances of heuristics are closer to that of the optimal solution. We note that the number of SRLGs and their non-failure probabilities have a significant impact on the average reliabilities of the paths. The reliabilities improve as the average non-failure probabilities of the SRLGs increase. In addition, when the number of SRLGs is small, a path is more likely to consist of links that belong to fewer SRLGs. Consequently, the path has higher reliability if the non-failure probabilities of the SRLGs do not differ significantly. As the number of SRLGs increases, the links of a path are more likely to belong to a larger number of SRLGs, which reduces the reliabilities of the paths.

We also note that, as the nodal degree increases, the average reliabilities of the paths increase. The reason for this behavior is that an increase in nodal degree results in higher average number of links belonging to each SRLG, thereby making a path more likely to consist of links that belong to fewer SRLGs. The network topology and the size of the network also affect the reliabilities of the paths. Larger networks with more nodes result in a higher average hop count for the paths; hence, for the same nodal degree and the same number of SRLGs, the links of the paths in a network with more nodes will belongs to a greater number of SRLGs than paths in a network with fewer nodes, which reduces the reliabilities of the paths in a larger network.

Connection Holding-Time Aware Survivable Service Provisioning

A great deal of research has been conducted on survivable service provisioning in WDM optical networks. Previous work has considered several types of traffic models, e.g., static traffic, dynamic random traffic, admissible set, and incremental traffic, where the connection holding-time of demands is not explicitly taken into account for service provisioning. While different traffic models are valid and useful in many circumstances, these models are not able to capture the traffic characteristics of applications that require resources during specific time intervals, for instance, circuit leasing on a short term basis, where a client company may request certain amount of scheduled bandwidth from a service provider to satisfy its communication requirements at a specific time, e.g., between headquarters and production centers during office hours or between data centers during the night when backup of databases is performed, and so on. Many applications require provisioning of scheduled dedicated channels or bandwidth pipes at a specific time with certain duration. These scheduled bandwidth demands (Kuri, 2003) are dynamic in nature. They are not static in the sense that the demands only last during the specified time intervals.

In the work of Kuri et al. (Kuri, 2003; Kuri, Puech, Gagnaire, Dotaro, & Douville, 2003a), a scheduled lightpath demand model was proposed. The routing and wavelength assignment problem is solved using a branch & bound algorithm and a tabu search algorithm. The issue of diverse routing of scheduled lightpath demands was addressed in another work from the same group (Kuri et al., 2003b). The problem was formulated as an optimization model, which is basically a two-step optimization approach and a simulated annealing based algorithm was proposed to find heuristic solutions to the optimization problem. The work of Tornatore et al.(2005) exploits the connection-holding-time information to dynamically provision shared-path-protected connections using heuristic algorithms. The work of Saradhi, Wei, & Gurusamy (2004) considers the provisioning of fault-tolerant scheduled lightpath demands based on a two-step optimization that uses a set of pre-computed routes for working and protection paths. In (Li & Wang, 2005), given a set of scheduled traffic demands, we provide a set of joint routing and wavelength assignment (RWA) ILP problem formulations that enable the maximum resource sharing in both space and time (i.e., use backup resource sharing and resource sharing that exploits time disjointness among demands). This problem has been shown to be *NP*-hard. We therefore have studied a two-step optimization approach which divides the joint RWA problems into a routing subproblem and wavelength assignment subproblems; and then solves them individually (Li & Wang, 2005; Li, Wang, Xin, & Zhang, 2005).

To provide time-efficient solutions, in (Wang & Li, 2009), we have proposed an iterative survivable routing (ISR) scheme that utilizes a capacity provision matrix and processes demands sequentially using different demand scheduling policies. The objective is to minimize the total network resources (e.g., number of wavelength-links) used by working paths and protection paths of a given set of demands while 100% restorability is guaranteed against any single failure. The ad-

ditional information on connection holding-time offers a service provider a better opportunity to optimize the network resources jointly in space (i.e., backup resource sharing) and in time (i.e., taking advantage of time-disjointness amongst demands). Since a demand is considered accommodated as long as it is provisioned during its holding time, time disjoint demands (working path and protection path alike) can therefore share network resources.

Specifically, we have considered a traffic model termed as scheduled traffic model in (Wang & Li, 2009). The network has an arbitrary topology $G = (N, L)$, where N, L are the set of nodes and the set of directed links, respectively. Each link, represented by a pair of ordered nodes, has a set of wavelengths K. A set of scheduled traffic demands, D, is given, each demand of which is represented by a tuple $(s_r, d_r, n_r, \alpha_r, \beta_r)$, where s_r and d_r are the source and destination nodes of demand r, n_r is the number of requested lightpaths, α_r and β_r are the setup and teardown time of the demand, respectively. The scheduled traffic model is different from the static and dynamic random traffic models generally assumed in the literature. In the static traffic model, all demands are known in advance and do not change over time, while the dynamic random traffic assumes that a demand arrives at a random time, the inter-arrival time and holding time of demands are random or conform to some probability distribution. The scheduled traffic model explicitly considers the time dimension of demands since many demands for bandwidth in ultra high-speed networks will be time-limited rather than permanent. The model is also dynamic in the sense that demands only last during the specified time intervals. Given a set of scheduled demands, some demands may not overlap in time. This motivates us to take into account the time disjointness (if any) among demands along both working and protection paths in addition to optimizing the spatial network resource sharing based on backup resource sharing, to achieve a higher degree of overall network resource shareability.

Our proposed algorithm strives to minimize the total network resources used by both working paths and protection paths of a given set of demands through exploiting network resource reuse in both space and time domains simultaneously. That is, both working capacity and protection capacity are minimized simultaneously. Given a demand, our approach tries to accommodate the demand by finding working and protection paths that will use the least amount of additional network resource after using sharable resources as much as possible. The time relationship among demands is represented by an interval graph. By exploiting the time disjointness among a given set of demands, we transform the problem of finding the minimum total capacity needed on a link to accommodate a set of the demands for their working capacity as well as protection capacity to the problem of finding a maximum weighted clique of the weighted interval graph corresponding to the demand set. Once a working path and a protection path are found for each demand in the given demand set, the total network resources used can be determined. The algorithm then runs iteratively to reduce the total resources used by rearranging the paths of demands. Because the algorithm iteration keeps reducing the objective function, ISR can converges quickly in a stable network (i.e., a fixed topology and a given fixed demand set).

The ISR scheme is evaluated against solutions obtained by integer linear programming (Li & Wang, 2005). The simulation results indicate that the ISR algorithm is extremely time efficient while achieving excellent performance in terms of total network resources used. The impact of demand scheduling policies on the ISR algorithm is also studied. It appears that no significant difference exists among various demand scheduling policies for small demand sets and networks. For large networks and demand sets, in many cases, the ISR algorithm employing the Most Conflicting Demand First (MCDF) policy achieves the best performance, and the one using the Least Con-

flicting Demand First (LCDF) policy needs more wavelength-links compared with other policies.

In the work of (Jaekel & Chen, 2006), a new ILP formulation for routing and wavelength allocation, under the scheduled traffic model, was presented that minimizes the *congestion* of the network. Two levels of service were proposed where idle backup resources can be used to carry low priority traffic, under fault-free conditions. When a fault occurs, and resources for a backup path need to be reclaimed, any low priority traffic on the affected channels is dropped. The results demonstrate that significant improvements can be achieved over single service level models. Optimal solutions can also be generated for moderate sized networks, within a reasonable amount of time. The authors also present a heuristic that can quickly generate good solutions for much larger networks.

In the work of (Wang, Li, Luo, Fan, & Xin, 2005), we propose a general scheduled traffic model called the *sliding scheduled traffic model*. In this model, the setup time α_r of a demand r, whose holding time is τ time units is not known in advance. Rather α_r is allowed to begin in a pre-specified time window $[t_s, t_e]$ subject to the constraint that $t_s \leq \alpha_r \leq t_e - \tau$. We then solve two problems: (1) how to properly place a demand within its associated time window to reduce overlapping in time among a set of demands; and (2) how to route and assign wavelengths (RWA) to a set of demands under the sliding scheduled traffic model in mesh reconfigurable WDM optical networks without wavelength conversion.

As follow-up work, the authors of (Jaekel & Chen, 2007; Jaekel, 2006) present a new ILP formulation for both fixed window model, and the sliding scheduled traffic model. They consider fault-free as well as survivable networks using path protection, and do not require any wavelength conversion. The ILP can jointly optimize the problem of scheduling the demands in time and allocating resources for the scheduled lightpaths. It is shown that the complexity of the ILP formulation for sliding scheduled traffic model, in terms of the number of integer variables, is less than previous ILP formulations for the simpler fixed window model. For very large networks, a fast two-step optimization process is also proposed. The first step schedules the demands optimally in time, such that the amount of overlap is minimized. The second step uses a *connection holding-time aware* heuristic to perform routing and wavelength assignment for the scheduled demands. The work of (Saradhi & Gurusamy, 2007) develops a time conflict resolving window division algorithm which places a given set of sliding scheduled light-path demands within their allowed intervals, and two routing and wavelength assignment (RWA) algorithms for routing sliding scheduled lightpath demands in WDM optical networks.

Survivable Service Provisioning in OBS Networks

In optical burst switched (OBS) networks, wavelengths or channels on a link are divided into control channel group (CCG) for the transmission of burst control packets (BCPs) and data channel group (DCG) for the transmission of data bursts. Network failures can manifested as data channel failure, control channel failure, fiber failure, link failure, and node failure. Survivability in OBS networks is challenging, and survivability approaches can be classified into link- and path-level restorations. Link-level restoration is performed at the local node, while path-level restoration is done globally between *OBS path switch LSRs* (OPSLs) and *OBS path merge LSRs* (OPMLs). The OBS restoration procedure is briefly summarized as follows. Every node in the OBS domain monitors the network status, and when a fault is detected, an appropriate restoration technique is applied according to the type of the fault detected. For channel failure and fiber failure, link-level restoration can be performed, whereas for link failure and node failure, path-level restoration can be attempted. For channel failure, survivability of a data channel and a control channel must

be considered together. The interested reader is referred to (Dixit, 2003) for additional details.

Survivable Service Provisioning in Light-Trail Networks

In a typical WDM optical network, the connection between end users is supported by establishing a lightpath. Once a lightpath is established, the entire wavelength is used exclusively by its source and destination node pair, and no wavelength multiplexing between multiple nodes along the lightpath is allowed. Similar to a lightpath, a light-trail is an all optical connection. However, a light-trail allows the nodes (other than the source and the destination) on the optical connection to access the optical path by either inserting or receiving data through the light-trail, which makes possible finer bandwidth granularity control and more efficient utilization of the optical bandwidth. A light-trail can be considered as a generalization of a lightpath – leading to multiple users being able to access an optical path.

Survivability is equally challenging in light-trail networks because one single link failure could cause failures of a set of light-trails, each of which carries multiple connections. In the work of (He, Fang, & Somani, 2004), two protection schemes are proposed, namely connection based protection and link based protection. In connection based protection, for each connection request, the resources are allocated to a primary connection in a light trail LT_1 and a backup connection in another light trail LT_2. LT_1 and LT_2 are link-disjoint. The primary connection is the working connection when there is no link failure. If a link on LT_1 fails, the failure information is propagated through the control channel. When the source node of the request receives the failure information, it starts to transmit the data on LT_2 to the destination through the backup connection. In link based protection, for each link on a light-trail, a backup sub-light-trail is provided. When a link on a light-trail fails, the light-trail will be rerouted around the failed link and use the backup sub-light trail. The restored light-trail in link based protection is longer than original light-trail. In the work of (Balasubramanian, He, & Somani, 2005), the authors study dedicated and shared protection for light-trail networks. It was determined that, with dedicated protection, about 200% redundancy may be required. Shared protection performs much better and full protection can be achieved in the presence of single link failures with less than 100% redundancy. A limit of this work is that light-trails are assumed to carry primary connections alone or secondary connections alone. An optimal solution would allow a light-trail to multiplex both primary and secondary connections.

FUTURE RESEARCH DIRECTIONS

As optical networking technologies have undergone dramatic development, the field of survivability provisioning in optical networks has also evolved significantly over the years. However, many avenues of research still remain open for future exploration. For example, more studies are needed to provide effective means of protection and restoration in relatively recent light-trail networks and OBS networks.

While a wide range of protection mechanisms against various types of failure have been studied, most of the efforts have been generally directed at single domain protection where it is assumed that each node in the network has complete knowledge of the physical topology of the entire network. A multi-domain network is a network composed of several single-domain networks, interconnected by inter-domain links. Each single domain can be regarded as an independent network that has its own local rules of operation and management to provide services. The internal topological details of a domain are usually not shared externally. As a result, no node in a multi-domain network can have complete information on the multi-domain network. Thus, the survivability provisioning

in multi-domain networks is more challenging than that of single domain networks. So far, only a handful of published works have focused on survivability in multi-domain optical networks. Many issues remain open such as effective topology aggregation with SRLGs, multi-domain survivability mechanisms, trade-offs between different concurrent objectives (for instance efficient use of backup resources, scalability, and fast recovery time).

Moreover, survivable service provisioning in optical network architectures, such as optical flow switching, optical split-and-direct switching, SMART (scalable multi-access reconfigurable transport), and others remains largely unexplored.

CONCLUSION

This chapter has presented the principles and state-of-the-art of survivability provisioning in optical networks, in particular, in optical networks that employ wavelength division multiplexing (WDM). Concepts of survivability provisioning in optical networks such as protection and restoration, dedicated versus shared survivability, path-based, link-based, segment-based, cycle-based survivability, and so on, have been covered to provide multiple classes of quality of protection against single failure, dual-failure, multiple simultaneous failures, or shared risk link group failures, in WDM mesh networks. Recent developments in survivable service provisioning have also been summarized, such as survivability provisioning that takes into account the connection holding-time, survivability in WDM light-trail networks and optical burst switched networks. Finally, the chapter briefly examined future research directions.

REFERENCES

Asthana, R., Singh, Y. N., & Grover, W. D. (2010). *p*-Cycles: An overview. *IEEE Communications Surveys & Tutorials*, *12*(1), 97–111. doi:10.1109/SURV.2010.020110.00066

Balasubramanian, S., He, W., & Somani, A. K. (2005). Light-trail networks: Design and survivability. In *Proceedings of IEEE Conference on Local Computer Networks* (pp. 174-181).

Bhandari, R. (1998). *Survivable networks: Algorithms for diverse routing*. Norwell, MA: Kluwer Academic Publishers.

Bouillet, E., Labourdette, J.-F., Ramamurthy, R., & Chaudhuri, S. (2002). Enhanced algorithm cost model to control tradeoffs in provisioning shared mesh restored lightpaths. In *Proceedings of OFC*.

Chen, Y., Qiao, C., & Yu, X. (2004). Optical burst switching (OBS): A new area in optical networking research. *IEEE Network Magazine, 18*(3), 16–23. doi:10.1109/MNET.2004.1301018

Datta, P., & Somani, A. K. (2004). Diverse routing for shared risk resource groups (SRRG) failures in WDM optical networks. In *Proceedings of Broadnets* (pp. 120-129).

De Patre, S., Maier, G., Pattavina, A., & Martinelli, M. (2002). Optical network survivability: Protection techniques in the WDM layer. *Photonic Network Communications, 4*(3/4), 251–269. doi:10.1023/A:1016047527226

Dixit, S. S. (2003). *IP over WDM: Building the next generation optical Internet*. Wiley-Interscience.

Doucette, J., & Grover, W. D. (2002). Capacity design studies of span-restorable mesh transport networks with shared-risk link group effects. In *Proceedings of OptiComm* (pp. 25-38).

Elie-Dit-Cosaque, D., Ali, M., & Tancevski, L. (2002). Informed dynamic shared path protection. In *Proceedings of OFC*.

Ellinas, G., Hailemariam, A. G., & Stern, T. E. (2000). Protection cycles in mesh WDM networks. *IEEE Journal on Selected Areas in Communications, 18*(10).

Gerstel, O., & Ramaswami, R. (2000). Optical layer survivability: A service perspective. *IEEE Communications Magazine, 38*(3), 104–113. doi:10.1109/35.825647

Gerstel, O., & Sasaki, G. (2001). Quality of protection (QoP): A quantitative unifying paradigm to protection service grades. In *Proceedings of SPIE OptiComm*.

Grover, W. D. (2003). *Mesh-based survivable transport networks: Options and strategies for optical, MPLS, SONET and ATM networking*. Prentice Hall PTR.

Grover, W. D., & Stamatelakis, D. (1998). Cycle-oriented distributed preconfiguration: Ring-like speed with mesh-like capacity for self-planning network restoration. In *Proc. IEEE International Conference on Communications* (pp. 537-543).

Gumaste, A., & Chlamtac, I. (2004). Light-trails: An optical solution for IP transport. *Journal of Optical Networking, 3*(4).

Guo, L. (2007). Heuristic survivable routing algorithm for multiple failures in WDM networks. In *Proceedings of 2nd IEEE/IFIP International Workshop on Broadband Convergence Networks*.

He, W., Fang, J., & Somani, A. K. (2004). On survivability design in light-trail optical networks. In *Proceedings of 8th Conference on Optical Networks Design and Modeling (ONDM)*.

Ho, P.-H., & Mouftah, H. T. (2002). A framework for service-guaranteed shared protection in WDM mesh networks. *IEEE Communications Magazine, 40*(2), 97–103. doi:10.1109/35.983914

Ho, P.-H., Tapolcai, J., & Cinkler, T. (2004). Segment shared protection in mesh communication networks with bandwidth guaranteed tunnels. *IEEE/ACM Transactions on Networking, 12*(6), 1105–1118. doi:10.1109/TNET.2004.838592

Hu, J. Q. (2003). Diverse routing in optical mesh networks. *IEEE Transactions on Communications, 51*(3), 489–494. doi:10.1109/TCOMM.2003.809779

Huang, C., Li, M., & Srinivasan, A. (2007). A scalable path protection mechanism for guaranteed network reliability under multiple failures. *IEEE Transactions on Reliability, 56*(2), 254–267. doi:10.1109/TR.2007.896739

Jaekel, A. (2006). Lightpath scheduling and allocation under a flexible schedule traffic model. In *Proceedings of IEEE Globecom.*

Jaekel, A., & Chen, Y. (2006). Routing and wavelength assignment for prioritized demand under a scheduled traffic model. In *Proceedings of Broadnets Workshop on Guaranteed Optical Service Provisioning.*

Jaekel, A., & Chen, Y. (2007). Demand allocation without wavelength conversion under a sliding scheduled traffic model. In *Proceedings of IEEE Broadnets.*

Kaminow, P., & Koch, T. L. (1997). *Optical fiber telecommunications IIIA*. Academic Press.

Kitamura, Y., Lee, Y., Sakiyama, R., & Okamura, K. (2007). Experience with restoration of Asia Pacific network failures from Taiwan earthquake. *IEICE Transactions on Communications. E (Norwalk, Conn.), 90-B*(11), 3095–3103.

Kodialam, M., & Lakshman, T. V. (2000). *Dynamic routing of bandwidth guaranteed tunnels with restoration* (pp. 902–911). Proc. IEEE INFOCOM.

Kodialam, M., & Lakshman, T. V. (2001). *Dynamic routing of locally restorable bandwidth guaranteed tunnels using aggregated link usage information* (pp. 376–385). Proc. IEEE INFOCOM.

Kuri, J. (2003). *Optimization problems in WDM optical transport networks with scheduled lightpath demands*. Unpublished doctoral dissertation, ENST Paris.

Kuri, J., Puech, N., Gagnaire, M., Dotaro, E., & Douville, R. (2003). Routing and wavelength assignments of scheduled lightpath demands. *IEEE Journal on Selected Areas in Communications, 21*(8), 1231–1240. doi:10.1109/JSAC.2003.816622

Kuri, J., Puech, N., Gagnaire, M., Dotaro, E., & Douville, R. (2003). Diverse routing of scheduled lightpath demands in an optical transport network. In *Proceedings of Fourth International Workshop on the Design of Reliable Communication Networks.*

Li, G., Wang, D., Kalmanek, C., & Doverspike, R. (2002). Efficient distributed path selection for shared restoration connections. In *Proceedings of IEEE INFOCOM* (pp. 40–149).

Li, T., & Wang, B. (2005). On optimal survivability design in WDM optical networks under a scheduled traffic model. In *Proceedings of the 5th IEEE International Workshop on Design of Reliable Communication Networks.*

Li, T., Wang, B., Xin, C., & Zhang, X. (2005). On survivable service provisioning in WDM optical networks under a scheduled traffic model. In *Proceedings of IEEE Globecom.*

Liu, C., & Ruan, L. (2005) p-Cycle design in survivable WDM networks with shared risk link groups (SRLGs). In *Proceedings of 5th International Workshop on the Design of Reliable Communication Networks.*

Liu, Y., Tipper, D., & Siripongwutikorn, P. (2001). Approximating optimal spare capacity allocation by successive survivable routing. In *Proceedings of IEEE INFOCOM* (pp. 699–708).

Luo, X., & Wang, B. (2005). Diverse routing in WDM optical networks with shared risk link group (SRLG) failures. In *Proceedings of IEEE International Workshop on Design of Reliable Communication Networks.*

Mohan, G., & Murthy, C. S. R. (2000). Light-path restoration in WDM optical networks. *IEEE Network, 14*(6), 24–32. doi:10.1109/65.885667

Ou, C., Zang, H., Singhal, N. K., Zhu, K., Sahasrabuddhe, L. H., MacDonald, R. A., & Mukherjee, B. (2004). Subpath protection for scalability and fast recovery in optical WDM mesh networks. *IEEE Journal on Selected Areas in Communications, 22*(9), 1859–1875. doi:10.1109/JSAC.2004.830280

Ou, C., Zhang, J., Zang, H., Sahasrabuddhe, H., & Mukherjee, B. (2004). New and improved approaches for shared-path protection in WDM mesh networks. *Journal of Lightwave Technology, 22*(5), 1223–1232. doi:10.1109/JLT.2004.825346

Qiao, C., & Xu, D. (2002). Distributed partial information management (DPIM) schemes for survivable networks—Part I. In *Proceedings of IEEE INFOCOM* (pp. 302-311).

Ramamurthy, S., & Mukherjee, B. (1999). Survivable WDM mesh networks – Part I: Protection. In *Proceedings of IEEE INFOCOM* (pp. 744-751).

Ramamurthy, S., & Mukherjee, B. (2003). Survivable WDM mesh networks. *Journal of Lightwave Technology, 21*(4), 870–883. doi:10.1109/JLT.2002.806338

Ramaswami, R., & Sivarajan, K. N. (2002). *Optical networks: A practical perspective* (2nd ed.). Morgan Kaufmann Publisher.

Saradhi, C. V., & Gurusamy, M. (2007). Scheduling and routing of sliding scheduled lightpath demands in WDM optical networks. In *Proceedings of OFC*.

Saradhi, C. V., & Murthy, C. (2002). Dynamic establishment of segmented protection paths in single and multi-fiber WDM mesh networks. In *Proceedings of OptiComm* (pp. 211-222).

Saradhi, C. V., Wei, L. K., & Gurusamy, M. (2004). Provisioning fault-tolerant scheduled lightpath demands in WDM mesh networks. In *Proceedings of Broadnets* (pp. 150-159).

Sebos, P., Yates, J., Hjalmtysson, G., & Greenberg, A. (2001). Auto-discovery of shared risk link groups. In *Proceedings of Optical Fiber Communication Conference*.

She, Q., Huang, X., & Jue, J. P. (2006). Maximum survivability under multiple failures. In *Proceedings of IEEE/OSA Optical Fiber Communication Conference*.

Shen, G., & Grover, W. D. (2005). Automatic lightpath service provisioning with an adaptive protected working capacity envelope based on *p*-cycles. In *Proceedings of 5th International Workshop on the Design of Reliable Communication Networks*.

Stamatelakis, D., & Grover, W. D. (2000). IP layer restoration and network planning based on virtual protection cycles. *IEEE Journal on Selected Areas in Communications, 18*(10), 1938–1949. doi:10.1109/49.887914

Strand, J., Chiu, A. L., & Tkach, R. (2001). Issues for routing in the optical layers. *IEEE Communications Magazine, 39*(2), 81–87. doi:10.1109/35.900635

Su, C., & Su, X. (2001). Protection path routing on WDM networks. In *Proceedings of OFC*.

Suurballe, J. W. (1974). Disjoint paths in a network. *Networks, 4*(2), 125–145. doi:10.1002/net.3230040204

Suurballe, J. W., & Tarjan, R. E. (1984). A quick method for finding shortest pairs of disjoint paths. *Networks, 14*(2), 325–336. doi:10.1002/net.3230140209

Tornatore, M., Pattavina, A., Zhang, J., Mukherjee, B., & Ou, C. (2005). Efficient shared-path protection exploiting the knowledge of connection-holding time. In *Proceedings of OFC*.

Wang, B., & Li, T. (2009). Survivable scheduled service provisioning in WDM optical networks with iterative routing. *Optical Switching Network and Networking, 7*(1).

Wang, B., Li, T., Luo, X., Fan, Y., & Xin, C. (2005). Routing and wavelength assignment under a scheduled traffic model in reconfigurable WDM optical networks. In *Proceedings of Broadnets*.

Xu, D., Xiong, Y., & Qiao, C. (2003). Novel algorithms for shared segment protection. *IEEE Journal on Selected Areas in Communications, 21*(8), 1320–1331. doi:10.1109/JSAC.2003.816624

Yoo, M., & Qiao, C. (1999). Supporting multiple classes of services in IP over WDM networks. In *Proceedings of IEEE Globecom* (pp. 1023–1027).

Yuan, S., & Jue, J. P. (2002). Shared protection routing algorithm for optical network. *Optical Networks Magazine, 3*(3), 32–39.

Yuan, S., & Jue, J. P. (2004). Dynamic path protection in WDM mesh networks under risk disjoint constraint. In *Proceedings of IEEE Globecom* (pp. 1770–1774).

Yuan, S., Varma, S., & Jue, J. P. (2005). Minimum color problem for reliability in mesh networks. In *Proceedings of IEEE INFOCOM* (pp. 2658-2669).

Yuan, S., Varma, S., & Jue, J. P. (2008). Lightpath routing for maximum reliability in optical mesh Networks. [OSA]. *Journal of Optical Networking, 7*(5), 449–466. doi:10.1364/JON.7.000449

Yuan, S., Wang, B., & Waller, W., & Delavina, E. (2010). Reliable path routing in mesh networks under multiple link failures. *IEEE Transaction on Reliability.*

Zang, H., & Mukherjee, B. (2001). Connection management for survivable wavelength-routed WDM mesh networks. *SPIE Optical Networks Magazine, 2*(4), 17–28.

Zang, H., Ou, C., & Mukherjee, B. (2003). Path-protection routing and wavelength assignment (RWA) in WDM mesh networks under duct-layer constraints. *IEEE/ACM Transactions on Networking, 11*(2), 248–258. doi:10.1109/TNET.2003.810313

Zhang, Y., Taira, K., Takagi, H., & Das, S. K. (2002). An efficient heuristic for routing and wavelength assignment in optical WDM networks. In *Proceedings of IEEE International Conference Communications* (pp. 2734-2739).

Zhang, Z., Zhong, W. D., & Bose, S. K. (2005). Dynamically survivable WDM network design with *p*-cycle based PWCE. *IEEE Communications Letters, 9*(8), 756–758. doi:10.1109/LCOMM.2005.1496606

Zhou, D., & Subramaniam, S. (2000). Survivability in optical networks. *IEEE Network, 14*(6), 16–23. doi:10.1109/65.885666

ADDITIONAL READING

Akyamac, A., Sengupta, S., Labourdette, J., Chaudhuri, S., & French, S. (2002) *Reliability in Single domain vs. Multi domain Optical Mesh Networks.* In *Proceedings of National Fiber Optic Engineers Conference.*

Chan, V. W. S. (2009). Optical flow switching: a new "green" transport mechanism for fiber networks. In *Proceedings of 11th International Conference on Transport Optical Networks.*

Drid, H., Cousin, B., Molnar, M., & Lahoud, S. (2010). A survey of survivability in multi-domain optical networks. *Computer Communications, 33*(8), 1005–1012. doi:10.1016/j.comcom.2010.02.003

Ghani, N., Liu, Q., Benhaddou, D., Rao, N. S. V., & Lehman, T. (2008). *Control plane design in multi-domain/multilayer optical networks.* *IEEE Communications Magazine, 46*(6), 78–87. doi:10.1109/MCOM.2008.4539470

Grover, W. D. (2004). The protected working capacity envelope concept: An alternate paradigm for automated service provisioning. *IEEE Communications Magazine, 42*(1), 62–69. doi:10.1109/MCOM.2004.1262163

Guo, L. (2007). LSSP: A novel local segment-shared protection for multi-domain optical mesh networks. *Computer Communications, 30*(8), 1794–1801. doi:10.1016/j.comcom.2007.02.010

Guo, L., Wang, X., Li, Y., Wang, C., Li, H., Wang, H., & Liu, X. (2009). A novel domain-by-domain survivable mechanism in multi-domain wavelength division-multiplexing optical networks. *Optical Fiber Technology*, *15*(2), 192–196. doi:10.1016/j.yofte.2008.10.001

Ho, P.-H. (2004). State-of-the-art process in developing survivable routing schemes in mesh WDM networks. *IEEE Communications Surveys*, *6*(4), 2–16. doi:10.1109/COMST.2004.5342295

Li, T., & Wang, B. (2006). On Optimal *p*-cycle based Protection in WDM Optical Networks with Sparse-partial Wavelength Conversion. *IEEE Transactions on Reliability*, *55*(3), 496–506. doi:10.1109/TR.2006.879650

Li, T., & Wang, B. (2007). Approximating optimal survivable scheduled service provisioning in WDM optical networks with shared risk link groups. In *Proceedings of IEEE Broadnets*.

Luo, X., & Wang, B. (2009). Integrated Scheduling of Grid Applications in WDM Optical Lighttrail Networks. accepted for publication in *IEEE/OSA. Journal of Lightwave Technology*, *27*(12), 1785–1795. doi:10.1109/JLT.2009.2020997

Maier, G., Busca, C., & Pattavina, A. (2008). Multi-Domain Routing Techniques with Topology Aggregation in ASON Networks. In *Proceedings of ONDM*.

Mouftah, H. T., & Ho, P.-H. (2003). *Optical Networks: Architecture and Survivability*. Kluwer Academic Publishers.

Qiao, C., & Yoo, M. (1999). Optical burst switching (OBS) – a new paradigm for an optical Internet. *Journal of High Speed Networks*, *8*(1), 69–84.

Shen, G., & Grover, W.D., W. D. (2003). Extending the *p*-cycle concept to path segment protection for span and node failure recovery. *IEEE Journal on Selected Areas in Communications*, *21*(8), 1306–1319. doi:10.1109/JSAC.2003.816598

Somani, A. K. (2006). *Survivability and traffic grooming in WDM optical networks*. Cambridge University Press. doi:10.1017/CBO9780511616105

Szigeti, J., Romeral, R., Cinkler, T., & Larrabeiti, D. (2009). p-Cycle Protection in Multi-Domain Optical Networks. *Photonic Network Communications*, *17*(1), 35–47. doi:10.1007/s11107-008-0141-2

Truon, D., & Thiongane, B. (2006). Dynamic routing for shared path protection in multi-domain optical mesh networks. *Journal of Optical Networking*, *5*(1), 58–74. doi:10.1364/JON.5.000058

Wang, B. (2008). Traffic Grooming under Scheduled Service. In Dutta, R., Kamal, K. E., & Rouskas, G. (Eds.), *Traffic Grooming for Optical Networks: Foundations and Techniques*. Springer. doi:10.1007/978-0-387-74518-3_11

Wu, T.-H. (1995). Emerging technologies for fiber network survivability. *IEEE Communications Magazine*, *33*(2), 58–59, 62–74. doi:10.1109/35.350377

Xie, X., Sun, W., Hu, W., & Wang, J. (2007). A shared sub-path protection strategy in multi-domain optical networks. In *Proceedings of Optical Fiber Communication and Optoelectronics Conference*.

Xin, C., Wang, B., Cao, X., & Li, J. (2006). Logical Topology Design for Dynamic Traffic Grooming in Mesh WDM Optical Networks. *Journal of Lightwave Technology*, *24*(6), 2267–2275. doi:10.1109/JLT.2006.874562

Zang, H. (2003). *WDM Mesh Networks: Management and Survivability*. Kluwer Academic Publishers. doi:10.1007/978-1-4615-0341-5

Zhang, X., Liao, D., Wang, S., & Yu, H. (2010). On segment shared protection for dynamic connections in multi-domain optical mesh networks. *International Journal of Electronics and Communications*, *64*(4), 366–371. doi:10.1016/j.aeue.2008.12.011

Zheng, S. Q. (2006). SMART: an optical infrastructure for future Internet. In *Proceedings of IEEE Broadnets*.

KEY TERMS AND DEFINITIONS

Backup Resource Sharing: A technique that allows multiple protection paths to share some common wavelength-links as long as their corresponding working paths are link disjoint.

Primary/Active/Working Path: The path used for carrying working traffic when the network has no failure.

Protection: Techniques that use pre-assigned capacity to ensure survivability.

Quality of Protection: The level at which survivable services are restored.

Restoration: Techniques that re-route affected traffic after a failure occurrence using available capacity to ensure survivability.

Secondary/Backup/Protection Path: The alternate path used for carrying detoured traffic when the network suffers from a failure.

Shared Risk Link Group: A group of resources such as links that fail simultaneously due to a common fault.

Chapter 2
Protection Survivability Architectures:
Principles and Challenging Issues

Péter Babarczi
Budapest University of Technology and Economics, Hungary

János Tapolcai
Budapest University of Technology and Economics, Hungary

ABSTRACT

Survivable routing serves as one of the most important issues in optical backbone design. Due to the high data rates enabled by the wavelength division multiplexing technology, any interruption in the service results in the loss of a large amount of application data. Thus, making efforts to calculate and signal the protection resources promptly after the failure occurred would lead to an unacceptable high delay. As the main purpose of this chapter, the principles of pre-planned protection approaches in mesh optical backbone networks are discussed. The Shared Risk Link Group (SRLG) concept is introduced modeling physical and geographical dependency among seemingly unrelated link failures. Finally, methods are presented for calculating the exact end-to-end availability of a connection.

INTRODUCTION

Communication network design serves as an important issue for service providers among the rapidly changing and emerging technologies. It is particularly critical when an all-optical back-bone is in place due to its high data rate along each fiber and transparency in the data plane. The transparency – lack of O/E/O conversion at the intermediate nodes – enables very high data rates exceeding 10 or even 40 Gbps. There has been an increasing interest in providing high data-rate services such as video-conferencing or multimedia internet access recently. The rapidly

DOI: 10.4018/978-1-61350-426-0.ch002

increasing thirst for bandwidth and the spread of multicast technology provide new challenges for engineers. The persistent change of the underlying technology (e.g. Wavelength Division Multiplexing (WDM) networks, wavelength conversion capability, dynamically switched multi-layer networks) always requires new design goals and methodologies. However, the main design goals and Quality of Service (QoS) requirements of the network are permanent: low capital expenditure (CAPEX) and operational expenditure (OPEX), throughput efficiency, and survivability. *Survivability* – the capability of a network to recover ongoing connections disrupted by a failure of a network component (Mouftah and Ho 2003) – has emerged to be the most important aspect in designing the control and management planes for next-generation networks. The techniques proposed for survivability in optical network infrastructures can be classified into two general categories: *pre-designed protection* and *dynamic restoration.*

In circuit switched and virtual circuit switched mesh networks, like the extensively deployed wavelength-division multiplexing networks, one of the key quantifiable properties of survivability is the end-to-end availability provided by the network to the connection during its lifetime. Thus, we give a special attention to this metric in this chapter. In WDM networks each optical fiber carries a large number of wavelength channels modulated at even 40 Gbps, thus a short transport level interruption may lead to an enormous loss of application data. *Availability* refers to the probability of a reparable system to be found in the operational state at some time t in the future. End-to-end connection availability refers to the case when the source and destination nodes are connected by at least one path of operating edges and nodes, given that the connection was established at time $t=0$ (Sterbenz et al. 2010). Faults possibly cause the disruption of a connection if the users' data is carried only along one path in the

network, which might not be sufficient to fulfill the required connection availability defined in the Service Level Agreement (SLA) contracted between the service provider and customers. Providing optical backbone network services high connection availability is essential for service providers, as they gain more profit from higher rates on reliable transfer. In the SLA, the operator declares the minimal service conditions able to carry the customer's data in the network for a given charge. The customer states his/her required bandwidth – if it is known – and chooses one of the connection availability classes offered by the provider.

The chapter gives insight into various pre-designed protection techniques, as one possible way of designing a survivable network. First, we discuss the pre-designed single failure resilient protection methods proposed for survivable mesh networks. In order to handle the protection methods independently from the underlying technology, the edges used for carrying users' data in the absence of failures are referred to as *working (or active) path.* To deal with unexpected interruptions caused by accidental events (such as rodent bite on communication fibers), pre-planning *protection (or backup) resources* during connection setup with sufficient bandwidth for each working path has been widely accepted as the most effective solution. The strategy is also known as survivable routing in the path selection stage. For each type of protection, we briefly explain its distinctive characteristics, such as cost versus efficiency tradeoffs, resource utilization (fibers, wavelength, switches, etc.), speed of recovery, and other features that set it apart from other types.

As the main focus of the chapter, the principles of multiple link failure resilient protection methods are investigated. Most of the methods proposed for optical layer protection can handle single link and node failures. As the network components (fibers, optical cross connects (OXC), transmitters, regenerators, etc.) have high availability, single link failure is a statistically dominant effect in optical networks, thus protecting single failures can fulfill the availability requirement of a wide range of applications. On the other hand, failures rarely occur independently and targeted attacks or large-scale natural disasters resulting in dual-link or equivalent failures cause the outage of network components in the same geographic region. These failures cause intolerable disruption in the service of mission-critical applications, and they are also worthy of our attention. Therefore, the notion of Shared Risk Link Group (SRLG) is introduced to support multiple dependent failure resilient network design.

Finally, different availability evaluation methods are discussed to evaluate the performance of protection methods. In a network composed of *reparable subsystems*, the quality of service is usually measured in terms of availability of an individual connection. From the service provider's perspective, it is important to know the availability of a single connection, in order to provide different services for different rates. *Reliability* refers to the probability of the connection being adequately operational for the period of time intended in the presence of network faults. An important difference between reliability and availability is that reliability refers to a failure-free operation during a time interval, while availability refers to failure-free operation at a given instant of time. Thus, the notions are really close to each other (they are the same in non-reparable systems) as a reliable system is being found in the operational state with high probability. Accordingly, mathematical formulations are, in general, the same for network reliability if we substitute reliability parameters to availability parameters. Therefore, we introduce general approaches for calculating two-terminal reliability of a probabilistic network. As availability evaluation methods have high computational complexity, the chapter gives an insight into the applied exact and approximation two-terminal reliability evaluation methods.

BACKGROUND

Evolution of Technologies

In the case of statically configured networks, the network was provisioned, configured, maintained and supervised through the management plane via a centralized management system. Such networks were mainly designed in a point-to-point manner, and the signal was converted to the electronic domain at each node. As a second step networks were designed in a shape of ring. In these synchronous digital hierarchy (SDH)/synchronous optical networks (SONET) networks survivability mechanisms like automatic protection switching (APS) between redundant links in a point-to-point manner or SONET self-healing rings (SHR) in a ring topology were implemented. Later, due to the limited connectivity and reliability potentials, networks were deployed in the shape of a mesh in the backbone and metro networks. Mesh topologies offer high connectivity which greatly improves network reliability and design flexibility. On the other hand, because of the greater number of routing and design decisions (Maier et al. 2002) it leads to a bunch of complex problems like signaling between the nodes or the availability calculation of a connection. In opto-electronical cross-connects (EXCs) the optical signal is first converted to electrical signal then electrical space-switching is performed and finally it is converted back to optical domain again to any wavelength. By using EXCs in the network the total transparency of bit rates and signal formats is lost.

In order to improve the transmission potentials of the fiber optic cables on the same cable topology, the wavelength division multiplexing and dense wavelength division multiplexing (DWDM) technology was introduced, offering tremendous amount of bandwidth by simultaneously transmit data of multiple connections on non-overlapped wavelength channels on a single fiber. The bandwidth of a fiber link can be divided into tens (or hundreds) of non-overlapped wavelength channels (i.e., frequency channels) and each cable contains many (e.g. 20 or more) fibers. Thus, the WDM technology is expected to play a significant role in next-generation networks.

As the technology evolved and optical cross-connects (OXCs), optical add/drop multiplexers (OADMs) and photonic switches were introduced the optical signal of a WDM channel could be switched from an input port to an output port without any optoelectronic conversion. Thus, the costly and time consuming operation of electronic processing was eliminated at the intermediate nodes. In DWDM networks OXCs are used to switch individual wavelengths optically and establish lightpaths between nonadjacent nodes. A *lightpath* is an optical path established between two nodes of the network, carrying only optical signals. Two lightpaths can use the same links if and only if they use different wavelengths. In these high capacity networks there was an even growing need for a dynamically change the optical layer connectivity within milliseconds, i.e. a whole lightpaths can be deployed and released with user initiated signals within milliseconds in a distributed manner. Thus, the *control plane* (CP) was introduced in the networks, which communicates with signals to perform dynamic behavior with the other layers through well-defined interfaces (Automatically Switched Optical Network, ASON). Later, to reach finer granularity, the horizontal diversification of the network was started (Cinkler 2003). Thus, multi-layer networks emerged, from which the most promising architecture at this time is the Automatically Switched Transport Network/Generalized Multi-Protocol Label Switching (ASTN/GMPLS) (Bernstein et al. 2003). In ASTN/GMPLS networks the communicating entities could be connected on fiber, waveband, wavelength, TDM frame or packed level granularity. In such dynamic networks, the connection requests are handled independently, they are arriving and getting served sequentially, without any knowledge of future incoming requests.

In dynamic networks it is important to develop a suite of inter-operable strategies that can, in real-time, find working path and protection resources upon the current load with efficient resource utilization. However, the trade-off has to be considered in optical backbone design between the network cost and operational complexity. Optical cross-connects may or may not be equipped with *wavelength converters*, i.e. devices that transform data streams coming in at one specific wavelength into an outgoing data stream at another specific wavelength. The price of an optical wavelength converter is high, thus, most of the all-optical networks have been built without any wavelength converter. In these networks lightpaths have to be routed along the same wavelength on each link it traverses, leading to a complex routing problem. However, with the application of wavelength converters minimal cost routes can be found rapidly.

Although the methods introduced in the chapter were designed for optical layer protection in WDM networks, most of the protection methods can also be implemented at a number of layers including IP, MPLS, ATM, SDH/SONET, Next Generation SDH/SONET, ASON, and ASTN/GMPLS. Although each layer could have its own recovery schemes, they all show a rather similar succession of phases, that is, the recovery cycle (Vasseur et al. 2004). In multi-layer networks the inter-working between the layers is a challenging issue. In the case the failure is reparable at the optical layer in milliseconds, a survivable network should not allow the upper layers to take their own recovery action as it could lead to an unacceptable long interruption in the service. The first solution for this problem is the application of a hold-off timer, i.e. upper layers are delay their recovery action to allow the lower layers to repair the failure. The second approach use a recovery token signal, that is, the layer which owns the signal is responsible to recover from the failure. In the lower layers, e.g. WDM layer operating with high capacity and carrying aggregated traffic it is essential to keep the recovery time as short as possible. The ideal recovery time is considered to be less than 50 ms in next generation optical networks (Haider and Harris 2007). In this scenario, the interruption perceived by higher layers can be managed in a graceful manner.

Principles of Pre-Designed Protection Schemes

In this section, the basic notations and objectives of the pre-planned survivable network design approaches is introduced if the connection needs to be single link and node failure resilient. As we have mentioned before, the survivable methods can be classified under two general categories: pre-designed protection and dynamic restoration. Dynamic restoration approaches, when the protection resources are computed and allocated after a failure occurred, are out of the scope of this chapter. However, we note that restoration schemes are utilizing network capacity more efficiently as they do not reserve any backup resources in advance (100% resource utilization). On the other hand, these methods have slower restoration time and 100% recovery can not be guaranteed after disruptions as the protection resources are computed (and allocated) after a failure event occurs. Based on the actual load of the network it could happen, that some connections are disrupted because of insufficient network resources. Pre-designed protection approaches guarantee 100% recovery from failures against which they were designed (e.g. single link and node failures). On the other hand, some extra resources are idle in the case of the network's operational state. As one can see, no universal method exists for survivable network design. Based on operational premises, the network operator can choose the appropriate survivability technique based on the current service classes he wants to provide, depending on its properties, e.g. resource consumption, recovery time, recovery percentage. If the operator needs to recover all the connections under the considered failure scenarios, then a pre-designed protection

scheme is the appropriate choice, discussed in this chapter.

Applying protection approaches the backup resources are pre-planned, i.e. calculated before connection setup. They can be classified as path, link and segment protection based on the number of edges are protected along the working path with a single protection structure, shown in Figure 1. If a pre-designed protection scheme computes a node disjoint protection path between the end nodes of the connection, it is referred to as *path protection*. Protecting each edge of the working path individually is called *link (or line) protection*. The case where multiple overlapping (or non-overlapping) edges of the working path are protected by a single protection path is referred to as *segment protection*. The source node of the protection path/segment is called the *upstream node*, while the terminating node is called the *downstream node*. In the presence of a link failure disrupting a working path, the upstream nodes are still reachable on operating edges from the source of the working path. Similarly, from the downstream nodes the destination is still reachable on operational edges of the working path.

All of these methods have two types. In *dedicated protection* the backup resources (wavelengths, switches etc.) are dedicated to a single working lightpath, thus, they can be reserved and configured at connection setup (hot stand-by) and used till the connection is torn down. Or they can be *shared*, in which case spare resources can be used to provide protection to multiple working paths. Dedicated and shared protection schemes have their main differences in the amount of spare resources reserved for a connection, the signaling complexity, and the recovery time of the traffic after a failure occurred. The service provider's goal is to maximize the number of customers to gain more income, while minimizing the total resources allocated for a single user but still maintaining the required service level. Obviously, allocating more dedicated protection paths and path segments for a connection, higher con-

Figure 1. Classification of pre-designed protection schemes in optical mesh networks

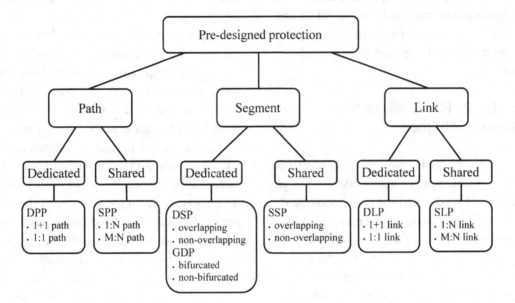

nection availability can be perceived by the customer at the expense of more reserved spare resources. Therefore, a trade-off has to be made between availability and resource consumption. In the case of single link failure resilient network design, a straightforward idea is to share the protection resources among users with disjoint working paths (a single link failure affects at most one of the working paths). However, after the failure has occurred, signaling is required between the upstream and downstream nodes of the path or the segments affected by the failure to reserve the protection resources. Thus, for the price of efficient resource utilization service recovery time is longer in comparison with that of dedicated protection. Dedicated protection is favored for its simplicity compared to shared protection. The complexity of shared protection lies firstly in the signaling efforts in case of a failure, and secondly, in computing the appropriate working and shared protection paths during connection setup.

In the following some properties are mentioned along the protection methods can be further divided into several groups. If the same protection resources are used regardless of the failure it is called *failure independent strategy*. Another possible approach is the *failure dependent protection scheme*, that is, if the working path consists of *n* links, then there are (at most) *n* different backup paths, which protection resources can be shared with themselves or with other protection paths. Finally, protection (and restoration) approaches can be differentiated on whether they release the still functioning segments of the working path (called *stub-release*) after it is affected by a failure or not. With the application of stub-release often considerable amount of network resources can be free up which can be used as spare capacity in the restoration process.

Single Failure Resilient Dedicated and Shared Protection Approaches

In most of the current networks dedicated protection is used, in which the protection paths are operational for the whole duration of the connection, avoiding the technical difficulties of setting up the protection routes promptly after the failure of the working path is perceived. In dedicated protection, each working and protection path pair is pre-configured, and is launched with the same copy of data between a source-destination pair at the same time during a normal operation. While dedicated protection provides a very fast restoration service, the resource efficiency is really poor, as protection paths are longer than working paths.

When using dedicated protection (since the allocated protection resources are dedicated) it is possible to send customer's data along all paths simultaneously; therefore, when the switching node (one of the downstream nodes of the path/ segment) senses the degradation of the signal on the working path it only needs to switch over listening to the protection path without any signaling. In the case of *dedicated path protection (DPP)* the entire working path is protected with a single node disjoint protection path (Ramamurthy et al. 2003). Because of its simplicity, *DPP is the most widely deployed protection method in current optical backbone networks.* If 1+1 protection is applied, an optical splitter is used at the source node in order to transmit user data on both the working and protection paths simultaneously. If a failure occurs along the working path, at the destination node the signal degradation on the working path is detected and only requires switching to the protection path without any signaling in the control plane. Thus, 1+1 DPP offers a rapid restoration. However, it does not use protection resources efficiently. To overcome this difficulty, 1:1 DPP is an option for use protection resources to carry low priority data while the working path is operational. If the working path is disrupted, signaling is needed for both the source and destination nodes (both are switching nodes in this case) to switch over to the protection path. Although dedicated path protection was designed to be single failure resilient, it is able to provide recovery from some multiple link failures, as only those multiple failures cause disruption which affects the working and protection paths simultaneously (Fumagalli and Valcarenghi 2000). As protection paths are usually much longer than primary paths, considering transmission and switching impairments

and the degradation of the transmitted signal at the protection path selection could achieve better dual failure restorability (Georgakilas et al. 2010) than the methods based on e.g. the hop count.

Although, in telecommunication networks the links and nodes are fairly reliable - their availability values are close to one -, the connection availability depends on all the single links the connection is traversing (discussed later), which is result in a lower perceived connection availability. If the provided availability for mission-critical applications like remote-surgery using only disjoint paths for protection does not fulfill the availability requirement stated in the SLA, the connection will be blocked. Thus, it is an obvious idea that the working and protection paths are not constrained to be disjoint to provide higher availability services and admit those demands in the network which would be blocked with DPP. Such an approach, called *Dedicated Segment Protection (DSP),* has been well investigated (Xue et al. 2007). In this protection method the working path is divided into segments, and each (overlapping or non-overlapping) segment is protected independently with one or multiple protection routes. Note, that the number of overlapping segments can be exponential in the number of the edges along the working path, thus, considering a large number of segments can lead to really poor resource utilization.

Dedicated link protection (DLP) refers to the case when each link of the working path is protected with its own protection path. It could be treated as the borderline case of DSP when each working path segment contains only one link. DLP reserves protection wavelengths between the end nodes of each link utilized by working paths. Similarly to DSP, dedicated link protection may require more spare capacity allocation than DPP. On the other hand, because the switching between the upstream and downstream node of the connection can be performed locally (the adjacent nodes to the failed link), 1:1 DLP can be even achieve better restoration time than 1:1 DPP.

In the case where the main objective of survivable network design is minimizing the cost of service (i.e. the reserved bandwidth), shared protection, which allows spare capacity to be shared by a number of working paths, is more appropriate than dedicated protection. However, the restoration involves more processing for signaling and establishing the cross-connections along the restoration path. There is thus an evident tradeoff between capacity utilization and recovery time.

Using shared protection for single failure resilience, also referred to as *1:N*, the spare capacity taken by backup links are only reserved but the switches are not configured during the normal operation. Therefore, the spare capacity can also be used by some best-effort traffic that can tolerate a service interruption. In case of a failure interrupting a working path (a fiber-cut or loss of signal due to the failure of any network element), the switching fabric structures in the nodes along the corresponding protection path are configured by prioritized signaling followed by a traffic switchover to recover the original service supported by the working path. After a failure occurred and the common pool of spare resources is occupied by a connection, the other working paths are unprotected until the original working path is reestablished or the protection is reconfigured.

In *Shared Path Protection (SPP)* (also termed *1:N* or *M:N* path protection) only the N working paths are launched with data flows (Lucerna et al. 2009) and a common pool of spare resources (1 or M) is calculated and reserved without the necessity of being pre-configured. These resources are shared with the other protection paths. It has been observed that the spare capacity resource sharing between different protection paths can substantially achieve better capacity efficiency required to achieve 100% restorability at the price of more complex control. If the source and destination nodes of the N connections is distinct, signaling is required between all nodes after a failure occurred, thus, this approach leads to a

longer restoration completion time (on the order of 100 ms) (Fumagalli and Valcarenghi 2000). The studies on SPP do not impose any limitation on the length of the backup paths, and may impair the overall performance. It is notable that the most of the SPP methods are Active-Path-First-based (Xu et al. 2002) (which means that the working path is derived first), in which little attempt has been made to jointly get the working and shared protection paths.

In addition to the end-to-end protection (similarly to the dedicated case), the study in (Ho and Mouftah 2002) suggest to segment each working path and find a protection path segment for each working path segment to reach higher connection availability. This is called *Shared Segment Protection (SSP)*. A framework called Short Leap Shared Protection (SLSP) along with a dynamic algorithm called CDR (Cascaded Diverse Routing) is proposed to perform segmented shared protection, in which the enumeration of *k*-shortest paths in each segment of the working path is performed. In addition, the segment-based shared protection takes extra signaling efforts and is imposed of high requirements on the hardware responsiveness.

Shared Link Protection (SLP) (also termed *1:N* or *M:N* link protection) applies the SSP technique locally to the failed link. In both SSP and SLP better resource sharing can be achieved due to the fact that shorter segments of the working paths are more likely disjoint. Thus, they can share the spare resources more likely. In this case, it is possible to use shared protection resources among different segments of the same working path too, thus yielding better resource utilization than achieved in DLP. The restoration time of SLP is generally faster than that of SPP because of the same reasons as DLP faster than DPP.

A promising way in survivable network design is deploying pre-configured cycles or *p-cycles* in the network. P-cycles are virtual rings formed in the spare capacity of the network in order to achieve ring-like restoration speed (50-150 ms) (Haider and Harris 2007). Originally, a set of rings that covers every link in the mesh at least once was searched for. On these rings the former SHR protection technique proposed for ring topologies was adopted. Later, the method was fully adopted in the mesh optical network design, and in addition to protecting on-cycle links, p-cycles can also protect straddling links (i.e. links between non-adjacent nodes of the cycle). Thus, it supports local recovery and high degree of node and multiple failures survivability.

PROTECTION SCHEMES AGAINST MULTIPLE FAILURES

The Corresponding Graph Representation of a DWDM Network

The aim of this section is to introduce the graph representations built up on optical networks, which serves as the input of resilient routing algorithms. Optical networks architecturally have two layers: the *physical layer* and the *optical layer*. The physical layer consists of fibers and OXCs, while the optical layer consists of optical links (lightpaths) and the corresponding nodes from the physical layer where lightpaths terminate. In contrast to static configured networks, assuming dynamically switched networks, lightpaths can be established within milliseconds between arbitrary pairs of nodes in the network. Thus, we introduce the graph representation of the physical layer (OXCs and fiber links), where an arbitrary path could be on optical link in the optical layer.

Figure 2 presents an example of the graph representation $G=(V,E)$ with a set of links E and nodes V for an optical network. The nodes of the graph represent Optical Cross-Connects or Optical Add-Drop Multiplexers, where connection demands can enter and leave the network. In most of the approaches, the network nodes are assumed to be fully reliable (have an availability equal to one). An undirected edge (representing bi-directional fiber links between adjacent nodes)

of the graph corresponds to an *Optical Multiplex Section (OMS)* of the network between two OXCs, and the cost function c_j on edge j corresponds to the cost of allocating a unit of demand flow (i.e. wavelength) on that particular edge. Cost function c_j may represent the length of the link (the number of optical amplifiers (or *Optical Transmission Sections (OTS)*), or signal quality degradation on long links. The connection between the first and last OXC in the optical domain is called the *Optical Channel (OCh)* section. An OCh is represented as a path in the graph.

In practical applications, often more network features are required. First of all, for certain communication network models, instead of bi-directional fiber links, we may need to consider directed links (arcs) and similarly, directed demands, and directed or bi-directed link capacities. Furthermore, in some practical applications the assumption of fully reliable nodes is not an appropriate model. Thus, node failures may have to be considered in addition to link failures. Node failures can be simulated by link failures in an auxiliary graph, where the node splitting technique is applied (Orlowski and Pioro 2009). First, each undirected edge is replaced by a pair of anti-parallel arcs. Secondly, every node v is split into two nodes v' and v'' connected by an arc $v' \rightarrow v''$. Each incoming edge of v is then directed to v', while each outgoing edge of v is directed from v'', as shown in Figure 3(b).

In dynamically switched networks, connection requests arrive one after the other without any knowledge of future arrivals. In such a scenario the general goal is to develop a suite of inter-operable strategies that has a superb overall performance with low blocking probability, short average, and maximal waiting time of establishing connections, and low network utilization. In the working path selection stage Dijkstra's shortest path finding algorithm (Dijkstra 1959) is the most commonly applied method, which uses the cost function on the edges of the underlying graph. Thus, setting the cost function on the edges prop-

Figure 2. (a) Optical channel, multiplex and transmission sections; (b) the corresponding graph representation

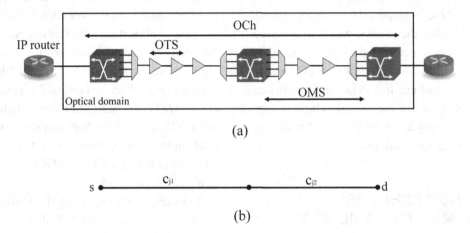

(a)

(b)

Figure 3. (a) Example network with cost function $c_{j1} = c_{j4} = 3$; $c_{j2} = c_{j3} = c_{j5} = 1$; (b) Node splitting technique on node v; (c) Edge contraction technique on edge j_3

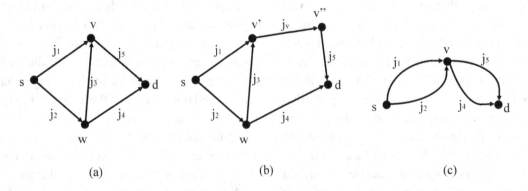

(a)   (b)   (c)

erly could be used for routing the traffic on those parts of the network where sufficient resources are available, while avoiding network components the free capacities of which are scarce. Applying this method (called *Traffic Engineering (TE)*) can lead to lower blocking probability and can increase the overall network performance. A very common idea in TE is to use load balancing functions, which set the weights on the links (c_j) according to the topology and traffic characteristics such that a good overall performance can be expected

using capacity-efficient routing algorithms for each connection request.

Finally, each edge has a capacity function corresponding to the available bandwidth on the given link (e.g. the free wavelength channels). The total capacity of link $j \in E$ of the graph G could be categorized into the following three types (see Figure 5(a) for details):

- **Working capacity** (denoted as q_j) which is the link capacity already taken by some working paths, and cannot be taken used until the corresponding working paths are torn down,

- **Spare capacity** (denoted as v_j), which is the link capacity reserved by some backup paths,
- **Free capacity** (denoted as f_j), which is the unreserved link capacity that can be reserved as either working or spare capacity.

In the single failure scenario the task is to find a working path (the set of edges along the working path is denoted by W) and a protection path (the set of edges along the protection path is denoted by P) between the source s and destination node d with the required bandwidth b. Depending on the applied protection scheme different constraints need to be satisfied by the solution. In the case of dedicated protection the working and protection paths can use any links with a sufficient free capacity. For shared protection, the constraint of spare capacity sharing must be investigated upon each network link before the best protection path can be derived for a working path. Whether or not a link has *sharable spare capacity* for a protection path depends on the physical location of the corresponding working path. This is also known as the *dependency* of the protection path on its working path.

Although different protection approaches require different algorithms and different auxiliary graphs to get a working and protection path pair, finding a disjoint pairs of paths between two nodes (often referred to as *diverse-routing*) in the network is the basis of the previously introduced single failure resilient schemes. In the diverse-routing problem the task is to find a link-disjoint pair of paths between two nodes of the graph. On the stipulation of resource availability and dependencies of the applied protection method, Suurballe's algorithm solves the diverse-routing problem in the graph with the modified cost function c_j. *Suurballe's algorithm*, first reported in the early 70's (Suurballe 1974) is famous for its polynomial computation complexity (originally $O(n^2 \cdot logn)$ time) in finding optimal disjoint pairs of paths in terms of cost sum of the two paths in a directed graph. It is notable that the algorithm uses the same suite of link-state to derive the two paths. Suurballe's algorithm finds the minimal cost disjoint pair of paths among all pairs of paths in the network (if exists). Finding a disjoint working and protection pair of paths with Suurballe's algorithm also avoids the *trap situation* which could happen due to greedily selecting the shortest path in the network as the working path, and as a second step a disjoint protection path is computed. For instance, a trap situation could occur in Figure 3(a) if the working path is selected with Dijkstra's algorithm ($s \to w \to v \to d$). In this situation, removing the edges from the topology (in order to get edge-disjoint working and protection paths) result in an s - d cut, thus, the connection is blocked as there is no disjoint pair of paths providing the required availability level. However, if Suurballe's algorithm is used for finding disjoint pairs of paths, it will return $s \to w \to d$ and $s \to v \to d$, and the connection can be established.

Note that the computational complexity of the diverse routing problem mainly depends on the wavelength conversion capability of the OXCs. If the nodes are capable of converting the wavelengths, the problem of finding a minimum cost edge and (except for the source and destination nodes of the connection) node disjoint working and protection paths is polynomial time solvable with Suurballe's algorithm. On the other hand, if the OXCs are unable to convert the wavelength i.e. the *wavelength continuity constraint* (Chlamtac et al. 1992) holds along the lightpath then the problem of finding two edge-disjoint lightpath in the network is NP-complete, both for the dedicated (Andersen et al. 2004) and the shared case (Ou et al. 2004).

Most of the restoration architectures are designed assuming statistically independent single failure cases, which is not adequate in present day networks. This simplification comes from the assumption that the probability of each physical conduit to be subject to a failure is small and thus can be regarded as independent events even under the single failure scenario. However, dual failures are the most significant effects of disruptions in a single failure resilient network. Modeling multiple failures purely at the graph representation, failure states (Orlowski and Pioro 2009) can be defined. At this representation we concentrate failures in a single layer of the network, e.g. in this example on the element failures in the physical layer. In this case, as the input of the routing problem a list of *failure scenarios* is given, and the connection needs to be resilient against all failures in the list. For this, we introduce a set $S \subseteq 2^E$ of *network states* each of which corresponds to a subset of failing links. Set S is called the failure scenario. It is assumed that S contains the normal, failure-less state $\emptyset$ in which all links are operational. The set $S^*=S\backslash\emptyset$ contains the *failure states (FS)* in which at least one link fails and each FS has a probability value that the corresponding failure state occurs. The number of states is exponential in the size of the network. In optical networks, the network elements have quite high availability. Therefore, in survivable network design failure states containing more than two or three elements are not worth our attention. Thus, the number of states is reduced to be polynomial with respect to the network size. If the states are assigned with the probability corresponding to the given dependent failure scenario measured by the network operator rather than the probability calculated from the independent individual link availability values; then the failure state approach can model failure dependencies, too. Single link failure resilience could be treated as the special case of the failure state model (i.e. each single network element in the network topology serves a failure state).

Shared Risk Link Groups

At the graph representation level, the layered structure of the WDM optical network, the topology lay-

out (e.g. physical location of the cables, common threats for multiple fibers) is lost. However, in an accurate network model, these properties have to be considered in a resilient network design. One of the possible ways of handling dependent multiple failures in optical networks uses *Shared Risk Link Groups (SRLG)* (or Shared Risk Resource Group or Shared Risk Group) (Strand et al. 2001; Ellinas et al. 2003; Papadimitriou 2002). SRLG expresses statistical dependencies between failures, that is, a group of network elements (i.e., links, nodes, physical devices, software/protocol identities, etc, or a mix of them) possibly subject to the same risk of single failure. In practical cases an SRLG may contain several seemingly unrelated and arbitrarily selected network elements. For instance, two links belong to the same SRLG if they share the same tunnel or conduit. Based on the observations at AT&T (Strand et al. 2001) a link may belong to over 100 SRLGs, each corresponding to a separate fiber group. In (Strand et al. 2001), SRLGs are characterized by 2 parameters. *Type of compromise* refers to the shared risk (e.g. shared fiber cable, shared conduit, etc.). The *extent of compromise* expresses the length of the sharing. The mapping of links and different types of SRLGs is in general defined by network operators based on the definition of each SRLG type. Links belong to the same SRLG type because they are in the same *physical hierarchy*, which is related to the fiber topology (more generally the physical resources) of the optical network including the lightpaths built on top of this physical topology, or *logical hierarchy*, which is related to the geographical topology of the network (Papadimitriou 2002).

The failures like cable cuts and OXC failures occur in the physical layer. However, in the physical hierarchy circuits are routed in the *optical layer* on optical links (lightpaths). Thus, an optical link failure could be affected by multiple link or node failures in the physical layer and belongs to those SRLGs. An example of possible SRLGs is defined in Figure 4. Since link failures in the optical layer are not mutually independent, the overall availability of a lightpath in the optical layer is lower than if assuming independent failures and leads to an inaccurate end-to-end availability value. In order to achieve a precise availability evaluation, these failure dependencies should be considered at the path selection stage and considering multiple failures and dependencies among failures allow us to develop efficient routing methods. In addition, since SRLG relationships are not necessarily self-discoverable and do not change dynamically, they don't need to be advertised by network elements. It can be configured in some central database and be distributed to or retrieved by the nodes. On the other hand, the information about link failure dependencies of SRLGs in the same logical hierarchy is inaccurate even at the service provider - who may have a long list of historical failure events, since they can only expect possible failures (e.g. disruptions in the same geographic region because of earthquakes, floods, etc.) in the future with the measured probability values. This makes the SRLG model hard to use in practice.

The presented SRLG model assumes that once an SRLG failure event occurs, all of its associated links fail simultaneously. However, this deterministic failure model can not describe e.g.

an event of a natural disaster, where some, but not necessarily all links in the vicinity of the disaster may be affected. There are promising ways of generalizing the notion of an SRLG to account for probabilistic link failure, called Probabilistic SRLG (Lee and Modiano 2009).

In contrast to single link failure resilience, when a general definition of the SRLGs is desired, a more complicated description and further elaborations are required to achieve an efficient implementation of any survivable routing algorithm for dedicated and shared protection. This is because an SRLG could contain a wide range of number and type of network elements. These elements are mainly overlapped and/or contained by other elements; thus these routing problems are mainly NP-complete. Therefore, most of the solutions proposed for this problem are either optimal (e.g. Integer Linear Programming (ILP) formulations) and slow, or fast, but do not give optimal solution for all problem instances (e.g. heuristic or approximation methods). The rest of the section provides a survey of these approaches. Throughout the remainder of the section, without loss of generality, we assume that *a single failure event corresponding to an SRLG arrives at a time*.

Figure 4. SRLGs defined on an example network; the two working paths W_1 (between source s_1 and destination d_1) and W_2 (between s_2 and d_2) are link disjoint, but they are involved in a common SRLG (namely $SRLG_4$)

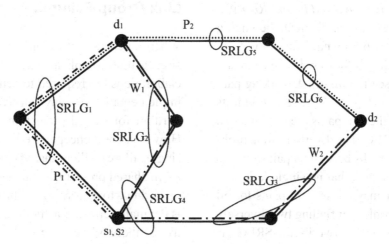

In the case when two simultaneous failure events (corresponding to two SRLGs s_1 and s_2) need to be considered, the two failure events will be redefined as a single failure event, and the links in s_1 and s_2 will be taken as a new SRLG (i.e., s_3), which is further considered in the approaches. For a connection demand between source node s and destination node d the SRLGs in the set could be categorized as follows (Tapolcai 2005):

- *Considered SRLG*: An SRLG belongs to this type if the network still remains $s - d$ connected after the failure occurs, that is the connection can be restored. In other words, the failed elements in the SRLG do not form a cut in the network topology; in this case, the working path affected by the failure is restorable.
- *Cut SRLG*: An SRLG belongs to this type if the source and destination nodes are in multiple isolated fragments when the network is attacked by a failure. In other words, the failed elements in the SRLG form a cut between the source and the destination node. Thus, the interruption upon the associate working paths can never be restored.

The cut SRLGs can not be protected with any survivable routing method, thus, the given SRLG list always contains considered SRLGs. We define that a *working path is involved in an SRLG* if it crosses any of the network elements belonging to that SRLG. Two working paths share the same risk of a single failure if they are involved in any common SRLG (see Figure 4.). A working path is said to be *SRLG disjoint (or diverse)* with its protection path if the two paths are not involved in any common SRLG. The diverse routing problem is to find two paths between a pair of nodes in the optical layer such that no single failure in the physical layer may cause both paths to fail (Hu 2003). The problem of finding two diversely routed paths in optical networks for SRLGs is much more difficult than the traditional edge/node disjoint path problem in graph theory. For the single link failure case, finding link and node disjoint path-pair with wavelength converters is polynomial time solvable (e.g. Suurballe's algorithm). However, if an arbitrary set of links can belong to the same SRLG, then the problem of finding an SRLG disjoint path pair between a pair of nodes in the network is NP-complete. Essentially, the difficulty of 1+1 SRLG-diverse routing arises because the architecture allows SRLGs to be defined in arbitrary and impractical ways which intuitively forces an algorithm to enumerate (a potentially exponential number of) paths in worst-case (unless $P = NP$). In (Ellinas et al. 2003), an auxiliary graph is used, in which each SRLG type is expressed as a subgraph. Applying these representations of the SRLGs considered in the input of the routing problem, the SRLG diverse routing problem could be solved with traditional edge/node disjoint path finding algorithms in the auxiliary graph. As expected from the general definition of SRLGs and from the high computational complexity of the SRLG diverse routing problem, for some complicated types of SRLG there is no feasible graph representation. Thus, some of the routing computations are not physically feasible.

Shared Path Protection (SPP) against Shared Risk Link Group Failures

When a general definition of the SRLGs is desired, a more complicated description and further elaborations are required to achieve an efficient implementation of any survivable routing algorithm for shared protection (Tapolcai 2005). However, the concepts are transferable for the single failure resilient shared protection schemes.

In shared protection schemes, the SRLG disjointedness for a working and protection path-pair is the major effort of achieving 100% restorability for the working data flows under the single

Figure 5. (a) An illustration of the categories of capacities along link j; (b) The possible situations of a different cost function defined in Eq. (2) (Ho et al. 2004)

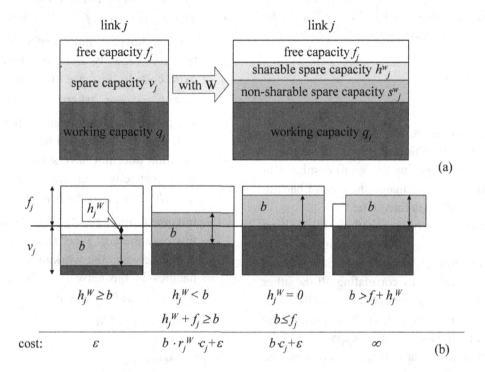

(a)

(b)

SRLG failure scenario. In addition, the SRLG-disjointedness must be kept for the working paths whose protection paths share spare capacity, such that the restoration will not be subject to resource contention when two working paths are interrupted by a single failure. In other words, whether or not two protection paths can share a spare capacity depends on the physical location of their working paths. A simple example is shown in Figure 4., where W_1, P_1 and W_2, P_2 form two working and protection path pairs. The protection path of W_2 should exclude the possibility of using any of the spare capacity taken by P_1 because both W_1 and W_2 are involved in $SRLG_4$. Thus, they share the same risk of a single failure. On the other hand, both W_1, P_1 and W_2, P_2 working and protection path pairs are SRLG disjoint.

Now we define the cost function of a connection for non-bifurcated demands (unsplittable flows). It represents the weighted capacity allocation and is composed of the cost of the working path and the protection path. The task is to find a working path and protection path between the source s and destination node d with bandwidth b. The cost of each link j in the network is denoted by c_j and is computed by the load balancing function.

In order to find a working path a cost function on the set of links is defined as follows:

$$c_j^W = \begin{cases} b \cdot c_j + \varepsilon & \text{if link } j \text{ is reservable} \\ \infty & \text{otherwise} \end{cases},$$

(1)

where c_j is the cost for each unit of bandwidth taken by working paths along link j, and ε is a small number defined as

$$\varepsilon = \frac{\min_{j \in E}(c_j)}{|E|}.$$

The fact that link j is not reservable by W can only be due to $b > f_j$, where f_j is the free capacity along link j (as illustrated in Figure 5(a)). The total cost of the working path is

$$\sum_{j \in W} \left\{ b \cdot c_j + \varepsilon \right\}.$$

Instead of using ∞ as a cost, it is more common to define feasible conditions. The purpose of additionally imposing the small number ε in the cost function is to match the cost of backup path, which will be defined later.

For finding the backup path of W, we first need to define the corresponding spare link-state and cost functions, which are specific to W, and can only be derived by correlating all the other working paths on the same SRLG with W. Spare link-state refers to the link-state assigned to the protection path (sharable spare capacity, and the non-sharable spare capacity), which depends on the corresponding working path.

With the presence of W, the spare capacity along link j can be further categorized into the following two types:

- ***Sharable spare capacity*** (denoted as h_j^W), which is the link capacity that has been reserved by some other backup paths, and is sharable with the backup path of W.
- ***Non-sharable spare capacity*** (denoted as s_j^W), which is the link capacity that has been reserved by some other protection paths, and is not sharable with the protection path of W due to the SRLG constraint. Note that $v_j = s_j^W + h_j^W$, which is the total spare capacity along link j.

The protection path may traverse link j in any one of the following three states:

- The case where the link has sufficient sharable spare capacity (i.e., $h_j^W \geq b$, in which

the backup path can take this link with the smallest cost (denoted as ε);
- The case where $f_j + h_j^W \geq b > h_j^W$, and the backup path must partly (or totally) take free capacity along this link with an extra cost. In this case the spare link-state is $b \cdot r_j^W \cdot c_j + \varepsilon$, where r_j^W is a $[0,1]$ scaling parameter determined by the location of W and will be defined later.
- The link does not have sufficient sharable spare capacity and free capacity (i.e., $f_j + h_j^W < b$, thus the backup path cannot traverse through this link by any means. In this case the cost is ∞. Due to the dependency between the working and spare capacities in the network, *the parameters r_j^W, h_j^W, and s_j^W cannot be defined until the presence of W*.

r_j^W is defined as

$$r_j^W = 1 - \frac{h_j^W}{b}$$

for any link $j \in E$. It is clear that r_j^W is 1 if there is no sharable spare capacity available along link j and it approaches 0 if h_j^W is close to b. In the former case (i.e., the case of $r_j^W = 1$), the cost for the backup segment to take this link is $b \cdot c_j + \varepsilon$ which is the same as that for the working path since all the reserved bandwidth has to be from the free capacity region as shown in Figure 5(b). The spare link-state for the protection path of W can be expressed as:

$$c_j^P = \begin{cases} \varepsilon & \text{if } h_j^W \geq b \\ b \cdot c_j \cdot r_j^W + \varepsilon & \text{if } h_j^W + f_j \geq b > h_j^W \\ \infty & \text{if } h_j^W + f_j < b \end{cases}$$

$$(2)$$

Figure 5(b) shows the three situations defined in Equation (2). In the left of Figure 5(b), the backup segment can have all b in the sharable spare capacity region; therefore, the cost is ε, as shown in the first condition in Equation (2). In the center of Figure 5(b), the backup path of W may partly take the free capacity region and the sharable spare capacity region; therefore, the link cost is $b \cdot r_j^W \cdot c_j + \varepsilon$, which is shown in the second condition in Equation (2). In the right side of Figure 5(b), the link cost is infinity because the backup path of W cannot be supported by the residual capacity of the link, which is shown in the third condition in Equation (2). Note that the protection path is assumed to take sharable spare capacity along a link whenever there is any sharable spare capacity available. If there is not enough sharable spare capacity along this link to cover the total bandwidth demand for protecting W (i.e., b), the backup path takes free capacity after considering all the sharable spare capacity.

Note that the adoption of the small constant ε is to keep the continuity between the first and second conditions in Equation (2). In this case, the cost of link j is set to ε as $h_j^W = b$ for both of the conditions. This is also the reason why we impose ε in the cost function for the working path shown in Equation (1), in which the costs of the working and backup path segments to take free capacity are the same.

Our objective is to determine c_j^P in Equation (2) - the spare link-state that defines the cost of the backup path of W passing through link j, in which h_j^W is the only variable that must be figured out (or equivalently, s_j^W since $v_j = s_j^W + h_j^W$). Note that h_j^W and s_j^W are network-wide link-state specific to the presence of W. The *spare provision matrix* is denoted as $\underline{\underline{S}}$, which is an $|E| \times |SRLG|$ matrix and the entry *(j,i)* of $\underline{\underline{S}}$ of (denoted as $s_{j,i}$), stores the amount of non-sharable spare capacity

along link j for the protection path if the corresponding working path is involved in the ith entry of SRLG. The most straightforward way of obtaining the matrix of $\underline{\underline{S}}$ is to simulate the failure of each SRLG and measure the amount of restoration traffic on each link. With the single failure scenario, only one SRLG could possibly be subject to an interruption at a moment. Thus, we can derive s_j^W for $j \in E$, by finding the maximum demand of spare capacity among all the SRLGs traversed by W, i.e.,

$$s_j^W = \max_{l \in W}\left(s_{j,l}\right).$$

The input for a new connection demand between arbitrary node pairs is the graph representation introduced for the topology with the cost functions calculated for the working and protection path selection stage, the list of SRLGs and the spare provision matrix $\underline{\underline{S}}$.

The following conditions have to hold for a solution in a network:

1. A working path (containing edges W) for the connection with capacity requirement b existing so that $\forall j \in W : f_j \geq b$ is feasible,
2. A protection path (containing edges P) for the connection with capacity requirement b existing so that $\forall i \in P : f_i + v_i \geq b + \max_{\forall j \in W} s_{i,j}$ is feasible,
3. Path W and P are SRLG disjoint.

The *objective* of the routing problem is to minimize the total cost of the used edges in the working or any protection path, formally:

$$\min\left(\sum_{j \in W}\left\{c_j^W \cdot b + \varepsilon\right\} + \sum_{j \in P}\left\{b \cdot c_j^P \cdot r_j^W + \varepsilon\right\}\right).$$

Generalized Dedicated Protection (GDP) against Shared Risk Link Group Failures

Dedicated protection schemes are less sensitive in the general definition of SRLGs than shared protection approaches. In dedicated protection, the whole spare capacity is dedicated (non-sharable, $v_j = s_j^W$). The only thing that needs to be satisfied is that at least one viable path is operational on the working and protection links between the source and destination node in a given set of SRLGs. In *Generalized Dedicated Protection (GDP)* (Babarczi et al. 2009) the primary goal is to assign an arbitrary subgraph dedicated for the connection using minimal bandwidth while all failures are survived in the predefined SRLG list *F*, instead of finding an SRLG disjoint working and protection path pair. However, the complexity of the GDP problem remains NP-complete similarly to the methods finding SRLG disjoint path pairs. Again, as the part of the input of the GDP problem a list of SRLGs is given, in which the SRLGs are defined according to operational premises, and the connection needs to be resilient against all failures in the list to satisfy the QoS requirements declared in the SLA.

In such an environment, *GDP builds up a working path and a protection structure in the form of an arbitrary subgraph, in which in all considered SRLG failure states the connection is operational*. GDP could be treated as a generalized dedicated segment protection combining the flexibility of protection structure design of DSP while keeping the amount of reserved spare resources on a considerable level like SSP or DPP. From another point of view, GDP is similar to a failure dependent shared protection approach, where the resource sharing is allowed only with its own spare resources, because contrary to shared protection, the spare resources will be calculated and signaled at connection setup (hot stand-by). Therefore, in GDP, the protection segments in

the auxiliary graph correspond to the subsequent failures (for each considered SRLG *a* an auxiliary graph $G_a=(V,E_a)$ is obtained by deleting the failed edges from *E*, where E_a denotes the links operational in SRLG *a*) are not computed independently. The already allocated dedicated spare resources to the connection can be shared without any additional cost at the protection of further segments (if the given link is operational in the corresponding SRLG to the given segment, i.e. it is in E_a). Thus, we can redefine the sharable spare capacity:

- ***Sharable own spare capacity*** (denoted as $h_j^{P(k)}$), which is the link capacity that has been reserved by the connection's working path, or by the previously computed backup segments *i*, $\forall i \leq k$, where *i=0* corresponds to the working path (failure-less state).

$r_j^{P(k)}$ is defined similarly to the shared case:

$$r_j^{P(k)} = 1 - \frac{h_j^{P(k)}}{b}.$$

The cost function for non-bifurcated demands for the $i+1^{th}$ segment $\forall j \in E_a$, for the corresponding SRLG to the segment:

$$c_j^{P(i+1)} = \begin{cases} 0 & if\ h_j^{P(i)} = b \\ c_j \cdot b \cdot r_j^{P(i)} & if\ f_j \geq b > h_j^{P(i)} \\ \infty & if\ f_j < b \end{cases}.$$

The input for a new connection demand between arbitrary node pairs in the graph representation is the same as we have seen at the previous algorithms, with the above cost function defined on the edges. We need that both the required capacity $b \in N$ and the free capacity $\forall j \in E : f_j \in N$ are natural numbers, in order to guarantee the convergence of the GDP approaches. In the generalized dedicated protection implementation in (Babarczi et al. 2009) wavelength conversion capable OXCs are assumed and wavelength continuity constraint is not considered. As the generalization of dedicated protection methods, each OXC is able to split an incoming wavelength into multiple outgoing channels (only the source node in DPP), and switch between two data fragments arriving on multiple incoming ports and sending out on one outgoing port (only the destination node in DPP).

As a future direction, we can introduce a new functionality for the OXCs called *network coding* capability. Network coding capability means that the OXC is able to perform primitive algebraic operations on the optical data, such as addition and multiplication of two optical data streams. The idea of network coding was first introduced in (Ahlswede et al. 2000) for single-source multicast. They showed that with network coding the achievable multicast capacity equals the minimum of the maximal unicast flows from the source to the receivers. In (Koetter and Medard 2003) robust network codes to a given set F of failure patterns (or failure states, or SRLGs) for multicast were proposed, i.e. the switching configurations at the network nodes remain unchanged even in the presence of failures. For each f in F an auxiliary graph G_f is constructed, that is obtained by deleting the failing links (similarly to the SRLG graphs are obtained). Network coding is widely investigated for wireless networks, and it gains an even growing interest in optical environment (Kamal 2010). The shared concept of *1:N* protection is generalized to *1+N* protection, where in contrast to *1:N* protection the spare resources are in hot-stand-by similarly to 1+1 protection. On the shared spare resource network coding is performed, and the linear combination of the input symbols of the N working paths is sent. In the following definition of the GDP method, algebraic operations of network coding are allowed on the working path besides the protection path, generalizing the concept of *1+N* protection:

1. A working path in $G=(V,E)$ (containing edges W) for the connection with required bandwidth b exists,
2. Protection path(s) (containing edges P) $\forall\, G_a = (V, E_a)$ for all f in the considered SRLG set F between s and d with larger minimal cut than b exist,
3. Each fragment of costumer's data must reach the destination in all considered SRLG state.

On the other hand, if only splitting and switching capability is available at the OXCs, and algebraic operations are not allowed, the routing problem described by condition (1)-(3) need to be supplemented with the following condition to get a valid solution (see Figure 6.):

4. Bandwidth is required to be non-bifurcated, i.e. b units of bandwidth are reserved on all edges among W and P.

The *objective* of the routing problem is to minimize the total cost of the used edges in the working or any protection path in all considered SRLG f:

$$\min\left(\sum_{j \in W,P} c_j \cdot h_j^{P(f)} \right).$$

Note, that the problems of selecting working and protection paths are closely coupled. If an optimal solution is sought both problems should be solved simultaneously. However, the protection structure assignment could take place after the working path is computed. By employing network coding capable OXCs we can solve the routing problem with the minimal consumed bandwidth. In the case of generalized dedicated protection with network coding instead of simply switching the optical signals at each OXC, some OXCs forward the linear combination of the incoming traffic (i.e. performs algebraic operation over a given Galois Field, e.g. exclusive or (*XOR*)

Figure 6. Condition (1)-(3) is satisfied in the example with network coding capable OXCs. However, without network coding Condition (3) is violated if either x or y is sent along the link (n_2, n_3) instead of x+y. Thus, Condition (4) is required to get a valid solution.

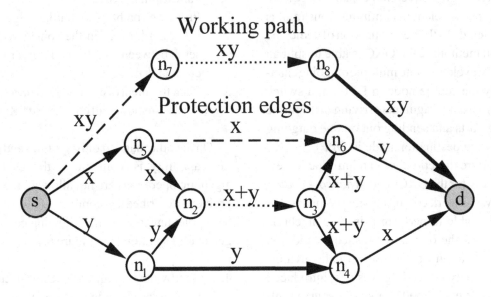

operation over *GF(2)*). Applying network coding, the destination node decodes the original traffic fragments by computing the inverse operation on each data fragment, if at least capacity *b* reaches it.

An example is presented in Figure 6. for network coding capable OXCs for a demand with 2 units of bandwidth request (*x* and *y*). The transmitted data is divided into two fragments on the protection resources, where node n_2, n_4 and n_6 perform network coding operations. In order to recover from the failure of the three SRLGs (dotted, dashed, bold), the destination node must receive enough pieces of information for final decoding in each auxiliary graph G_f. As one can easily check, the presented solution with network coding, the destination node is able to reproduce both data fragments in all SRLGs. However, without network coding capable OXCs either *x* or *y* is sent along the link (n_2, n_3) instead of *x+y*. If *x* or *y* is configured, in the bold and dashed SRLG respectively, *b* bandwidth arrives at the destination, but either *x* or *y* cannot be decoded without signaling in the control plane and OXC

reconfigurations, which are expensive and time-consuming operations. Note that the sub-graph on nodes *s, n_1, n_2, n_3, n_4, n_5, n_6* is called a butterfly network and is widely used as a simple topology to illustrate the benefits of network coding.

Note, that with the application of more complex OXCs (even in the single failure case) the resource efficiency with network coding can be lowered under the resource efficiency of DPP, while the rapid restoration time is maintained. For this, if three disjoint paths exist in the network, than sending only *b/2* data along each path the approach is still resilient, but uses with b/2 less bandwidth than DPP in the best scenario.

In the availability-aware implementation of GDP if the set of SRLGs is chosen properly, the connection only needs to be operational in the considered SRLG states in order to fulfill the given availability requirements. Thus, the complex problem of availability evaluation is moved to the SRLG selection stage, as we could be sure without evaluating the availability of the subgraph (which is a time-consuming process in the general case, see later) that the end-to-end connection availability meets the required level stated in the SLA.

CONNECTION AVAILABILITY EVALUATION METHODS

Series-Parallel Graphs

This section is an attempt to give an enumeration of the most important exact and approximation methods for computing the achieved availability value by a connection assuming mutually independent link and node failures in the network. However, in (Spragins 2002), link failure independence was investigated, and it was shown that such an assumption could be dangerously inaccurate. In telecommunication networks, failures occur because of cable cuts, Denial of Service (DoS) attacks or unpredictable events like earthquake or other natural disasters. Such failure events make more elements simultaneously unavailable in the same network region. In the previous section the SRLG concept was introduced for modeling failure dependencies in survivable network design. Note, that some methods in Figure 8. (e.g. Markov-models, state enumeration) are also applicable for computing the availability of dependent failure approaches.

The availability of some families of graphs can be evaluated with polynomial time algorithms. One family of these graphs is called series-parallel graphs. A graph is called *series-parallel*, if it can be turned into a graph of single link connecting two nodes s and d by a sequence of the following two operations, (1) replacing a pair of parallel edges with a single edge that connects their common endpoints, (2) replacing a pair of edges incident to a vertex of degree 2 other than s or d with a single edge. Series-parallel graphs play an important role in the availability evaluation of path and link protection schemes. Thus, we give a special attention of these graphs.

The evaluation of the connection availability A_{s-d} is based on the single link availability A_j (and unavailability $U_j=1-A_j$) values used by the connection. The basic equation for the availability of link j is

$$A_j = \frac{MTTF}{MTTF + MTTR},$$

where A_j is the availability, $MTTF$ is the mean time to failure, and $MTTR$ is the mean time to repair of the link under consideration. The mean time between failures ($MTBF$) is defined as $MTBF = MTTF + MTTR$. The $MTBF$ value of a link represents the average time elapsed between two subsequent failures. For the long-distance links the operator has cable-cut recordings, and they know how many cable cuts they can expect in a year approximately. Typical MTBF values for optical links range between 50 and 200 days per 1000 km of cable, while for an OXC is about 10^5 - 10^6 hours (Vasseur et al. 2004). Throughout the section, we assume fully reliable nodes ($A_j=1$), and *only link failures are considered*. Modeling node failures apply the node splitting technique in Figure 3(b).

A connection on a single working path is operational if all the links and nodes along the path are functional, in other words if any of the links along the path is cut the connection is disrupted. Thus, it is straightforward to have the following series rule (or Lusser's Law) for a working path containing n links:

$$A_{s-d} = \prod_{i=1}^{n} A_i. \qquad (3)$$

Applying the series rule, the elements on the working path can be replaced with a single link having the availability of the connection (Figure 7(a)).

The connection on parallel edges between the source and destination node of the connection is operational if any of the edges are operational. Thus, the availability value of a graph with *m* parallel elements can be computed with the parallel rule:

$$A_{s-d} = 1 - \prod_{i=1}^{m}(1 - A_i). \qquad (4)$$

Applying the parallel rule, the redundant links could be replaced with a single link having the availability of the connection (Figure 7(b)).

Now, let us see the availability of a connection with dedicated path protection, that is, a link and node disjoint path pair between the source and the destination node. It is easy to reduce the two paths for a parallel edge-pair with the series rule, but now we face with the problem of parallel edges. After the parallel rule is applied, the parallel edges can be replaced with a single edge between the source and the destination nodes having the overall availabil-

Figure 7. Each link j in the examples has $A_j = 0.9$; (a) Series graph (e.g. working path); (b) Parallel graph; (c) series-parallel graph (e.g. path protection); (d) a non series-parallel graph (e.g. GDP)

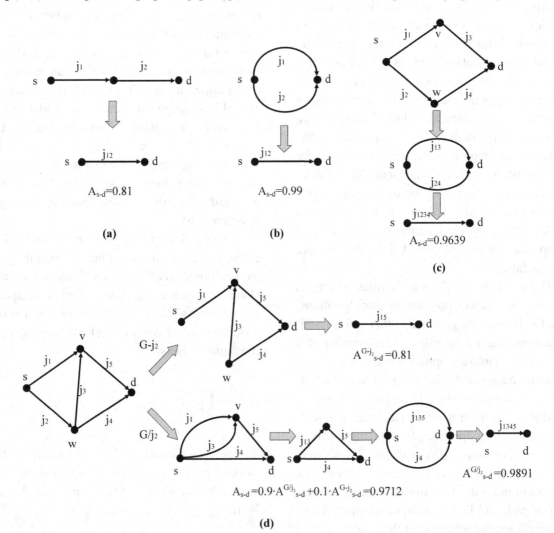

Figure 8. Classification of two-terminal reliability evaluation schemes

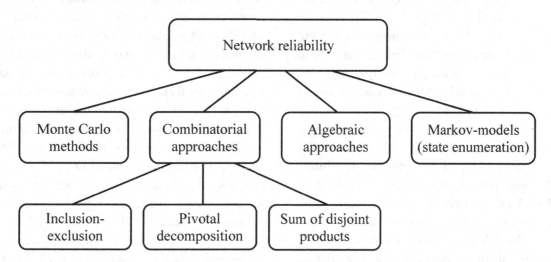

ity of the connection (Figure 7(c)). By definition, for series-parallel graphs the series and parallel rules can be applied one after the other, until the graph is reduced to a single link labeled with the availability of the connection. For example, the subgraphs on the working and protection edges as the result of dedicated path protection, link protection or non-overlapping segment protection are series-parallel graphs. However, by adding a new edge to a parallel graph as in Figure 7(d), the connection availability evaluation turns to be a complex problem, investigated in the next section.

As we discussed previously, the trade-off between resource utilization and connection availability have to be considered in the selection of the protection method. In the case of shared path protection, the perceived connection availability by the disjoin path pair is lower than in the case of DPP, because the protection resources can be used by other failed connections, make the restoration of the connection impossible. The connection routed with SPP is operational if its working path is operational, or the working path failed and the spare resources along the protection edges are not used by other connections (Zhang and Mukherjee 2004).

General Graph Topologies

The end-to-end availability of a connection is defined to be the probability that at least one viable path is operational between the source and destination node according to the given link availability values. Evaluating the network reliability for a general input is an NP-hard problem even for planar directed acyclic graphs (DAG) with a maximal nodal degree of 3 (Provan 1986). The problem of network reliability and survivability is defined in complexity theory as follows (Garey and Johnson 1979):

1. ***Network Reliability Problem:***
   - INSTANCE: Graph $G=(V,E)$, subset $V' \subseteq V$, a rational failure probability $p(e)$, $0 \leq p(e) \leq 1$,, for each $e \in E$, a positive rational number $q \leq 1$.
   - QUESTION: Assuming edge failures are independent of one another is the probability q or greater that each pair of vertices in V' is joined by at least one path containing no failed edge?
2. ***Network Survivability Problem:***
   - INSTANCE: Graph $G=(V,E)$, a rational failure probability $p(x)$, $0 \leq p(x)$

≤ 1, for each $x \in V \cup E$, a positive rational number $q \leq 1$.

◦ QUESTION: Assuming all edge and vertex failures are independent of one another, is the probability q or greater that for all $\{u,v\} \in E$ at least one of u, v or $\{u,v\}$ fail?

Considering single connection availability rather than evaluating the networks' reliability, $V' \subseteq V$ contains only the source and destination nodes of the connection, and the input graph contains only the edges used by the connection. This evaluation is called *two-terminal reliability problem*. It has been shown that the exact computation of two-terminal reliability is #P-complete (Valiant 1979; Provan 1986), thus a wide range of different approaches were proposed for bounding the problem (Shier 1991; Rai et al. 1995). Exact availability evaluation approaches are well investigated but the high computational complexity of the problem allows only small instances to be solved efficiently. Thus, pseudo-optimal solutions must be obtained, using enumerative approaches.

Several algorithms have been developed to solve the reliability problem for non series-parallel networks, shown in Figure 8. However, the calculation of these methods is, in general, quite involved.

There are approaches using algebraic formulation of the two-terminal reliability problem (Shier 1991). By contrast, the sum of disjoint products (SDP) and inclusion-exclusion techniques are based on a given enumeration of minpaths or mincuts (Rai et al. 1995). Monte Carlo methods are also applicable for the two-terminal reliability problem. Other methods are based on state space enumeration. For instance, if we have the considered failure states with a given probability of that failure event, in order to get the availability of the connection we have to sum up the probabilities of the failure states in which the connection is operating. Other possibility for the state space enumeration is a continuous time Markov model,

where the steady state probabilities of the states in which the connection is operating is summed up (Mello et al. 2005). All of *state enumeration methods need some simplification to avoid the exponential nature of the problem*, e.g. assume that not more than two simultaneous link failures occur in the network. This approximation, well accepted in the literature, is valid for most networks.

The studies in (Shier 1991; Provan 1986) assume mutually independent link failures, and the exact availability value is assigned by graph decomposition, called factoring. The assumption of mutually independent link failures allows us for a most effective way of viewing state space enumeration. Rather than fully specifying the states of all m edges at once, we can instead select a particular edge j as a condition. Instead of the reliability problem of $R_{s-d}(G)$, we obtain a new system, in which j is operational, and a new system, in which j is failed. This produces the *pivotal decomposition formula* (Shier 1991):

$$R_{s-d}(G) = A_j R_{s-d}(G | j \text{ is operational}) + (1 - A_j) R_{s-d}(G | j \text{ is not operational}). \qquad (5)$$

The recursive formula in Eq. (5) shows how reliability computations for a given system can be decomposed into those for two smaller systems with conditioning (or factoring). If the conditioning is done on all edges pivotal decomposition reproduces state-space enumeration, as all possible states of the network were considered. Fortunately, there are circumstances in which not all edges need to be considered for factoring. In fact, by judiciously selecting the edges for factoring, and using the series and parallel rule for appropriate subgraphs, significant savings on computational time is possible in practical cases. For a directed graph $G=(V,A)$, factoring on an edge j out of s, or into d, is especially helpful (see Figure 7(d), where edge j_2 is contracted) and always equivalent to the original network. However, *if the edge for pivotal decomposition is not properly selected, spurious paths could occur in the network*, and the availability of the new system can be higher than the original system (e.g. Figure 3(c) path j_1, j_4 is not realizable in the original graph).

In the following, the summary of the most important network reductions is shown for a possible implementation of the pivotal decomposition. We follow the method published in (Page and Perry 1989), which offers an efficient implementation of the pivotal decomposition in directed networks. As a result of factoring on a given edge we face two new problems, like in Figure 7(d). In the first case, when j is failed, we simply erase it from the graph *(G-j)*. The second case is more complicated, as the head and the tail of j need to be contracted *(G/j)* into a single node, also shown in Figure 3(c). In the former case, some false starts (i.e. a vertex other than the source having no edge directed into the vertex) and dead ends (i.e. a vertex other than the destination having no edge directed outwards from the vertex) could be in the network (e.g. the tail of j_3 and j_5 is false start in Figure 7(d) after j_2 is deleted). In the latter case, after the nodes are contracted some edges directed into the source or out of the destination are possible. In both cases these edges could be removed from the graph, as they do not affect the overall availability of the connection. Apply these and similar edge removing methods and use the series and parallel rule in Equation (3) and Equation (4), respectively, if applicable in an iterative manner. In case we get a single link after the iterations, the availability of the graph is the availability value the link is labeled with. If no more reduction is possible on the graph, and we still have a subgraph with multiple links, then we choose an edge and apply the pivotal decomposition formula in Equation (5) recursively.

Note, that the method outperforms state-space enumeration approaches in most cases (e.g. for most generalized dedicated protection solutions). However, where only few edge deletions and series-parallel reductions are performed during the computation, factoring is slower than state-space enumeration due to the extra effort made on eliminating some edges.

FUTURE RESEARCH DIRECTIONS

From a technological point of view restoration strategies may play an important role in future survivable optical backbone design, as soon as fast failure localization is solved in the all-optical environment that can meet the stringent timing requirements (50 ms) of optical layer recovery. At the same time, protection approaches will remain for situations where fast recovery is crucial, while resource efficiency is less important. As the technology evolving and more complex all-optical equipments will be used in practice, (e.g. optical buffering of the signal is solved), near optical packet switching (OPS) and optical burst switching (OBS) optical circuit switching with traffic grooming and network coding could be a competitive technology. We have presented a possible protection method for network coding capable OXCs (generalized dedicated protection) which leads an efficient utilization of resources. In all-optical environment the current architectures proposed for network coding use fiber delay lines to perform coding operations on the signals. The OXCs are equipped with these devices could lead to a finer granularity than current coarse circuit switched WDM networks.

From the view point of survivability, customers are mainly interested in QoS parameters of their connection (e.g. availability) rather than in the applied protection techniques, which is the responsibility of the network operator to choose the proper techniques to provide the QoS the customer had paid for. In order to provide well-defined service classes in the SLA for the customer, an availability-aware routing method is crucial. This could be achieved via e.g. an availability-aware implementation of dedicated or shared protection. Availability of the connections can be ensured by building up a proper SRLG list in a way that if all SRLGs are protected in the list, the network operator could be sure that the required service availability is provided for the customer without evaluating the availability of the connection. Note that in this case the complex problem of availability evaluation is by-passed, i.e. with the application of this indirection level the problem would be simpler, which could not be generally true. As a result, finding a proper SRLG list for a given availability level is still an NP-hard problem. Moreover, the information about link failure dependencies of SRLGs in the same logical hierarchy is inaccurate even at the service provider - who may have a long list of historical failure events. Updating the SRLG database manually could lead to a slow convergence to the current state of the network. Auto-discovery of the SRLG is also an option (Sebos et al. 2005), and could lead to a better network and failure model than manually configured databases. Even with the application of automatic SRLG discovery method, the accuracy of the SRLG list at the network operator remains the key aspect of the availability-aware methods and pre-planned protection approaches.

CONCLUSION

In this chapter the principles of pre-designed protection schemes were discussed. After the basic concepts of single failure resilient dedicated and shared protection methods were shown, we introduced the notion of shared risk link groups, the most widespread approach in multiple failure resilient network design for modeling failure dependencies between network components. Based on the SRLG concept, we gave a thorough analysis of a shared path protection approach and a generalized dedicated protection method. The SPP and GDP methods face with most of the problems of shared and dedicated protection, respectively. Thus, understanding the principles of these approaches provides us a good look into the basics of other protection approaches too. Finally, the main techniques for evaluating the two-terminal reliability of a connection were enumerated. We have seen that, the more sophisticated protection approach requires a more complex availability evaluation process.

The chapter focuses on the tradeoffs in an efficient backbone network design, among resource efficiency, complexity, availability and the price of the equipments for each protection scheme. Each of the previously mentioned approaches has different metrics. Therefore, network operators have to choose always the appropriate one for their network based on operational premises.

REFERENCES

Ahlswede, R., Cai, N., Li, S. Y. R., & Yeung, R. W. (2000). Network information flow. *IEEE Transactions on Information Theory*, *46*(4), 1204–1216. doi:10.1109/18.850663

Andersen, R., Chung, F., Sen, A., & Xue, G. (2004). On disjoint path pairs with wavelength continuity constraint in WDM networks. *In IEEE INFOCOM, 1*, (pp. 524–535).

Babarczi, P., Tapolcai, J., & Ho, P. H. (2009). Availability-constrained dedicated segment protection in circuit switched mesh networks. *In Reliable Networks Design and Modeling. RNDM, 09*, 1–6.

Bernstein, G., Rajagopalan, B., & Saha, D. (2003). *Optical network control: Architecture, protocols, and standards*. Boston, MA: Addison-Wesley Longman Publishing Co., Inc.

Chlamtac, I., Ganz, A., & Karmi, G. (1992). Lightpath communications: An approach to high bandwidth optical WANs. *IEEE Transactions on Communications*, *40*(7), 1171–1182. doi:10.1109/26.153361

Cinkler, T. (2003). Traffic- and λ-grooming. *IEEE Network*, *17*(2), 16–21. doi:10.1109/MNET.2003.1188282

Dijkstra, E. W. (1959). A note on two problems in connextion with graphs. *Numerische Mathematic*, *1*(1), 269–271. doi:10.1007/BF01386390

Ellinas, G., Bouillet, E., Ramamurthy, R., Labourdette, J. F., Chaudhuri, S., & Bala, K. (2003). Routing and restoration architectures in mesh optical networks. *Optical Networks Magazine*, *4*(1), 91–106.

Fumagalli, A., & Valcarenghi, L. (2000). IP restoration vs. WDM protection: is there an optimal choice? *IEEE Network*, *14*(6), 34–41. doi:10.1109/65.885668

Garey, M. R., & Johnson, D. S. (1979). *Computers and intractability. A guide to the theory of NP-completeness. A Series of Books in the Mathematical Sciences*. San Francisco, CA: WH Freeman and Company.

Georgakilas, K. N., Katrinis, K., Tzanakaki, A., & Madsen, O. B. (2010). *Impact of dual-link failures on impairment-aware routed networks*. In 12th International Conference on Transparent Optical Networks.

Haider, A., & Harris, R. (2007). Recovery techniques in next generation networks. *IEEE Communications Surveys & Tutorials*, *9*(3), 2–17. doi:10.1109/COMST.2007.4317617

Ho, P. H., & Mouftah, H. T. (2002). Allocation of protection domains in dynamic WDM mesh networks. *In 10th IEEE International Conference on Network Protocols*, 2002, (pp. 188–189).

Ho, P. H., Tapolcai, J., & Cinkler, T. (2004). Segment shared protection in mesh communications networks with bandwidth guaranteed tunnels. *IEEE/ACM Transactions on Networking (TON)*, *12*(6), 1105–1118. doi:10.1109/TNET.2004.838592

Hu, J. Q. (2003). Diverse routing in optical mesh networks. *IEEE Transactions on Communications*, *51*(3), 489–494. doi:10.1109/TCOMM.2003.809779

Kamal, A. E. (2010). 1+ N network protection for mesh networks: network coding-based protection using p-cycles. *IEEE/ACM Transactions on Networking*, *18*(1), 67–80. doi:10.1109/TNET.2009.2020503

Koetter, R., & Medard, M. (2003). An algebraic approach to network coding. *IEEE/ACM Transactions on Networking*, *11*(5), 782–795. doi:10.1109/TNET.2003.818197

Lee, H. W., & Modiano, E. (2009). *Diverse routing in networks with probabilistic failures*. IEEE INFOCOM '09.

Lucerna, D., Tornatore, M., Mukherjee, B., & Pattavina, A. (2009). Availability target redefinition for dynamic connections in WDM networks with shared path protection. *Design of Reliable Communication Networks DRCN 2009*, (pp. 235–242, 25-28).

Maier, G., Pattavina, A., De Patre, S., & Martinelli, M. (2002). Optical network survivability: protection techniques in the WDM layer. *Photonic Network Communications*, *4*(3), 251–269. doi:10.1023/A:1016047527226

Mello, D. A. A., Schupke, D. A., & Waldman, H. (2005). A matrix-based analytical approach to connection unavailability estimation in shared backup protection. *IEEE Communications Letters*, *9*(9), 844–846. doi:10.1109/LCOMM.2005.1506722

Mouftah, H. T., & Ho, P. H. (2003). *Optical networks: Architecture and survivability*. Kluwer Academic Publishers.

Orlowski, S., & Pioro, M. (2009). *On the complexity of column generation in survivable network design with path-based survivability mechanisms*. In International Network Optimization Conference (INOC).

Ou, C., Zhang, J., Zang, H., Sahasrabuddhe, L. H., & Mukherjee, B. (2004). New and improved approaches for shared-path protection in WDM mesh networks. *Journal of Lightwave Technology*, *22*(5), 1223. doi:10.1109/JLT.2004.825346

Page, L. B., & Perry, J. E. (1989). Reliability of directed networks using the factoring theorem. *IEEE Transactions on Reliability*, *38*(5), 556–562. doi:10.1109/24.46479

Papadimitriou, D. (2002). *Inference of shared risk link groups*. Internet Draft. Retrieved September 1, 2010, from http://tools.ietf.org/html/draft-many-inferencesrlg-02

Provan, J. S. (1986). The complexity of reliability computations in planar and acyclic graphs. *SIAM Journal on Computing, 15*(3), 694–702. doi:10.1137/0215050

Rai, S., Veeraraghavan, M., & Trivedi, K. S. (1995). A survey of efficient reliability computation using disjoint products approach. *Networks, 25*(3), 147–163. doi:10.1002/net.3230250308

Ramamurthy, S., Sahasrabuddhe, L., & Mukherjee, B. (2003). Survivable WDM mesh networks. *Journal of Lightwave Technology, 21*(4), 870–883. doi:10.1109/JLT.2002.806338

Sebos, P., Yates, J., Hjalmtysson, G., & Greenberg, A. (2005). Auto-discovery of shared risk link groups. *Optical Fiber Communication Conference and Exhibit 2001*, (p. 3).

Shier, D. R. (1991). *Network reliability and algebraic structures*. New York, NY: Clarendon Press.

Spragins, J. (2002). Dependent failures in data communication systems. *IEEE Transactions on Communications, 25*(12), 1494–1499. doi:10.1109/TCOM.1977.1093787

Sterbenz, J. P. G., et al. (2010). *ResiliNets: Resilient and survivable networks*. Retrieved September 1, 2010, from https://wiki.ittc.ku.edu/resilinets

Strand, J., Chiu, A. L., & Tkach, R. (2001). Issues for routing in the optical layer. *IEEE Communications Magazine, 39*(2), 81–87. doi:10.1109/35.900635

Suurballe, J. W. (1974). Disjoint paths in a network. *Networks, 4*, 125–145. doi:10.1002/net.3230040204

Tapolcai, J. (2005). *Routing algorithms in survivable telecommunication networks*. Doctoral dissertation, Budapest University of Technology and Economics, Hungary. LAP Lambert Academic Publishing AG & Co KG. ISBN 978-3-8383-9297-4

Valiant, L. G. (1979). The complexity of enumeration and reliability problems. *SIAM Journal on Computing, 8*, 410. doi:10.1137/0208032

Vasseur, J. P., Pickavet, M., & Demeester, P. (2004). *Network recovery: Protection and restoration of Optical, SONET-SDH, IP, and MPLS*. Morgan Kaufmann Publishers.

Xu, D., Qiao, C., & Xiong, Y. (2002). An ultra-fast shared path protection scheme - Distributed partial information management, part II. *In IEEE ICNP: International Conference on Network Protocols*, (pp. 344–353).

Xue, G., Zhang, W., Wang, T., & Thulasiraman, K. (2007). On the partial path protection scheme for WDM optical networks and polynomial time computability of primary and secondary paths. *Management, 3*(4), 625–643.

Zhang, J., & Mukherjee, B. (2004). A review of fault management in WDM mesh networks: Basic concepts and research challenges. *IEEE Network, 18*(2), 41–48. doi:10.1109/MNET.2004.1276610

ADDITIONAL READING

Ball, M. O. (1980). Complexity of network reliability computations. *Networks, 10*(2), 153–165. doi:10.1002/net.3230100206

Barla, I. B., Rambach, F., Schupke, D. A., & Carle, G. (2010). Efficient Protection in Single-Domain Networks using Network Coding. In *IEEE Global Telecommunications Conference, GLOBECOM 2010*, pages 1–6.

Bhandari, R. (1999). *Survivable networks: algorithms for diverse routing*. Kluwer Academic Pub.

Colbourn, C. J. (1987). *The Combinatorics of Network Reliability*. NY, USA: Oxford University Press.

Demeester, P., Gryseels, M., Autenrieth, A., Brianza, C., Castagna, L., & Signorelli, G. (1999). Resilience in multilayer networks. *IEEE Communications Magazine*, *37*(8), 70–76. doi:10.1109/35.783128

Grover, W. D., et al. (2010). TRLab: papers on p-cycles. Retrieved January 24, 2011 from http://www.trlabs.ca/trlabs/research/library?new_url=people/edm-grover.html

Inkret, R., Lackovic, M., & Mikac, B. (2003). 266 Case Study Topologies. In *Proceedings of ONDM03* (pp. 3–5). WDM Network Availability Performance Analysis for the COST.

Jaggi, S., Sanders, P., Chou, P. A., Effros, M., Egner, S., Jain, K., & Tolhuizen, L. M. G. M. (2005). Polynomial time algorithms for multicast network code construction. *Information Theory. IEEE Transactions on*, *51*(6), 1973–1982.

Li, S. Y. R., Yeung, R. W., & Cai, N. (2003). Linear network coding. *Information Theory. IEEE Transactions on*, *49*(2), 371–381.

Maesschalck, S. D., Colle, D., Lievens, I., Pickavet, M., Demeester, P., & Mauz, C. (2003). Pan- European optical transport networks: an availability-based comparison. *Photonic Network Communications*, *5*(3), 203–225. doi:10.1023/A:1023088418684

Markopoulou, A., Iannaccone, G., Bhattacharyya, S., Chuah, C. N., & Diot, C. (2004). Characterization of failures in an IP backbone. *Proceedings - IEEE INFOCOM*, *4*, 2307–2317.

Mohandespour, M., & Kamal, A. E. (2010). 1+N protection in polynomial time: a heuristic approach. In *IEEE Global Telecommunications Conference (GLOBECOM 2010), pages 1–6*.

Molisz, W. (2004). Survivability function-a measure of disaster-based routing performance. *IEEE Journal on Selected Areas in Communications*, *22*(9), 1876–1883. doi:10.1109/JSAC.2004.829644

Rai, S., & Agrawal, D. P. (1990). *Distributed computing network reliability*. Los Alamitos, CA, USA: IEEE Computer Society.

Tornatore, M., Carcagni, M., & Pattavina, A. (2010). Availability formulations for segment protection. Communications. *IEEE Transactions on*, *58*(4), 1031–1035.

Tornatore, M., Maier, G., & Pattavina, A. (2005). Availability Design of Optical Transport Networks. *Selected Areas in Communications. IEEE Journal on*, *23*(8), 1520–1532.

Verbrugge, S., Colle, D., Demeester, P., Huelsermann, R., & Jaeger. M. (2005). General availability model for multilayer transport networks. In *Proceedings of Design of Reliable Communication Networks (DRCN 2005)*, page 8.

Wolsey, L. A. (1998). *Integer Programming*. Wiley-Interscience.

Zhou, D., & Subramaniam, S. (2002). Survivability in optical network. *IEEE Network*, *14*(6), 16–23.

KEY TERMS AND DEFINITIONS

Connection Availability: In a network with reparable links and nodes ~ refers to the probability that the source and destination nodes are connected by at least one path of operating edges at some time t in the future, given that the connection was established at time $t=0$.

Dedicated Protection: It is a pre-designed protection approach in which the spare resources can be used by a single connection.

Generalized Dedicated Protection: It is a pre-designed protection approach protecting all SRLGs in the list strategically defined according to specific operational premises by allowing all intermediate nodes to split an incoming signal or switch between incoming signals.

Non-Sharable Spare Capacity: It is the link capacity that has been reserved by some other protection paths j, and is not sharable with the protection path of the connection i due to the fact that the working paths of i and j share at least a common SRLG.

Pivotal Decomposition (or Factoring): It is an exact availability calculation method, which is turned to be efficient in the evaluation of the connection availability of dedicated protection approaches in communication networks, e.g. generalized dedicated protection.

Shared Protection: It is pre-designed protection approach in which a common pool of spare resources is available for a set of working connections.

Shared Risk Link Group: A group of links subject to the same risk of failure because of physical of geographical dependency, e.g. the duct is cut in which the links are buried.

Spare Provision Matrix: It is an $|E| \times |SRLG|$ matrix and the entry (j, i) stores the amount of non-sharable spare capacity along link j for the protection path if the corresponding working path is involved in the ith SRLG.

Chapter 3
Protection Architectures for WDM Passive Optical Networks

Anusha Sivakumar
Indian Institute of Technology Madras, India & India UK Advanced Technology Center of Excellence in Next Generation Networks, Systems and Services, India

Ganesh C. Sankaran
Indian Institute of Technology Madras, India & India UK Advanced Technology Center of Excellence in Next Generation Networks, Systems and Services, India

Krishna M. Sivalingam
Indian Institute of Technology Madras, India & India UK Advanced Technology Center of Excellence in Next Generation Networks, Systems and Services, India

Gerard Parr
University of Ulster, UK & India UK Advanced Technology Center of Excellence in Next Generation Networks, Systems and Services, UK

ABSTRACT

Passive Optical Networks (PON) support subscribers with bandwidth requirements more than 10 Mbps. Fiber and node failures in a PON network can lead to large amounts of data loss, while isolating the central office from the subscribers. Hence, high network availability is desired when a PON is used for business enterprises and for providing mobile backhaul services. To maximize network availability, several protection architectures have been proposed in literature. In this chapter, we critically analyze and compare novel WDM PON protection architectures amongst those proposed in the literature. The comparison is done from topology, resource utilization and power budget perspectives. We also discuss protection mechanisms that are typically used in the architectures and their impact on restoration.

DOI: 10.4018/978-1-61350-426-0.ch003

INTRODUCTION

Passive optical networks (PONs) are used to provide last mile connectivity, thereby fulfilling the rising bandwidth (BW) needs of home users and the industry (Lam, 2007). A PON is a point-to-multipoint network in which a single optical fiber supports high data rates while connecting multiple users. The components used along the transmission path are passive in nature, i.e. they do not consume any electrical power. This passive property of the network is considered as its biggest asset when compared to other networks which use active components. This significantly reduces the operational cost of the overall network by eliminating the need for running power cables alongside transmission cables. Current generation PONs support up to 64 customers spread over a distance of 20 Km from the central office. They support data rates of the order of Gbps. Factors such as energy efficiency, increased number of customers per central office port, unused bandwidth capacity of optical fibers and low-cost optical components employed in PON make it a better access network compared to others like WiMAX, Wi-Fi and traditional DSL based networks.

There are different PON types such as ATM PON defined in ITU-T G.983.1 (ITU-T, 1998, 2000), Broadband PON defined in ITU-T G.983.3 (ITU-T, 1998, 2001, 2002), Ethernet PON defined in IEEE 802.3ah (IEEE 802.3 Ethernet in the First Mile Study Group, 2001), Gigabit PON defined in ITU-T G.984 (ITU-T, 2003), 10G-EPON defined in IEEE 802.3av and Wavelength Division Multiplexed (WDM) PON (Banerjee et al., 2005). A-PON refers to ATM PON wherein ATM is used as the signaling protocol. B-PON refers to broadband PON. Ethernet PON makes use of IEEE 802.3 Ethernet frames for data transmission and G-PON refers to gigabit PON where in the data rates are of the order of Gbps. APON, BPON, EPON and GPON use Time Division Multiple Access (TDMA) protocol for data transmission between the Optical Line Terminal (OLT) and the Optical Network Units (ONU). In TDM PON a power splitter (PS) acts as a remote node (RN), where each ONU is allocated a part of the total bandwidth available for the PON network. All the ONUs in the TDM PON network need to be time synchronized and the data rate of all the ONUs should also be the same. Such constraints are overcome by WDM PONs.

Further in this chapter, we concentrate on WDM-PON (Mukherjee, 2006), a next generation broadband access technology which has high potential because of its high bandwidth capacity and ability to provide better security to the transmitted data. When compared to Time Division Multiplexed (TDM) PON, WDM PON does not require time synchronization among ONUs for upstream and downstream data transmission which greatly reduces the complexity of the network nodes. Also, each wavelength channel can have its own data rate, can operate using different protocols irrespective of other ONUs and every end user has to pay only based on his/her own usage and upgrade. Thus the channel access in this case neither requires coordination nor involves contention among the participating ONUs. At any instant of time the channel can be accessed by the ONUs thereby reducing the overall network latency and control overhead. WDM PON assigns static wavelengths to communicate between OLT and ONU with an Arrayed Waveguide Grating (AWG) acting as the RN. AWG acts as a wavelength router and routes the data traffic on dedicated wavelengths to the ONUs. There are a set of Hybrid PON architectures which use both AWG and PS as remote nodes. In these architectures wavelength and timeslot are allocated dynamically to ONUs. Hence, WDM PON is considered as a logical point-to-point network, where data sent on one wavelength is independent of the other. This reduces chances of being overheard which in turn increases security. WDM PONs have attracted research interest worldwide due to the enormous bandwidth unleashed by applying WDM to PONs.

With the access network supporting high data rates and using multiple wavelengths, the need for network survivability (Lam, 2007) has become increasingly evident. In case of failures, data transmission is disrupted thus causing heavy revenue loss for the service providers. Survivability plays a major role in such a high capacity access network as a single fiber cut can lead to tremendous amount of data loss. Fiber cut maintenance and subsequent time for restoration of the network typically takes hours and sometimes days. It is very important to prevent such failures and adequate protection mechanisms are essential in such phenomena. In case of WDM PON, the network topology is extensively modified in the various protection architectures, especially since the number of ONUs and in turn the number of end users that can be connected with the central office can be high. As fiber cuts and node failures can completely isolate the OLT and the ONU from each other, the network has to be protected to the required extent for faster restoration during failures.

Protection is one of the critical issues for Optical Metropolitan and Core Networks too. Most of these network architectures are based on mesh and ring topologies which inherently provide multiple paths (Hill et al., 2002) (Rubin & Ling, 2002) (Maier, Herzog, Scheutzow, & Reisslein, 2005). However PON which is based on the tree topology is sparsely connected. With WDM PON providing higher bandwidth, survivability of the PON network becomes even more critical.

In this chapter, we present a comprehensive study of WDM based PON protection architectures that have been proposed in literature. We have classified them based on their layout and explained in brief the architectures which fall under various categories like protection by fiber duplication, alternate path protection. The prominent source of failure in most cases, was observed to be a fiber cut situation, mainly caused due to underground maintenance activities. The chapter considers protection architectures specific to WDM PON

in detail. However, in general, any tree based network architecture such as TDM PON and Hybrid PON can adapt the protection strategies explained in the chapter.

The rest of this chapter is organized as follows. Section 2 presents a brief overview of Passive Optical Networks. Section 3 presents the ITU-T recommended protection architectures. Section 4 presents an overview of the protection methodologies available for PON. Section 5 describes the various proposed WDM PON protection architectures classified in to two broad categories. Section 6 presents a comparison of the different architectures proposed in literature. Section 7 summarizes the paper and presents possible future work in this topic.

BACKGROUND

This section presents a brief description of Passive Optical Networks. A typical PON network consists of an OLT (Optical Line Terminal) at the central office of the service provider, as shown in Figure 1. The OLT mostly consists of transmitters, receivers, couplers, Arrayed Waveguide Gratings (AWG), filters and broadband light sources (BLS). Lasers are most commonly used as transmitters and photodiodes are used as receivers. The transceivers can be either fixed or tunable. Fixed transceivers can tune to only one wavelength while tunable transceivers can tune themselves to any of the other available wavelengths thereby providing added flexibility.

The OLT is further connected to a Remote Node (RN) by means of a feeder fiber (FF). The RN, by means of distribution fibers (DF), is connected to a number of Optical Network Units (ONU) which in turn connects to the end nodes (EN). The Remote node consists of passive splitters, couplers, and AWGs to route the data to the destined ONUs through the respective distribution fibers. The ONUs also contain transceivers for communicating with the OLT. The access network

Figure 1. Architecture of a typical Passive Optical Network (PON)

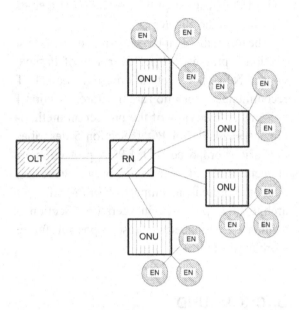

between the OLT and the ONU as a whole is referred to as the Optical Distribution Network (ODN). Using a dynamic bandwidth allocation algorithm, data transmission takes place between the OLT and the ONUs. The OLT broadcasts the downstream data to the different ONUs through the ODN. The ONUs identify the packet destined to them based on the packet headers, and drop the others. The OLT also allocates slots for the ONUs over which the ONUs send their upstream data. The data transmission is considered to be secure here as the data is always encrypted and sent from both ends of the network. The OLT acts as an interface between the backbone network and the PON, while the ONU provides the interface between the end users and the PON network

WDM-PON refers to wavelength division multiplexed (WDM) PON where the data is carried by multiple wavelength channels over a single fiber to different ONUs, as shown in Figure 2. This increases the data carrying capacity thereby providing high bandwidth to the users. In a WDM PON network, the OLT consists of wavelength routers, typically Array Waveguide Gratings (AWG), to route the data on the assigned wavelengths to the ONUs. The remote node also contains AWGs to route the data over the designated wavelengths to the destined ONUs.

Figure 2. Architecture of a typical WDM-PON

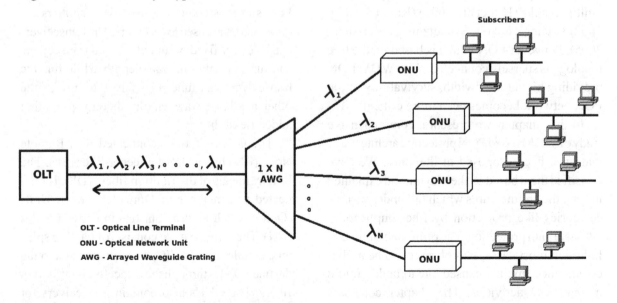

A 1×N AWG is a commonly used AWG configuration in WDN PON. The spectral periodicity property and the cyclic property of the AWGs play an important role for routing the data on various wavelengths through its I/O ports to the destined ONUs. For efficiently using the total number of available wavelengths, wavelength assignment plans are made. The upstream data and the downstream data are assigned to two different bands of wavelengths. WDM filters like Blue/Red filters are used to differentiate between the upstream and downstream bands of wavelengths and to transmit the data accordingly over the destined paths to reach the ONUs.

Wavelength/optical couplers are used to differentiate the upstream and downstream wavelengths at the ONUs. The ONUs should be kept simple, as making them complex further increases the network cost and hence the cost of the subscribers associated with them. Usage of fixed lasers at the ONUs, constrains them to tune to only a fixed set of wavelengths. Such ONUs are referred to as wavelength specific ONUs. However tunable lasers can be used at the ONUs which enable them to tune to multiple wavelengths. Such ONUs are referred to as colorless ONUs or non wavelength specific ONUs. In case of colorless ONUs, broadband light sources (BLS) are used at the OLT (WDM Technologies, 2008). The upstream and downstream data are divided into two bands of wavelengths namely the C-band and the L-band. The C-band stands for conventional band and has a wavelength range between 1530 nm and 1565 nm while the L-band, referred to as long band has the wavelength range between 1565 nm and 1625 nm.

ITU-T RECOMMENDED PROTECTION ARCHITECTURES

In this section, we describe the ITU-T recommended protection architectures and a classification of commonly used protection mechanisms.

The ITU-T recommendation on PONs G.983.1 (ITU-T, 1998) has suggested four possible fiber duplication and protection switching architectures. These protection architectures are relevant to WDM PONs and are referred to as Type A, B, C and D. All four types are based on tree topology to support point to multi-point links. Figure 3 depicts the architectures which are based on different levels of protection

In type A, only the feeder fiber between the OLT and the ONU is duplicated. In type B, along with feeder fiber duplication, the optical transceiver at the OLT is also duplicated. In case of a FF failure, the control is switched to the backup fiber link and hence to the backup optical transceiver. In type C, the optical transceivers not only at the OLT but also at the ONU are doubled. In addition to it the optical power splitter at the remote node along with the distribution fibers are also duplicated. In order to reduce the complexity and total network cost, the type D architecture apart from doubling the optical transceivers at the OLT, FF, DF and power splitter, optical power splitter at the RN was included additionally to avoid the need of duplicating the optical transceiver's in the ONU. In all the above schemes, in case of fiber cut or component failure, protection switching is done by switching the data to the backup fiber or component, thereby increasing the overall network survivability.

In case of WDM PON with tree topology, the splitter at the remote node is replaced by a wavelength router and the distribution fibers carry data to ONUs on specific wavelengths destined for each ONU. Since the ITU-T based 4 schemes also follow tree topology, they can also be employed for WDM PONs. In that case, the splitter is replaced with an Arrayed Waveguide Grating (AWG) wavelength router and the distribution fibers carry upstream and downstream data on dedicated wavelengths.

It can be observed that, to overcome feeder or distribution fiber failure, the fibers are often duplicated. However in some cases the nodes

Figure 3. ITU-T G.983.1 Recommended protection switching architectures for PON (ITU-T, 1998). (© 1998, [ITU-T]. Used with permission.).

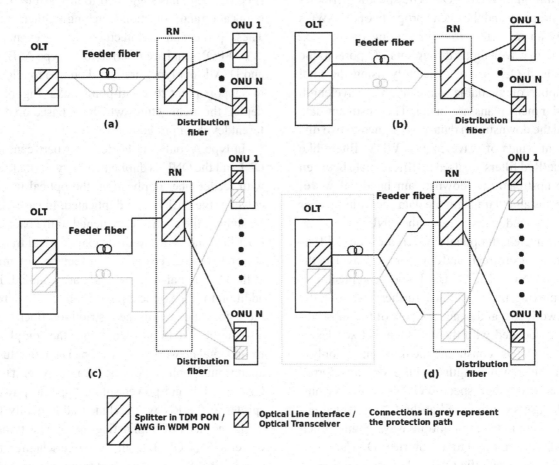

have been duplicated, by which alternate paths are established between the source and the destination in case of any failure.

METHODS OF PROTECTION

This section presents a classification of various protection methods which have been adapted for increased network survivability. Figure 4 presents this classification scheme.

Using Duplication

Fiber failures are overcome with the help of fiber duplication. The working fiber is duplicated to create a protection fiber. Efforts should be taken, however, to minimize fiber duplication in order to reduce the overall network cost. In most cases, the links and the nodes are duplicated for protection. These strategies have been further elaborated below.

Link Protection

Failures of feeder and distribution fibers lead to tremendous amount of data loss in the network. Hence, the fiber links are often duplicated such that an alternate path is made available for the data traffic to reach the destination. In most cases the feeder and distribution fibers are duplicated and in case of a feeder/distribution fiber failure, the fault is detected at the OLT or ONU. Once the fault is detected, the optical switches present at the

Figure 4. Classification of network protection methods

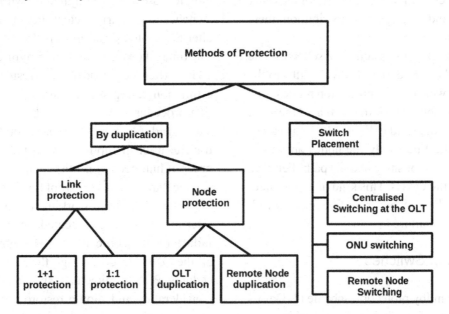

OLT/ONU change their switching states such that the signals now flow through the protection path. Under link duplication, there are two methods by which network is protected. They are referred to as 1 + 1 and 1:1 protection schemes.

1+1 Protection

In this scheme, the data traffic is sent both on the working and the protection fiber simultaneously to the remote node. During normal mode, the remote node/the ONU receives its data from the working fiber. In case of a fiber failure, the RN/ONU alter their switching states and receive traffic through the protection fiber. The same logic works for the DF failure also. In this approach, the time to restore the traffic only corresponds to the protection switching time. Thus, the restoring time is minimal and is of the order of fewer milliseconds (ms). This protection mechanism is followed by the Type A architecture shown in Figure 3(a).

1:1 Protection

In this scheme, the working fiber carries the data traffic during the normal operation while the protection fiber remains redundant. As illustrated in Figure 3(b), the Type B architecture follows this principle. Along with the feeder fiber, the optical transceiver at the OLT is also duplicated. In case of a feeder fiber cut, the optical switch present in the OLT is reconfigured after fault detection so that it transmits the data through the protection fiber. The restoration time in this scheme is higher as it includes protection switching time and the propagation delay.

Node Duplication

To handle node failures, node duplication architectures have been proposed. In this case the fibers are not duplicated but alternate paths are made available by duplicating the OLT and the remote node as illustrated in Figure 5. In case of feeder fiber failure, the neighboring WDM PONs, acting as protection pairs, protect each other. In case of FF1 failure, the state of the optical switch OS1 is changed and the downstream data of WDM OLT1 flows on FF2 and reaches remote node 2 and then to the respective ONUs as shown in Figure 5. Similarly in case of a remote node failure, an alternate path is always available for the traffic to reach the destination with the help of the other remote node and the respective distribution fibers to the ONUs.

Using wavelength assignment schemes, the signals travel on the destined wavelengths to the ONUs in the working and protection modes. The layout of this method is similar to that of the Type C architecture depicted in Figure 3(c). Instead of duplicating the fibers only the nodes are duplicated, thereby establishing disjoint paths between the OLT and the ONUs. This kind of an architecture protects the network from fiber cuts as well as remote node failure.

Placement of Switches

The active/standby protection scheme (1:1) uses an optical switch for switching from working path to protection path. Optical switches (Sivalingam & Subramaniam, 2004) are power driven and due to the passive nature of RNs, these optical switches can be placed either in the OLT or in the ONU or in both places. Architectures that use optical switches only at the OLT (Z. Wang, Sun, Lin, Chan, & Chen, 2005) provide centralized protection switching as an operational advantage in addition to keeping the cost of ONUs low. Power

monitors are used in conjunction with the optical switches to primarily detect the failure and then alter the switch position to restore the network. The time taken for restoration is typically around 18 ms. After the physical link is restored, ranging estimation, re-registration and time synchronization procedures are triggered before the data is exchanged. The total time for restoration of data transfer after disruption due to failure is in the order of hundreds of milliseconds.

Such architectures disrupt traffic to all ONUs when switching from working to protection path after a DF failure is detected. In this case, the failure of a specific RN-ONU segment causes all the ONUs to go through the restoration procedures. PON architectures should take this into consideration and should restore traffic without disrupting the unaffected segments.

Architectures with optical switching elements at the ONU (Chan, Chan, Chen, & Tong, 2003) make the ONU costlier. Optical switches are active elements such as MEMS, whose mean life time affects the overall availability of the network. Decentralization of switching elements makes the architectures more robust and helps in confining the traffic disruption to affected segment(s).

Figure 5. Node duplication mechanism

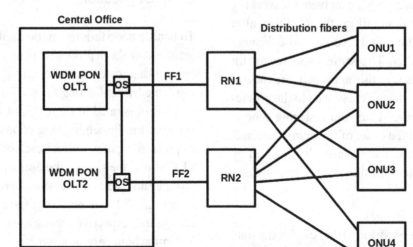

In (J. H. Lee et al., 2009), the architecture proposed constitutes an optical switching element at the remote node to provide protection. However this makes the remote node active, thereby bringing the need for laying additional power cable along the fiber.

Architectures with optical switching elements at both OLT and ONU offer advantages of both centralized and decentralized switching but at a higher cost. Switching element synchronization overhead is added to the restoration time where switches at both the ends will have to react to a specific failure scenario. In the next section, we present a subset of the existing protection architectures described in literature.

EXISTING PROTECTION ARCHITECTURES

A number of protection architectures for WDM PON have been proposed in the literature. This section presents a comparative study on these resilient architectures proposed. We have classi-

Figure 6. Classification of protection architectures

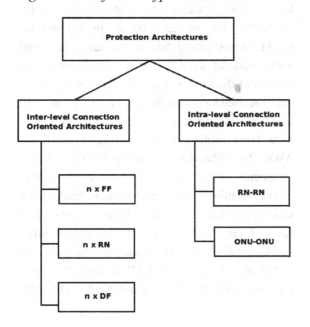

fied the protection architectures into two broad categories namely interlevel and intralevel connection oriented architectures as shown in Figure 6 and they have been explained further in the following sections.

In the study, it was observed that the tree topology has been adapted extensively in most of the architectures, with the advantage of the topology being the ability to connect the OLT with multiple ONUs and hence end users, thereby being more convenient and scalable than other access networks like WiMAX, Wi-Fi, and copper networks. Since fiber cut is most critical in PON, most of the protection architectures explained will be focused on fiber link failure.

Inter-Level Connection Oriented Architectures

This section describes those architectures which have multiple inter level connections in the network to provide protection. This is normally done by connecting the OLT with a small number of remote nodes (usually 2), which are in turn connected to multiple ONUs via multiple disjoint feeder and distribution fibers respectively. Figure 3 presents such an approach. This helps in establishing parallel data flows between the central office and the ONUs, thereby providing increased network survivability. As an alternative to fiber duplication, alternate paths can be established between the OLT and the ONUs such that in case of fiber failure, the data can reach the destination via the alternate path established. For example, in case a connection between the OLT and the RN fails, the traffic is sent via the other link existing between the OLT and RN. The same logic hold good for the connections between the RN and the ONU.

Thus we have classified these approaches into three sub-categories, with the protection aspect of every approach elaborated with respect to the architectures proposed in literature. The various approaches under the inter-level oriented connection category are as described below:

n x FF

In an n x FF architecture, there may be 'n' feeder fibers (usually 2 to 3) running in parallel between the OLT and the remote node. One feeder fiber acts as a working fiber while the other acts as a backup link referred to as the protection fiber as shown in Figures 3(a) and 3(b). In the normal mode, based on the link protection mechanisms (1+1 or 1:1) used, the data is carried by the working and/or the protection fiber. In case of a working feeder fiber cut, the data reaches the remote node through the protection fiber from the OLT. The layout of the working and protection fibers should be such that they do not run in the same conduit in order to overcome simultaneous working and protection feeder fiber cuts. These constraints have to be taken care of during the network planning phase to efficiently protect the network and prevent data loss. The architectures falling under this category have been described below.

Protection Architectures for Bidirectional WDM PON

In (K. Lee et al., 2007), a self restorable architecture for bidirectional WDM PON with the help of colorless ONUs (Iwatsuki & Kani, 2009) has been proposed. This architecture is similar to the Type B architecture shown in Figure 3(b). The wavelength bands of the two different Broadband Light Sources (BLS) used in the OLT are separated by an integer multiple of the free spectral range (FSR) of the 2XN AWG. Wavelength bands A-BLS and B-BLS are assigned to upstream and downstream traffic respectively. The 1:1 link protection mechanism has been used here. In case of working FF failure, the fault monitor present at the OLT detects the signal loss and alters the state of the switches such that the AWG at the OLT routes the data through the backup feeder fiber to the remote

node and then to the ONUs. The wavelengths of the signals in the protection mode are shifted by 100 GHz to the longer wavelength.

An 8-channel WDM PON with 16 Fabry Perot Laser Diodes (FP-LD) wavelength locked with Amplified spontaneous emission (ASE) (H. D. Kim, Kang, & Le, 2000) and 20 Km long feeder fiber were used to experimentally verify the architecture proposed in (K. Lee et al., 2007). The optical powers of the upstream and downstream signals were monitored by a 90:10 tap coupler and an optical spectrum analyzer. The restoration switching time was found to be 8 ms with negligible power penalty. However, this architecture has emphasized only on protection against feeder fiber failure and does not support distribution fiber failure. Hence in case of DF cut the ONUs may get completely isolated from the OLT.

In (K. Lee, Mun, Lee, & Lee, 2008), a similar bidirectional WDM PON architecture surviving both feeder fiber and distribution fiber failure has been proposed. The layout of this architecture is similar to that of the previous architecture except that, the DFs are also duplicated here. The architecture is as shown in Figure 7.

Among the broadband light sources present at the OLT, the L-band and C-band BLS are used for the down- stream and upstream signals respectively. The wavelengths in the L-band are used for transmitting the downstream signals and the unmodulated optical carriers for the upstream signals to the ONUs; the wavelengths in the C-band are used for transmitting the upstream signals. The 1+1 link protection scheme has been used here. The remote node here consists of an N × N AWG. The Nth and N^{*} port of the NXN AWG are connected to the working and protection feeder fiber respectively. In case of working feeder fiber failure, the data is retrieved from the protection feeder fiber. The remaining N-1 input and output ports of the AWG are connected through two distribution fibers to every ONU. One DF acts as the working fiber while the other acts as the pro-

Figure 7. Architecture proposed for bidirectional WDM PON in (K. Lee, Mun, Lee, & Lee, 2008). (© 2008, [IEEE]. Used with permission).

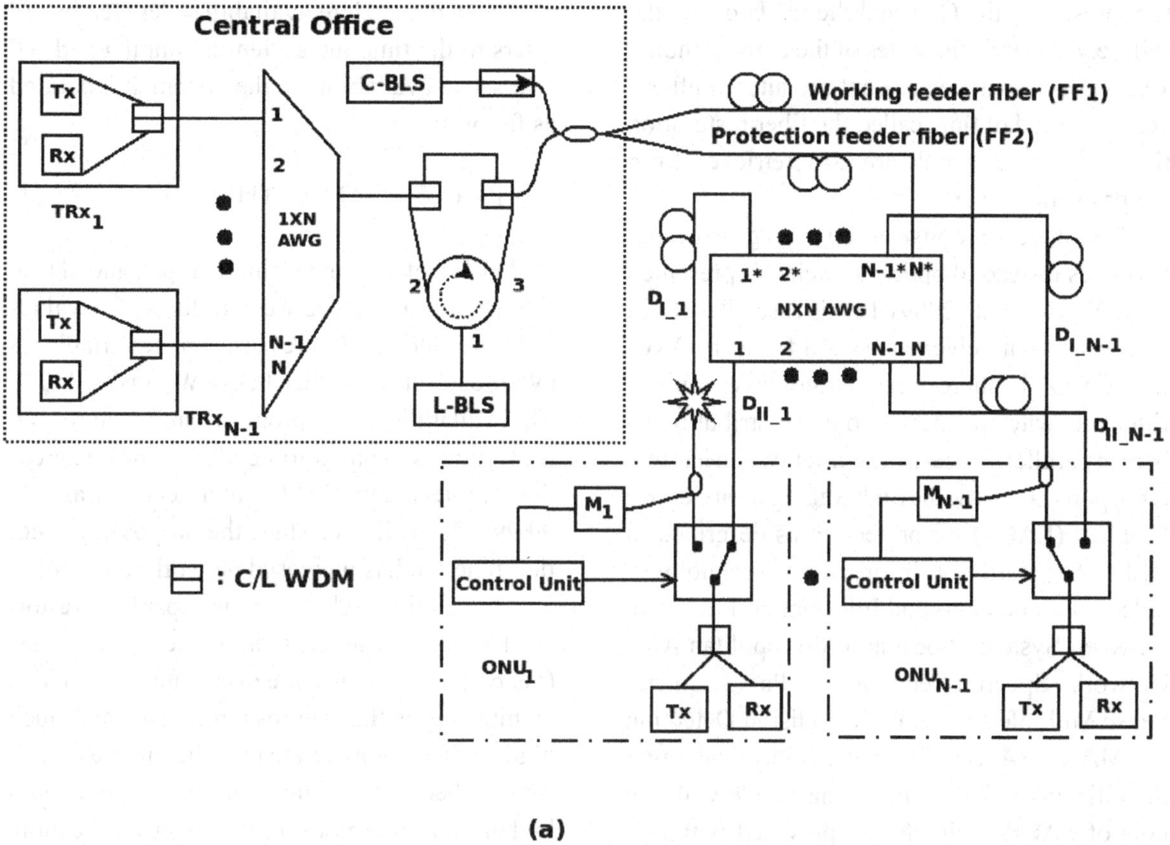

(a)

tection fiber. This architecture supports N-1 colorless ONUs.

In case of a FF cut or a DF cut, the power monitor at the ONU detects the failure and alters the state of the optical switch present at the ONU such that it transmits and receives signals through the protection Feeder fiber or distribution fiber. Thus this architecture can survive both FF and DF cuts, since there is always an automatic protection function existing and the signals can always reach the destination via the alternate path by duplicating only the feeder and the distribution fibers.

To experimentally demonstrate the above self protecting architecture, a 16 X 16 AWG with 100 GHz channel spacing and FSR of 31 nm were deployed at the RN. The experimental setup included an OLT connected to 2 ONUs. The length

of the FF and DF were set to be 15 Km and 5 Km respectively. One FP-LD is connected to the AWG at the remote node was modulated at 1.25 Gbps with $2^{31} - 1$ PRBS data. Only the DF link cut has been demonstrated. The switching time from the working fiber to the protection fiber was reported to be 4 ms. The receiver sensitivities measured at a BER (Bit Error Rate) of 10^{-10} were around -25.5 dBm both in normal and protection states.

Protection Architecture for WDM PON with Colorless Optical Source

An architecture similar to (K. Lee et al., 2008) has been proposed in (J.-Y. Kim, Mun, Lee, & Lee, 2009). This architecture includes N transceivers, optical switches and fault monitors at the OLT for every ONU. Under normal operation, the signals travel through the working feeder and distribution fiber. In case of a fiber failure, fault monitors present at the OLT and the ONU detect the failure. After this, the states of the corresponding 1X2 optical switches present are altered after a certain period of time called the fiber restoration time. Thus, the signals are now retrieved from the protection fiber.

Architecture constructed using Ethernet Switches instead of optical switches is presented in (J.-Y. Kim et al., 2009). In this case, there exist two optical transceivers at the OLT for every ONU. In every ONU, there are two transceivers and an Ethernet switch connected to two distribution fibers in parallel. Since the Ethernet switch has two input ports, it can use the Link Aggregation Control Protocol (LACP) for protection as described in (IEEE Standard for Information Technology-Telecommunications and Information Exchange Between Systems-Local and Metropolitan Area Networks-Specific Requirements Part 3: Carrier Sense Multiple Access With Collision Detection (CSMA/CD) Access Method and Physical Layer Specifications, 2006). In normal mode with the help of LACP, each ONU is provided with two times the bandwidth assigned to each channel. This condition prevails even in case of a feeder fiber failure. However, when the working DF fails the bandwidth is reduced by half for that particular affected ONU. The link aggregation mechanism being widely used in Ethernet Switches seems to be advantageous for this scheme. Thus only in case of fiber failure, the system switches the affected signal path to the protection path while others maintain their normal operation. The advantage with this scheme is that the signal fault is localized with respect to the transmission path while the other transmission channels remain unaffected.

FF and DF fiber cuts were experimentally demonstrated with a 32 channel 1.25 Gbps WDM PON connected to 16 ONUs, using optical switches. It was observed that the fiber break restoration time was 10 ms. The author's BER performance analysis shows that in most cases the power penalty

was less than 1 dB. Apart from the above experiments, a detailed system availability analysis has also been done. Here availability denoted by 'A' refers to the time the system is running and 'U' refers to unavailability of the system. It is defined as follows:

$$A = 1 - U; U = MTTR/MTBF \qquad (1)$$

MTTR refers to mean time to repair and MTBF refers to mean time between failures. The values of MTTR and MTBF are based on the variables in (Verbrugge et al., 2006; Chen & Wosinska, 2007). The availability of the proposed architecture with protection is claimed to be 99.99876% whereas the unprotected WDM PON architecture was only 99.9917% available. Thus, the proposed protection function has increased network reliability.

Though this architecture is capable of restoring the network against simultaneous fiber cuts (i.e. both FF and DF), the disadvantage with this architecture is that the resources used are much higher when compared to the other architectures. This is because the number of components is higher and there is excessive fiber duplication. Hence, the cost of network protection doubles itself thereby making the architecture expensive.

Novel Group Protection Architectures

In the previously explained architectures, the number of wavelengths has been equal to twice the number of ONUs. Each ONU requires 2 wavelengths; one for upstream and the other for downstream data transmission. A novel 1: N protection scheme based on group protection has been proposed in (Sue, 2006b) where N ONUs require only N wavelengths for data transmission. FF failure is overcome by means of alternate path switching. The layout of this architecture is similar to the one shown in Figure 3(a). For simplicity, a 1:2 protection scheme has been explained further, which can be expanded to a 1: N protection scheme. In case of a 1:2 protection scheme, the first two feeder fibers (FF1 and FF2) (as shown in Figure 8) are considered as working fibers which carry downstream and upstream traffic respectively while the third FF (FF3) acts as the protection fiber. The 8 DFs carry 8 working wavelengths and 4 protection wavelengths in both directions. Three ONUs form a group and are connected to each other by means of interconnection fibers and every ONU in a group shares a transmitting or receiving wavelength with another ONU in that group. In this scheme, a total of 2S interconnection fibers (IF) and S wavelengths are used for S ONUs. Each ONU is assigned distinct upstream and downstream wavelengths.

Two wavelength assignment schemes have been proposed based on which either the leftmost four DFs or the rightmost four DFs are protected. To retrieve the upstream and downstream signals in case of failure, FF3 uses two fiber couplers to tap the signals from FF1 and FF2. At the time of FF1 failure the OLT routes the downstream signals

through FF3. In the normal mode, the power monitors are not activated and the optical switches are set to OFF state. In the protection mode, the decision circuit switches the state of the optical switch to send the signals via the B/R filter to the ONUs. The affected ONUs also detect the DF cut and they in turn send their signals to the OLT via FF3. Since the distribution fiber failure is detected independently at both the OLT and the ONU, synchronization between them is not required.

For supporting the proposed wavelength assignment scheme, each ONU requires 2 input and output ports. Each pair of ports protect the leftmost and rightmost group of DFs. In case of a DF cut, the power monitor (MO) triggers the decision circuit to set the rfn (receive from neighbor) and ttn (transmit to neighbor) ports to ON state. The affected ONU sends the backup upstream signals from the ttn port to the ttf (transmit to fiber) port of the unaffected ONU in the same group, which

Figure 8. (a) OLT design; (b) ONU design presented in (Sue, 2006b). (©2006, [IEEE]. Used with permission.).

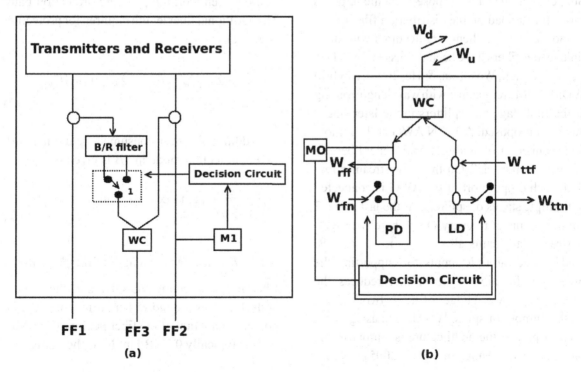

FF1   FF3   FF2

(a)   (b)

in turn sends the affected ONU's signals to the OLT via FF3. Also the rff (receive from fiber) port of the unaffected ONU sends the backup downstream signals to the rfn port of the affected ONU. Thus, both the upstream and downstream signals are transmitted and received by the affected ONU. Though this architecture does not support simultaneous DF cut at the rightmost and leftmost group of fibers, this scheme is capable of surviving more simultaneous faults than the previous schemes. But on the other hand, the number of interconnection fibers and the complexity at the ONU is higher thereby increasing the overall network cost.

This 1:2 scheme explained above can be extended to an 1: N protection scheme, but this will need 3S IFs and some more network components added to the present architecture, where N is less than or equal to S-1. As an extension of (Sue, 2006b) a detailed traffic load analysis has been done in (Sue, 2006a) which shows that even though the number of IFs are greater and the number of wavelengths are lesser than (Chan et al., 2003; C. Lee, Chan, Chan, Chen, & Lin, 2003; Z. Wang, Sun, et al., 2005), the proposed scheme imposes lower traffic load on the distribution fibers.

So far only resilient architectures with multiple feeder fibers have been discussed for WDM PON. In (Chen, Wosinska, & He, 2008), a hybrid WDM/TDM architecture with wavelength routing at the first stage and splitters in the later stages has been proposed. A 4 X N AWG at the remote node connects to 4 unidirectional feeder fibers all the way to the OLT. On the downstream direction, each output port of the AWG is connected to multiple stages of splitters. At the last stage, 2 splitters connect to every ONU thereby providing redundant paths all the way to the ONUs. ONUs have multiple inputs for supporting the protection function. Each splitter reduces the operational power budget of the network based on the number of splits. Wavelength assignment used as part of the architecture is similar to the one used by (Z. Wang, Sun, et al., 2005). 4 unidi-

rectional links increase the fiber installation cost. Power variation here is dependent on the number of splitters encountered and the length of fiber encountered along the working/protection paths. A detailed availability analysis has been done in this paper and the architecture proposed here has been compared with (Sue, 2006b; Z. Wang, Sun, et al., 2005). It has been observed that the number of wavelengths used is reduced by 50% and with additional FFs it increases connection availability. Other variations of this architecture can be found in (J. H. Lee et al., 2009; Chan et al., 2003; Z. Wang, Zhang, Lin, & Chan, 2005; Z. Wang, Sun, et al., 2005).

It is observed that in the architectures where 1+1 protection is used, the signals are carried by the protection path but they are not used by the ONU when the working path is in use. Thus in such cases, the restoration time of those architectures is limited to fiber fault detection and the local switching times at the respective OLT/ONU to retrieve the data from the protection path.

Further we analyze some of the architectures with respect to protection cost and power variation between working path and protection path. The additional fiber length required for protection is given by:

$$\Delta f_{nFF} = \sum_{i=1}^{N_1} \sum_{j-1}^{n} l_{RN_{ij}}^{u} - \sum_{i=1}^{N_1} \min_{j}(l_{RN_{ij}}^{u})$$

(2)

Additional power consumed by the network in the protection mode in dB is given by:

$$\Delta p_{nFF}^{\max} = \max_{j}[\max_{i}(l_{RN_{ij}}^{u}) - \min_{i}(l_{RN_{ij}}^{u})]P_{fiber}$$

(3)

where $l_{RN_{ij}}^{u}$ denotes the length of the jth upstream fiber link connecting the OLT and the remote node, typically around 15 Km. Pfiber denotes the power attenuation of the fiber per km. For SMF-28 it is typically 0.2 dB/km. N_1 is the number of remote nodes used in the protection PON architecture. n is the number of duplicate resources used in the protection PON architecture.

n x RN

In an inter-level connection oriented tree architecture with n remote nodes (RN), the feeder fibers connecting the OLT and the remote nodes are link disjoint. The layout of this kind of architecture is similar to the Type C and D architectures shown in Figures 3(c) and 3(d). Remote Nodes are essentially AWGs which route the data carried over multiple wavelengths to the destined ONUs. Unlike the n x FF architecture, an alternate path is established between the OLT and ONU, without duplicating the fibers. Hence here the network protection cost corresponds to the extra remote node deployed and fiber laid for protection. This kind of architecture not only supports fiber cuts but also provides protection against remote node failure. With the help of a novel wavelength assignment plan, the network can be restored from fiber and remote node failure thereby increasing the availability of the network. Architectures presented in (Sun, Chan, & Chen, 2006; Chowdhury, Huang, Chien, Ellinas, & Chang, 2008; Son, Han, Lee, & Chung, 2005) use this protection technique.

Protection Architecture Based on Node Duplication

In (Son et al., 2005) there exist two remote nodes connected to two WDM PON OLTs in parallel. This architecture is similar to the one in Figure 5 and adopts the node duplication strategy for network protection. The fiber links of the neighboring WDM PON are used to restore the disrupted signals. Two different wavelength bands are used for the neighboring WDM PONs and the cyclic property of AWG has been effectively utilized for routing the wavelengths appropriately in the normal and protection modes of operation.

Each WDM PON consists of N transceivers and N wavelength couplers for the N ONUs present in each of the two groups. The 3-dB couplers at the OLT divide the optical path of the WDM signals from the respective WDM PON OLTs to the destined remote nodes. The remote nodes in turn connect to all the ONUs via mutually exclusive distribution fibers. The RN consists of a 1X N AWG and N Blue/Red filters. ONUs consist of power monitors and optical switches (OS) for fault detection and to switch to the protection mode respectively in case of failure. The blue bands of the B/R filters in both RN1 and RN2 are connected to the ONUs present in group1 while the red bands of the same are connected to the ONUs in group2. Hence each ONU now has 2 mutually exclusive optical paths to transmit and/or receive their upstream and downstream signals. Thus in case of fiber failure, the signals from the ONUs will reach the OLT via the alternate path (i.e. via the alternate DF/FF path) since the power monitors trigger the optical switches to change their state in the ONU.

This architecture supports simultaneous distribution fiber cuts as long as at least one path exists to a given ONU from the OLT. Also, the DF cut of one ONU does not affect the operation of the other ONUs. Since the FFs and all the distribution fibers are active throughout, the time to restore the network in case of a fiber or node failure is less compared to other schemes, since only optical switching delay is incurred.

Protection Architecture Based on Centralized Alternate Path Protection

An architecture similar the one shown in Figure 5 is proposed in (Sun et al., 2006), except that it contains a single WDM PON OLT. The layout of this architecture is similar to that of the Type C architecture shown in Figure 3(c). This architecture survives FF, DF and RN failures and employs a simple centralized alternate path switching scheme based on a novel wavelength assignment plan. The

OLT consists of N transceivers, each connected to an optical switch and B/R filter, which are in turn connected to an NX 2 AWG. The OLT is further connected to the B/R filters and the 2X N AWG at the two RNs via 2 disjoint FFs. Wavelengths in the blue band and the red band of the B/R filters at the remote node are transmitted through the first and the last N/2 output ports of 2X N AWGs at the RNs. The upstream signals from the ONU reach the OLT via both the RNs while the downstream signals from the OLT reach the ONU via a single DF. The wavelength assignment utilizes the spectral periodicity property of the AWG.

In case of a DF cut, the monitoring unit present at the ONUs detects the failure and toggles the state of the corresponding 2X 2 optical switch at the OLT. As a result, the downstream signals of the affected ONU will be received via an alternate path while all the transceivers at the OLT will still receive a copy of the upstream signal based on the wavelength assignment plan. In this way, the architecture protects ONUs from FF and AWG failures.

2.5 Gbps directly modulated DFB laser diodes were used at the OLT and the ONU. A 16X 16 AWG with 100 GHz channel spacing and FSR=12.8 nm was used at the OLT and two 1X 16 AWGs with 50 GHz channel spacing and FSR=6.4 nm were used at the Remote Node. A 22 Km long single mode FF was used. B/R filters used at the OLT and the RN have a pass band of 18 nm for both blue and red bands each. The receiver sensitivity was reported to be between -20 dBm and -21 dBm. The switching time of the optical switches at the OLT in case of simulated distribution fiber cut was measured to be 3 ms.

It is observed that though the OLT complexity is high in this architecture, the ONU has been kept simple and all the protection switching is performed only at the OLT without disrupting the operation of the unaffected ONUs. With the OLT cost being shared equally among all the subscribers and easy network management, OLT complexity is justified.

Protection Architecture Based on Cyclic Wavelength Assignment Scheme

In most of the protection architectures, different wavelengths are assigned for the upstream and downstream traffic. This can prove to be expensive especially when the number of ONUs in a network increase. Hence an approach which overcomes this problem has been proposed in (Chowdhury et al., 2008), where the same wavelength is reused i.e. the working wavelength of one ONU acts as a protection wavelength for the clock-wise adjacent ONU as both the wavelengths travel in disjoint paths. This way the total number of wavelengths is reduced considerably and the available wavelengths are efficiently utilized, thereby reducing the overall network cost. The layout of this architecture is similar to that of the Type C architecture shown in Figure 3(c).

For N ONUs, N wavelength channels are used for bidirectional transmission. Centralized light sources have been used to reduce the system cost. Each wavelength λ_I is subdivided into two sub-wavelength channels, namely λ_I^u and λ_I^d, for the upstream and downstream traffic of Ith ONU using optical carrier suppression technique. The clock wise wavelength sharing scheme is such that in the working mode for I = {1, 2... N}, ONU_1 will be served by λ_1^u and λ_1^d while ONU_2 will be served by λ_2^u and λ_2^d and so on. Similarly for N ONUs, ONU_I is served in the protection mode by λ_{I-1} i.e. (λ_{I-1}^u and λ_{I-1}^d) and ONU_1 is served by λ_N (i.e. λ_N^u and λ_N^d).

The wavelengths generated by the OCS unit are fed in to ONU specific Network Unit Controllers (NUC) using an AWG and 3 dB filters. The NUCs perform transceiver and protection switching operation with respect to their destined ONUs. The NUC decides the wavelength channel to be used based on the mode of operation (working or protection mode). The fault monitor present in the NUC detects fiber failure. The interleaver filters are used to separate the upstream and downstream carriers. The wavelength channels from the NUC are in turn connected to the respective feeder fibers via AWGs based on the mode of operation. EDFA (Erbium Doped Fiber Amplifiers) amplifiers at the OLT are used for amplifying the upstream and downstream signals. The AWGs at RN1 and RN2 route the respective wavelengths to the designated ONUs in the working and protection modes.

In case of any FF, DF, remote node or any transmitter failure, the power monitor in the NUC detects it and changes the state of the optical switch to protection state. The signals are sent over the wavelengths in the protection mode. The overall power loss for a downstream signal is 13 dB while for the upstream signal including the modulator loss (4dB), it is 30dB as it travels twice the distance of a downstream signal. To demonstrate the survivability of the proposed architecture, experiments were conducted with the distance between the OLT and ONU being 20 km. Each upstream and downstream channel carries 10 Gbps data with PRBS (Pseudo Random Binary Sequence) length equal to $2^{31}-1$. At BER equal to 10^{-10} the power penalty in the working and the protection mode downstream and upstream channel was observed to be less than 0.7 dB and 1.2 dB respectively. However, the upstream transmission suffers an additional power penalty of 1.5 dB when compared to the downstream transmission.

In this architecture, though a number of components have been used at the OLT, they are shared by multiple subscribers and the number of wavelengths used are reduced by half when compared to other architectures. Also this architecture supports FF, DF, remote node and laser failure.

Further we analyze the architectures from different perspectives. When there are 'n' remote nodes, the variation in fiber length, i.e. the extra feeder and distribution fiber length laid for protection is given by:

$$\Delta f_{nrN} = \Delta f_{nFF} + \Delta f_{nDF} \tag{4}$$

The power variation in case of a protected WDM PON network with 'n' remote nodes upon simultaneous feeder and distribution fiber cut is as below:

$$\Delta p_{nRN}^{\max} = \Delta p_{nFF}^{\max} + \Delta p_{nDF}^{\max} \tag{5}$$

where Δf_{nDF} and $\Delta p_{nDF}^{\max}$ refer to the additional fiber length laid for protection and power variation of architectures adapting the n x DF protection scheme.

n x DF

In this protection scheme, more than one fiber runs to the ONU from the same remote node, i.e. the distribution fibers are duplicated. Either 1+1 or 1:1 link protection mechanism can be adapted for link restoration. In most cases 1:1 link restoration is used and the fiber fault is detected by the fault monitors present at the OLT or ONU. In case of subscriber side switching where the optical switches are placed at the ONU, as and when fiber cut is detected, the control unit triggers the optical switches to change their state. Now the data is received and transmitted by the ONU via the protection path. The architectures which follow this principle include (K. Lee et al., 2008; J.-Y. Kim et al., 2009; J. H. Lee et al., 2009). Protection architectures other than (J. H. Lee et al., 2009) have been explained earlier, hence only the architecture proposed in (J. H. Lee et al., 2009) has been explained here.

A Hybrid TDM/WDM PON Architecture

An architecture providing a smooth evolution method from TDM PON to Next Generation PON with protec- tion capability has been proposed and demonstrated in (J. H. Lee et al., 2009). Here the WDM PON is referred to as Next Generation (NG) PON and is realized using the same legacy

PON infrastructure. The Central Office consists of a legacy PON OLT and a WDM PON OLT with each of them providing their services to the subscribers. Here the signals from the WDM PON OLT and the TDM PON OLT are combined using a wavelength band cou- pler across the feeder fibers. For protection, feeder fibers are duplicated and based on the mode of operation, the data is sent through the working or the protection fiber, which is in turn controlled by the optical switches present at the OLT and at the RN. This architecture makes the remote node a power driven network node, as an electrically controlled photovoltaic converter is used for triggering the optical switch to change its state.

At the RN, signals are again split and sent to the optical power splitter and the AWG for serving the legacy subscribers with TDM PON services and the NG subscribers with WDM PON services respectively. From the re- mote node, two distribution fibers are connected to the ONU. Through one distribution fiber the legacy subscribers are provided with TDM PON service while the other distribution fiber provides WDM-PON service to the next generation subscribers. With the help of the optical switches present at the RN and the OLT, the subscribers have the flexibility to switch between the two types of services as per their needs. Hence, the NG-PON subscribers will have two distribution fibers running in parallel through every output port of the AWG thereby protecting them from distribution fiber failure.

In case of FF failure, the fiber fault is detected at the central office, based on the loss of signal (LOS) output, and following which the state of the optical switches at the OLT and RN are changed. Now the optical power with control information at the central office is sent to the remote node through the protection fiber. In case of DF failure, the fault is detected at the CO and at the subscriber side, after which the state of the optical switches at the RN are reconfigured remotely and simultaneously the switch at the subscriber side is reconfigured. Hence the data travelling through the failed fiber

gets rerouted via the alternate path made available to the subscribers. The remote node here consists of a control unit which controls the state of the optical switches present at the OLT, RN and ONU to enable alternate path protection switching.

With respect to legacy subscribers, who are provided with both TDM and WDM PON services simultaneously, only a specific service with higher priority is protected while the rest of the services are left unprotected. This paper also proposes a hybrid TDM/WDM PON architecture with color-less ONUs along with different switching block configurations at the remote node to protect both TDM PON and WDM PON signals in case of DF failure.

The architecture explained above was experimentally demonstrated using color free sources. A 2 X 32 splitter and a 2 X 40 AWG were used for TDM and WDM PON demonstration. The length of the FF and DF was set to 10 Km. The protection mechanisms for both FF and DF were demonstrated. In the protection mode, the signals in the back up fiber were shifted 200 GHz towards the longer wavelength based on the routing property of the AWGs. It was observed that the restoration time for feeder fiber failure was less than 1ms. In case of DF failure, it took 12ms to switch the specific optical switch at RN to send the signals in the protection mode. The total restoration time measured along with switch reconfiguration of the remote node was recorded as 48ms.

Though this architecture provides both TDM and WDM PON services, the cost of the network is high as it uses many optical switches. More fiber has to be laid for supporting both the services simultaneously which increases the overall network cost.

In certain cases, as explained earlier, when multiple remote nodes are there, the fibers are not duplicated but an alternate path is established. In such cases, multiple fibers terminate at the ONU from one or more Remote Nodes. Such architectures require ONUs to be equipped with an optical switch/monitor or additional transceivers to be able to switch between the fiber paths leading to OLT. The Type C architecture in Figure 3(c) represents one such architecture variant where ONU is connected to 2 RNs. In this topology the power variation will depend primarily on the path length difference. Since the connections to a given ONU originate from different RNs, the possibility of working/protections fibers running along the same path or same conduit is reduced. The architecture requires additional transceivers or an optical switch which increases the cost of the ONU marginally. (Chen et al., 2008; Son et al., 2005; Sun et al., 2006; Chowdhury et al., 2008) are some variants of the architecture which have been explained earlier.

With respect to this classification, when the distribution fibers are duplicated the excess fiber length laid for protection is expressed as:

$$\Delta f_{nDF} = \sum_{i=1}^{N_2} \sum_{j=1}^{n} l_{ONU_{ij}}^u - \sum_{i=1}^{N2} \min_j (l_{ONU_{ij}}^u)$$
(6)

The power variation in terms of dB for a network protected with 'n' distribution fibers is given by:

$$\Delta p_{nFF}^{max} = \max_j [\max_i (l_{ONU_{ij}}^u) - \min_i (l_{ONU_{ij}}^u)] P_{fiber}$$
(7)

where $l_{ONU_{ij}}^u$ denotes the length of the jth upstream fiber link connecting ONU_i and the remote node (typically around 5 Km); and N_2 denotes the number of ONUs supported by the protection PON architecture.

Unlike the n x FF architecture, fiber length addition for n x DF architecture is dependent on the number of ONUs to be connected. Taking a typical 15 Km FF length and 5 Km DF length, supporting more than 3 ONUs will require more fiber to be installed for n x DF scheme than for n x FF scheme. From the protection perspective, DF is more vulnerable to fiber cuts than FF and

so protecting the network from DF failures will improve the overall network availability. Power variation depends on working/protection path length difference and this should be minimized in order to reduce the impact on network coverage.

Intra-Level Connection Oriented Architectures

This section describes architectures based on different types of intra level connections in the network to provide protection. The connection is formed either at the remote node level or at the ONU level. These architectures have interconnection fibers between a pair of ONUs or RNs and form intra-level connections with respect to the layout of the network. Most of them use an ONU to ONU interconnection, while some of the architectures also have a RN to RN interconnection for providing protection. These have been illustrated in Figures 9 and 10 and have been explained further in the following sections.

These architectures are intended to provide group protection such that each node provides protection to its peer level node. In case of ONUs, as shown in figure 9, an ONU-ONU interconnection provides protection against the distribution fiber failure. Similarly as shown in figure 10, a RN-RN interconnection provides protection against feeder fiber cut. These intra-level connection oriented architectures are further described in the following sections.

ONU-ONU Interconnection

Most papers under this classification have two ONUs forming a group, connected to each other by means of an interconnection fiber. The data of the two ONUs in a group is sent to both the ONUs of the group. Hence when the distribution fiber of one ONU fails, the affected ONU changes the state of its optical switch such that it receives its downstream data from the ONU that is paired to it via the interconnection fiber between them.

Figure 9. ONU-ONU interconnection

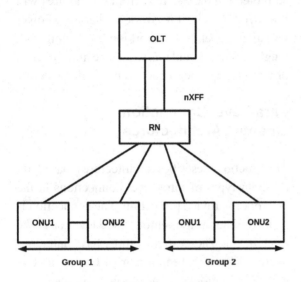

interconnection fiber length is much less than the length of the distribution fiber with regard to the network protection cost. In places where presence of obstacles between the ONUs increases the path length beyond the DF length, this architecture tends to increase the capital expenditure of the network. Further these architectures need an even number of ONUs to be deployed which is one of the constraints when it comes to deployment. When the current topology is balanced and there is a new customer to be added, two ONUs will need to be added to accommodate protection. Moreover when active components are involved along the protection path irrespective of whether they are being used or not, ONUs will have to be in the ON state always for the sake of protecting the node paired to it, thereby increasing the overall network power consumption. Figure 9 depicts an ONU based on intra-level oriented connection. The architectures which follow the same principle as this are described below.

Similarly, the affected ONU can send its upstream data through the interconnection fiber to the ONU paired to it and hence reach the OLT via the remote node. The interconnection fiber laid for protection adds value to the architecture only when the

Figure 10. RN-RN interconnection

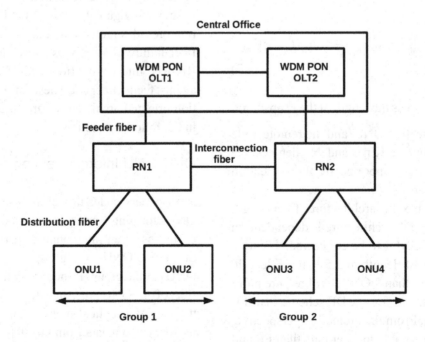

Group Protection Architecture with 1:1 Protection

A self protected WDM PON architecture providing bidirectional 1:1 protection against distribution fiber link failure has been proposed in (Chan et al., 2003). A backup fiber link exists between OLT and RN for automatic protection switching in case of any FF failure. Every output port of the 1X N AWG at the remote node connects to two adjacent ONUs through a fiber coupler. Two adjacent ONUs form a group and they are connected to each other by means of a single piece of fiber for alternate path protection. The Blue/Red filters in the ONUs are used to differentiate between the upstream and the downstream wavelengths destined for itself and its adjacent ONU, while wavelength couplers separate the downstream and the upstream wavelengths within an ONU. The spectral periodicity property of the AWG enables it to support 2 ONUs in each output port, thereby supporting 4 wavelength (upstream and downstream traffic) channels for 2 adjacent ONUs. The A, B, C and D wavelength bands are used for transmission of data. The A and C bands carry downstream traffic while B and D carry upstream traffic. The bands A and B are grouped as Blue band while the Red band comprises C and D wavelength bands.

In normal mode, the signals from the OLT reach the respective ONUs via the RN based on their dedicated upstream and downstream wavelength channels. In case of any DF failure, the power monitors at the ONUs detect loss of signal and reconfigure their optical switches to restore physical link. Now the signals from the affected ONU are routed through the adjacent ONU via the interconnection fiber installed between them. In this way adjacent ONUs in a group protect each other. Thus the ONU is able to support its own subscribers and simultaneously transmit the signals of the adjacent ONU to the OLT in protection mode, while the OLT remains unaware of DF failure.

The above architecture was experimentally demonstrated with the length of the DF being 20 Km and that of the interconnection fiber being 2 Km. An EDFA was used in the OLT to achieve the required transmitter power and to compensate for the insertion loss. In normal mode the signals travelled 20 Km while in the protection mode they travelled 22 Km. 2.5 Gbps $2^{23}- 1$ PRBS data was used to observe the downstream and upstream transmission performance under normal and protection modes. A 16-channel AWG with 100 GHz channel spacing and free spectral range (FSR) equaling 12.8 nm was used at the RN. The pass band of each Blue/Red filter was equal to 18 nm. The switching time or the restoration time in case of DF cut was reported to be 18 ms. The Bit Error Rate performance was also measured. The receiver sensitivities at 2.5 Gbps varied from -25 dBm to -26.5 dBm.

An architecture similar to (Chan et al., 2003) has been proposed in (C. Lee et al., 2003), which supports only half the number of ONUs when compared to (Chan et al., 2003). It has greatly reduced the number of optical couplers employed at the remote node and has faster restoration time. The wavelength assignment plan and the working principle are similar to that of (Chan et al., 2003). In this case, the actual traffic restoration time in case of DF cut was recorded as 9 ms while the measured receiver sensitivities at 2.5 Gbps varied from -31.5 dBm to -32.6 dBm.

Protection Architecture for Broadcast and Select WDM PON

A WDM PON architecture that supports broadcast-and-select services with the help of Coarse Wavelength Division Multiplexing (CWDM) technology at low cost and increased network reliability is presented in (Z. Wang, Zhang, et al., 2005). Here, the OLT consists of a CWDM mux and demux. Feeder fiber is protected with the help of additional back up fiber link. For N ONUs, N transceivers are placed at the OLT to provide dedicated services and one extra transmitter is deployed for providing broadcast services. The CWDM standard defines 18 wavelengths. The 1550 nm wavelength is allocated for broadcast service while the 1310 nm wavelength is located at the water peak of the standard fiber. The remaining 16 wavelengths are assigned for the dedicated upstream and downstream traffic of the 8 ONUs. A 2X N splitter acts as a remote node, connecting the OLT with N ONUs.

The ONUs consist of a Control Circuit (CC) for failure detection and an optical switch (OS) to switch the signals to the alternate protection path. A set of band filters have been used to separate the broadcast signals and the dedicated bidirectional signals at the ONU. A 50/50 coupler is placed in the broadcast path of one of the band filters to drop half of the broadcast signals at the broadcast receiver while the other set of broadcast signals are sent to the adjacent ONU in the same group via the interconnecting fiber so that they can be retrieved in the protection mode. Two cascaded 3 dB thin film filters (TFF) with 1550 nm as the center wavelength are used to transport the up and downstream signals to the transceivers present in the ONU. Each TTF acts as a unidirectional OADM where the downstream wavelength is dropped and the upstream wavelength is added for an ONU. In normal mode, the signals from the OLT reach the respective ONUs through the RN followed by the distribution fibers. In case of distribution fiber failure, the control circuit present in the ONU detects it and switches the data to the alternate path i.e. via the interconnection fiber present between the ONUs. Hence, the signals still reach the affected ONUs in spite of a DF failure. Similarly, with the help of the adjacent ONU, the affected ONU communicates with the OLT in the reverse manner.

The architecture was experimentally demonstrated with a 1X 16 splitter constructed using 2 stages of 1 X 4 splitter with an insertion loss of 12.5 dB for serving 16 ONUs. The length of the interconnecting fiber equaled 1 Km while the total distance between the OLT and the ONU was 20 Km. Different standard CWDM wavelength bands were used for the broadcast signal transmission and dedicated upstream and downstream signal transmission of the ONUs. With 1.25 Gbps PRBS data, the restoration time in case of DF failure was measured as 3 ms. BER was also measured and it was observed that at 10^{-9} BER, the receiver sensitivities were around -28 dBm. The System margin for dedicated traffic was 8-10 dBm and that of broadcast traffic in normal mode was 6.3 dBm and 2.7 dBm in protection mode.

Protection Architecture Based on Centralized Protection

Another ONU group protection architecture providing centrally controlled protection for traffic restoration in case of distribution fiber failure has been proposed in (Z. Wang, Sun, et al., 2005). To simplify ONU design and to reduce the total amount of required network resources, an optical isolator is placed in the OLT as a protection equipment. Here one FF carries downstream data to the ONUs and the other carries upstream data to the OLT. There are 4 groups of ONUs, with each group having a pair of ONUs, connect to the OLT. Here ONU_i^1 and ONU_i^2 refer to the first and second ONU in the ith group respectively.

In this architecture, the blue band wavelengths are allocated to ONU_i^1 where i = {1, 2, 3, 4} and the red band wavelengths are allocated to ONU_i^2. A fiber coupler present in ONU_i^1 combines its blue band output with the output from ONU_i^2. Similarly ONU_i^2 combines its red band output from ONU_i^1. In normal mode the optical switch in the OLT is in the bar state.

Due to the periodic and cyclic properties of the AWG and the proposed wavelength assignment plan, all the downstream traffic is connected to the left most ONUs present in every group. These downstream signals are then separated by the blue/red filter and passed on to the destined ONUs. The upstream signals are power split by the fiber coupler and transmitted to the OLT. Due to the presence of an optical isolator, only the upstream signals from the rightmost ONUs will be blocked. Thus under normal operation, only half the DFs and the interconnection fibers between the ONUs in the network are active. The rest of the fibers are meant for protection. In case of fiber failure, the decision circuit present in the OLT detects the failure and triggers the optical switch at the OLT to switch its state such that data is rerouted through the protection fibers. Now the upstream and downstream signals are sent in FF1 and FF2 respectively. Thus, all the bidirectional data wavelengths are switched from working fibers to their respective protection fibers. In a similar manner, failure of the interconnection fiber is protected to restore the traffic.

A 16-channel AWG with 100 GHz channel spacing and FSR = 12.8nm was used at the remote node. Instead of wavelength couplers, 50/50 couplers were used for simplicity. Length of FF and DF were 20km and 2km. The interconnection fiber was 2 kms long. In case of DF link failure receiver sensitivities were -24.5 dBm to -26dBm. The 1.5dB power penalty was due to fiber chromatic dispersion. The time to switch to the protection mode was reported to be 9ms.

Though this centralized protection scheme has reduced ONU complexity, it cannot support simultaneous fiber cuts of a working fiber or a protection fiber in a group, whose probability of occurrence is $N^2p^2/4 + O(p^2)$. Here, p represents the probability of a working or protection fiber being cut. Also the drawback with this proposed architecture is that the restoration of one DF failure interrupts the data transmission of all the other ONUs as the data of the unaffected ONUs also get switched to the protection path.

Another architecture has been proposed in (Sue, 2006b), where 3 feeder fibers have been used and 3 ONUs form a group. A detailed explanation of the same can be found in detail in Section 5.1.1. Though this architecture reduces the unavailability to a great extent, the cost of adding more interconnection fibers between ONUs adds to the overall network cost.

The additional interconnection fiber laid for protection in an architecture with ONU-ONU interconnection when N is even, is given by:

$$\triangle f_{ONU-ONU} = \sum_{t=1,3..}^{N_2} l^c_{ONU_i,ONU_{i+1}} = \frac{N_2}{2} \overline{l^c_{ONU,ONU}} \tag{8}$$

The power variation in terms of dB for a Intra-level connection oriented architectures with ONU-ONU inter- connection is given by:

$$\triangle p^{max}_{ONU-ONU} = 2P_{OSW} + P_{fiber}[\max_{i=1,3..}^{N_2-1} l^c_{ONU_i,ONU_{i+1}} \tag{9}$$

where $l^c_{ONU_i,ONU_{i+1}}$ is the length of the cross-stream fiber link connecting ONU_i to ONU_{i+1}; typically around 2 Km. P_{OSW} is the power attenuation of the optical switch; typically this value is around 1 dB, P_{fiber} is the power attenuation of the fiber per km. For SMF-28 the value is 0.2 dB and N2 is the number of ONUs supported by the protection PON architecture.

RN-RN Interconnection

Remote Node interconnection architectures can be used when there is a high possibility of fiber break in the feeder fiber segment. The primary objective of such architectures is to provide an alternate path when a feeder fiber fails.

As shown in Figure 10, RNs are interconnected to each other, to provide an alternate path for the data to flow from the OLT to the ONU in case of a FF failure. In some architecture variants like (X. Wang, Wang, Zhang, & Wang, 2009; Son et al., 2005), RNs are connected to different OLTs thus backing each other upon an OLT failure. These architectures provide management advantage for OLT maintenance without disrupting the services.

The wavelength allocation scheme is such that the adjacent ODNs operate using mutually exclusive wavelengths or wavelength bands.

In the architecture presented in (Son et al., 2005), only feeder fiber failure is supported. Here, the remote nodes are interconnected by a fiber and in case of failure, the fiber links of the neighboring WDM PON are used to restore the disrupted signals. Two different wavelength bands are used for the neighboring WDM PONs and the cyclic property of the AWG has been effectively utilized. Under normal operation, the signals from the OLTs are routed via respective feeder fibers and distribution fibers to the destined ONUs. In case of an FF failure, the power monitors at the OLT detect the loss of signal condition and reconfigure the optical switches at the OLT to restore the network via the RN-RN interconnection link. The Blue/Red filters present in the OLT and the RNs identify and differentiate the blue and red band signals (i.e. upstream and downstream) and route them accordingly to their respective ONUs.

The experimental setup for feeder fiber protection included 20 WDM signals for each ONU group. The length of FF, DF and IF were 20 Km, 5 Km, and 5 Km respectively. A 1 X 16 AWG with 100 GHz channel spacing and FSR equal to 12.8 nm was employed at the OLT and the RN. Hence the distance travelled by the signals in the normal mode and the protection mode is 25 Km and 30 Km respectively. For simulation Acoustic Optic Modulator (AOM) had been used for the FF cut scenario. Restoration time was observed to be 9 ms. Extinction ratio for the best and worst signal was 14.2 dB and 9.2 dB respectively.

The same paper proposes another architecture for higher reliability by including a System Management Module. This module checks whether the link partners are alive by active messaging to trigger the restoration process during failure. In case of failures, the optical switch units at the OLT are reconfigured by the system management module to restore connectivity to the ONU. This scheme uses TFF (Thin Film Filters) in place of normal wavelength filters, thus reducing the cost of the overall network protection architecture. The polling frames sent include a 4 bit slot number and 3 bit protection pair number to notify the ONU about the protection pair that is being used actively by the ONU. OLT maintains the working table of the ONUs and periodically polls each of the ONUs using the polling frame. Once an ONU receives the frame it responds to the polling frame and thus the working table is updated. Polling frame is sent every 5ms and restoration time was experimentally reported to be less than 16 ms. However the best case restoration time is 9.3 ms.

The extra fiber laid for protection when N is even can be expressed as:

$$\Delta f_{RN-RN} = \sum_{t=1,3..}^{N_1-1} l^c_{RN_i,RN_{i+1}} = \frac{N_1}{2} \overline{l^c_{RN,RN}} \qquad (10)$$

Power budgets of these architectures have to be carefully examined for variations due to additional fiber length and any optical component attenuation encountered when the operational path is switched from the working path to the protection path. The power consumed in dB with respect to this architecture is expressed below.

When the remote node is active:

$$\Delta p^{max}_{RN-RN} = 2P_{OSW} + P_{fiber}[\max_{i=1,3..}^{N_1-1} l^c_{RNi,RNi+1}] \qquad (11)$$

When the remote node is passive:

$$\Delta p^{max}_{RN-RN} = P_{OSW} + P_{\lambda Filter} + P_{combiner} + P_{fiber}[\max_{i=1,3..}^{N_1-1} l^c_{RNi,RNi+1}] \qquad (12)$$

where is the length of cross-stream fiber link connecting RN_i to RN_{i+1} which is typically around 3 to 5 Km, $P_{\lambda filter}$ is the power attenuation of the wavelength filter typically this value is around 1 dB, $P_{combiner}$ is the power attenuation of the splitter/combiner; typically this value equals $1.5 + 10\log_{10}M$ dB where M is the number of ports,

N1 is the number of remote nodes used in the protection PON architecture.

For failure protection these networks operate at much less than the peak bandwidth that can be supported by WDM PON. This scheme requires much less fiber overhead to support a feeder fiber failure for twin optical distribution networks. Power budget variations due to RN-RN interconnection need to be considered carefully, as typically the fiber length between the remote node and OLT is usually more than 75% of the total fiber length laid between the OLT and the ONU.

COMPARATIVE ANALYSIS

This section presents a comparison of the inter-level and intra-level connection oriented protection architectures based on the resources they have deployed for protecting the network. We assume that a typical WDM PON consists of transceivers and a 1 X N AWG at the OLT, a 1 X N AWG at the remote node and transceivers at the ONU for data transmission to take place between the OLT and the ONU. The components which have been deployed in the proposed protection architectures at the OLT, ONU and remote node, in addition to the typical WDM PON architecture assumed, have been listed in the comparison tables.

Tables 1-4 compare the total number of feeder, distribution, interconnection (if any) fibers and the total number of wavelengths used in the resilient architectures. We assume that for a 20km PON network, the feeder, distribution, and interconnection fibers are 15km, 5km and 2km long respectively. Given 'N' as the number of ONUs, the total number of ONUs that can be supported by the respective architecture (represented by 'K') has also been enumerated. For example, the architecture proposed in (K. Lee et al., 2007) supports N number of ONUs which is equal to the variable K i.e. K = N and hence requires 'K' distribution fibers to connect to the N ONUs. However in the architecture proposed in (Sun et al., 2006), the number of ONUs supported is N, which is equal to K i.e. K = N, but it requires 2K distribution fibers to connect to the N ONUs. In (Sun et al., 2006), if N = 8, K becomes 8, while the number of distribution fibers required to connect to the 8 ONUs is 2K which becomes 16. In order to provide protection, the additional fibers present in these architectures when compared to a typical unprotected WDM PON have been explicitly listed, in terms of the number of extra fibers deployed for protection, and in terms of the fiber lengths in kilometers.

For (Chen et al., 2008), the number of additional distribution fibers added for protection has been denoted by 'x', where 'x' refers to the number of stages at which splitters have been placed. The total number of ONUs that this architecture will support again depends on the same.

The architectures have also been compared based on their level of protection. Though few architectures support only FF failure, most of them are susceptible to a combination of FF, DF and remote node failures. These protection levels have also been recorded. The switching/restoration times mentioned in the table with respect to every architecture have been obtained from the experimental setup that the respective architecture has considered. In most cases, the architecture proposed and the experimental setup assumed differed i.e. if the architecture had been proposed for N ONUs, the experiments were performed only with 2 ONUs. The receiver sensitivities have been mentioned for upstream signal in the normal mode (NU) and in the protection mode (PU) at BER = 10^{-9}. It can be seen that the receiver sensitivities are either equal or have very low variation.

The power variation in dB has been calculated as the difference between the power consumed by the upstream or downstream signal in the working path and the protection path. Hence the values mentioned in the table basically state the extra power that would be consumed by the network in case of failure i.e. when the signal traverses through the protection path from the OLT to ONU

Table 1. Comparison of inter-level connection oriented architectures

	(K. Lee et al., 2007)	(Sun et al., 2006)	(Son et al., 2005)	(Sue, 2006b)	(K. Lee et al., 2008)	(J.-Y. Kim et al., 2009)	(Chowdhury et al., 2008)	(Chen et al., 2008)
Metrics								
Number of ONUs supported (K)	N	N	2N	N	N-1	N	N	–
No. of Feeder fibers	2	2	2	K + 1	2	2	2	4
No. of Distribution fibers	K	2K	4K	K	2K	2K	2K	–
No. of Interconnection Fibers	No	No	No	No. of ONUs * No. of wavelengths	No	No	No	M
No. of wavelengths	K upstream and K downstream	K upstream and K downstream	K upstream and K downstream	(1/2)K upstream and (1/2)K downstream	K upstream and K downstream	K upstream and K downstream	(1/2)K upstream and (1/2)K downstream	M
Additional fiber added for protection	1 FF	1 FF and K DFs	1 FF and K DFs	2 FF and IF = (No. of ONUs * No. Of wavelengths)	1 FF and K DFs	1 FF and K DFs	1 FF and K DFs	2 FF, x DF and M IF
Total length of additional fiber laid for protection (km)	15	15 + K*5	15 + K*5	30 + 2*(No. of ONUs * No. Of wavelengths)	15 + K*5	15 + K*5	15 + K*5	–
Failures Supported	FF only	FF, DF and RN	FF, DF and RN	FF and DF	FF and DF	FF and DF	FF, DF and RN	FF and DF
Switching time / Restoration time (ms)	8	3	9	–	4	10	–	–
Receiver Sensitivity (dBm)	NU = -38.5; PU = -39	NU = PU = -20.5	–	–	NU = -25.8; PU = -26	–	–	–
Power Variation (Working path vs. Protection Path) dB	$\Delta L_{FF} = 0.4$	$\Delta L_{FF} + \Delta L_{DF} = 0.6$	$\Delta L_{FF} + \Delta L_{DF} = 0.6$	$OS + \Delta L_{FF} = 1.4$	$\Delta L_{FF} + \Delta L_{DF} = 0.6$	$\Delta L_{FF} + \Delta L_{DF} = 0.6$	$\Delta L_{FF} + \Delta L_{DF} = 0.6$	$\Delta L_{FF} + \Delta L_{DF} = 0.6$
Power Penalty (WDM PON assumed vs. Working path of proposed architecture) dB	2	5	7	5	8	6	12 (compensated by EDFA)	14 dB (3 stage 1x2 splitter)

or vice-versa. For calculating this power variation, we assume the following:

- Feeder fiber length difference between normal mode and protection mode = 2km

- Distribution fiber length difference between normal mode and protection mode = 1km
- Length of the interconnection fiber = 2km
- Power penalty = 0.2dB/km

Table 2. Comparison of inter-level connection oriented architectures

Components additionally present for protection with respect to the normal PON architecture assumed									
	(K. Lee et al., 2007)	(Sun et al., 2006)	(Son et al., 2005)	(Sue, 2006b)	(K. Lee et al., 2008)	(J.-Y. Kim et al., 2009)	(Chowd-hury et al., 2008)	(Chen et al., 2008)
OLT	2XN AWG 2 optical switches 1 power monitor	N X 2 AWG N optical switches N WDM filters	N wave-length cou-plers 1 optical switch 1 3-dB coupler 1 WDM filter 1 power monitor	1 decision circuit 1 optical switch 2 fiber couplers 1 WDM filter 1 wavelength coupler 1 power monitor	1 fiber cou-pler 1 circu-lator N WDM filters	2 X 2N AWG N + 2 optical switches 2 WDM filter 2 circula-tors N + 1 power monitors	1 OCS unit[1] N 3-dB splitters N NUCs[2] 2 AWGs 2 EDFAs	default
ONU	default	N 2X2 couplers N 3-db fiber couplers	1 WDM filter 1 power monitor 1 optical switch	1 decision circuit 4 fiber couplers 1 power monitor 1 wavelength coupler	1 WDM filter 1 power monitor 1 optical switch 1 decision circuit	N power mon-itors N optical switches	1 Inter-leaver Filter	1 optical switch 1 WDM filter
Re-mote-Node	2 X N AWG	2 remote nodes - each having the following com-ponents extra: 2 X N AWG 1 B/R filter	2 remote nodes - each having the following components extra: N WDM filters	3 X N AWG	N X N AWG	2 X 2N AWG	1 remote node with 1 X N AWG	4 X M AWG

The power variation with respect to additional fiber traversed is calculated as the product of fiber length difference (FF or DF) and power penalty. We denote (ΔL_{FF}) and (ΔL_{DF})) as the extra power consumed by the signal when it traverses through the feeder fiber and distribution fiber in the protection mode. As per our assumptions of fiber length difference and the power penalty, (ΔL_{FF}) = 0.4dB; (ΔL_{DF}) = 0.2dB and (IF) = 0.4dB. The power penalty metric specifies the extra power consumed by the proposed resilient architectures when compared to the typical WDM PON archi-tecture assumed, which is mainly constituted by the power consumed by the additional compo-nents as mentioned in the table. The values that we have assumed for power penalty calculation are as follows:

- Optical coupler = 3 dB
- Wavelength coupler = 1 dB
- Arrayed Waveguide Grating = 4 dB
- Optical Switch = 1 dB
- WDM Filter or B/R Filter = 1 dB
- Splitter = $1 + \log_2 N$

Table 3. Comparison of intra-level connection oriented

	(Chan et al., 2003)	(Z. Wang, Sun, et al., 2005)	(C. Lee et al., 2003)	(Son et al., 2005)	(Z. Wang, Zhang, et al., 2005)	(X. Wang et al., 2009)
Metrics for Comparison						
Number of ONUs supported (K)	2N	N	N	2N	N	2N
No. of Feeder fibers	2	2	1	2	2	2
No. of Distribution fibers	2K	K	K	2K	K	2K
No. of Interconnection fibers	ONU-ONU (K)	ONU – ONU (K/2)	ONU – ONU (K/2)	RN – RN (1)	ONU – ONU (K/2)	RN – RN (1)
No. of wavelengths	K upstream and K downstream	K upstream and K downstream	K upstream and K downstream	K upstream and K downstream	(K upstream and K downstream) + 2	K upstream and K downstream
Additional fiber added for protection	1 FF + K IF	1F F + (K/2)IF	(K/2)IF	1FF and 1 IF between remote nodes	1 FF and K/2 IF	1 FF and 1 IF
Total length of additional fiber laid for protection (km)	15 + K*2	15 + 2*(K/2)	2*(K/2)	15+2 = 17	15 + 2*(K/2)	15 + 2
Failures Supported	FF and DF	FF and DF	DF only	FF only	FF and DF	FF and RN
Switching time / Restoration time (ms)	18	9	9	10	3	< 16
Receiver Sensitivity (dBm)	NU = -26; PU = -25	NU = PU = -24.5	NU=-31.7; PU=-31.8	NS = PS = -23.5	NU = PU =-28	–
Power Variation (Working path vs. Protection Path) dB	$2OS + \Delta L_{FF}$ =2.4	$IF + \Delta L_{DF}$ = 0.6	$2OS + IF + \Delta L_{DF}$ = 2.6	2 WDM filters + IF + ΔL_{FF} = 2.8	1 filter + IF = 1.4	$IF + \Delta L_{FF}$ 0.8
Power Penalty (WDM PON assumed vs. Working path of proposed architecture) dB	8	4	5	7	$2.5 + \log_2 N$	7

CONCLUSION AND FUTURE WORK

In this chapter, we have presented a study of WDM PON based protection architectures that have been proposed in literature so far. The protection architectures mostly support both distribution and feeder fiber failures, though some of the architectures can survive only one of the fiber cuts. Protection is provided either by duplicating the resources or by creating an alternate path, for the data to reach the destination in case of failure. Node failures such as Remote Node (RN) failures are overcome by duplicating them. Each architecture has adapted different strategies to restore the network from failure. The various protection schemes have been briefly described and then compared based on the number of components used and the time taken to restore the network from failures.

Table 4. Comparison of intra-level connection oriented architectures

Components additionally present for protection with respect to the normal PON architecture assumed						
	(Chan et al., 2003)	(Z. Wang, Sun, et al., 2005)	(C. Lee et al., 2003)	(Son et al., 2005)	(Z. Wang, Zhang, et al., 2005)	(X. Wang et al., 2009)
OLT	default	2 optical isolators 1 decision circuit 1 optical switch	default	Two WDM PON OLTs - each having the following components extra: 1 optical switch N wavelength couplers 1 WDM filter 1 fault monitor	CWDM mux/demux	Two WDM PON OLTs - each having the following components extra: 1XN AWG 1 (C + L f liter) 1 WDM filter 1 optical switch
ONU	1 B/R filter 1 wavelength coupler 2 power monitors 2 optical switches	1 B/R filter 1 wavelength coupler 1 fiber coupler	1 B/R filter 1 wavelength coupler 2 optical switches	1 wavelength coupler	2 band filters + 2 Thin film filters 1 50/50 coupler 1 control circuit 1 optical switch	1 CWDM Tunable laser Modulator
Remote Node	N 3-dB fiber couplers	2 X N AWG	2 X N AWG 1 3-dB fiber coupler	Extra remote node with every remote node having the following components extra: 1 3-dB fiber coupler 2 WDM filters	1 2 X N splitter 1 3-dB fiber coupler	1 3-dB fiber coupler 1 DCTF WDM

With respect to overall network availability, resilient PON architectures have to be further analyzed. The context of associated cost involved in implementing resilience, probability of availability and the impact on the network due to failure can be used to decide on the type of resiliency that can increase the availability of the network. Given that one of the primary factors adding to the cost of PON is the fiber installation cost, realizing desired availability using minimal overall length of fiber is one of the problems that can be studied further. Some of the PON architectures while trying to provide resiliency, reduce the maximum range of the Optical Distribution Network. Means to prevent this is another aspect that can be explored. A detailed quantitative analysis of the protection architectures can also be done in the future, to identify the best fit protection architecture for a given scenario.

REFERENCES

Banerjee, A., Park, Y., Clarke, F., Song, H., Yang, S., & Kramer, G. (2005). Wavelength-division-multiplexed passive optical network (WDM-PON) technologies for broadband access: A review [Invited]. *Journal of Optical Networking*, *4*(11), 737–758. doi:10.1364/JON.4.000737

Chan, T.-K., Chan, C.-K., Chen, L.-K., & Tong, F. (2003, November). A self-protected architecture for wavelength-division-multiplexed passive optical networks. *IEEE Photonics Technology Letters*, *15*(11), 1660–1662. doi:10.1109/LPT.2003.818657

Chen, J., & Wosinska, L. (2007). Analysis of protection schemes in PON compatible with smooth migration from TDM-PON to hybrid WDM/TDM-PON. *OSA Journal of Optical Networking, 6*(5), 514–526. doi:10.1364/JON.6.000514

Chen, J., Wosinska, L., & He, S. (2008, March). High Utilization of wavelengths and simple interconnection between users in a protection scheme for passive optical networks. *IEEE Photonics Technology Letters, 20*(6), 389–391. doi:10.1109/LPT.2007.915655

Chowdhury, A., Huang, M.-F., Chien, H.-C., Ellinas, G., & Chang, G.-K. (2008, February). A self-survivable WDM-PON architecture with centralized wavelength monitoring, protection and restoration for both up- stream and downstream links. In *Conference on Optical Fiber Communication/National Fiber Optic Engineers* (pp. 1-3).

Hill, A., Brierley, M., Percival, R., Wyatt, R., Pitcher, D., & Pati, K. (2002). Multiple-star wavelength-router network and its protection strategy. *IEEE Journal on Selected Areas in Communications, 16*(7), 1134–1145. doi:10.1109/49.725184

IEEE 802.3 Ethernet in the First Mile Study Group. (2001). *Ethernet passive optical networks* (EPONs).

IEEE. (2006). Standard for Information Technology-Telecommunications and information exchange between systems- Local and metropolitan area networks-Specific requirements part 3: Carrier sense multiple access with collision detection (CSMA/CD) access method and physical layer specifications.

ITU-T. (1998). *ITU-T recommendation G.983.1: Broadband optical access systems based on passive optical networks.* PON.

ITU-T.(2000). ITU-T recommendation G.983.2: The ONT management and control interface specification for ATM PON.

ITU-T.(2001). ITU-T recommendation G.983.3: A broadband optical access system with increased service capability by wavelength allocation.

ITU-T.(2002). ITU-T recommendation G.983.5: A broadband optical access system with enhanced survivability.

ITU-T.(2003). ITU-T recommendation G.984.1: Gigabit-capable passive optical networks (GPON): General characteristics.

Iwatsuki, K., & Kani, J. (2009, September). Applications and technical issues of wavelength-division multiplexing passive optical networks with colorless optical network units [Invited]. *OSA Journal of Optical Networking, 1*(4), 17–24.

Kim, H. D., Kang, S.-G., & Le, C.-H. (2000, August). A low-cost WDM source with an ASE injected Fabry-Perot semiconductor laser. *IEEE Photonics Technology Letters, 12*(8), 1067–1069. doi:10.1109/68.868010

Kim, J.-Y., Mun, S.-G., Lee, H.-K., & Lee, C.-H. (2009, November). Self-restorable WDM-PON With a Color- Free Optical Source. *OSA Journal of Optical Networking, 1*(6), 565–570.

Lam, C. (2007). *Passive optical networks: Principles and practice.* Elsevier/Academic Press.

Lee, C., Chan, T., Chan, C., Chen, L., & Lin, C. (2003). A group protection architecture (GPA) for traffic restoration in multi-wavelength passive optical networks. In *Proc. Of European Conference on Optical Communications* (ECOC) (vol. 2).

Lee, J. H., Choi, K.-M., Moon, J.-H., Mun, S.-G., Lee, H.-K., & Kim, J.-Y. (2009, October). A seamless evolution method with protection capability for next-generation access networks. *Journal of Lightwave Technology, 27*(19), 4311–4318. doi:10.1109/JLT.2009.2023608

Lee, K., Lee, S., Lee, J., Han, Y., Mun, S., & Lee, S. (2007, April). A self-restorable architecture for bidirectional wavelength-division-multiplexed passive optical network with colorless ONUs. *Optics Express, 15*, 4863–4868. doi:10.1364/OE.15.004863

Lee, K., Mun, S.-G., Lee, C.-H., & Lee, S. B. (2008, May). Reliable wavelength-division-multiplexed passive optical network using novel protection scheme. *IEEE Photonics Technology Letters*, *20*(9), 679–681. doi:10.1109/LPT.2008.919445

Maier, M., Herzog, M., Scheutzow, M., & Reisslein, M. (2005). Protectoration: A fast and efficient multiple-failure recovery technique for resilient packet ring using dark fiber. *Journal of Lightwave Technology*, *23*(10), 2816. doi:10.1109/JLT.2005.856165

Mukherjee, B. (2006). *Optical WDM networks*. New York, NY: Springer-Verlag Inc.

Rubin, I., & Ling, J. (2002). Failure protection methods for optical meshed-ring communications networks. *IEEE Journal on Selected Areas in Communications*, *18*(10), 1950–1960.

Sivalingam, K. M., & Subramaniam, S. (2004). *Emerging optical network technologies: Architectures, protocols and performance*. Secaucus, NJ: Springer-Verlag, Inc.

Son, E. S., Han, K. H., Lee, J. H., & Chung, Y. C. (2005, March). Survivable network architectures for WDM PON. In *Conference on Optical Fiber Communication/National Fiber Optic Engineers, Technical Digest* (vol. 5).

Sue, C. C. (2006a, November). 1: N protection scheme for AWG-based WDM PONs. In *IEEE Global Telecommunications Conference* (GLOBECOM) (pp. 1-5).

Sue, C. C. (2006b, July). A novel 1: N protection scheme for WDM passive optical networks. *IEEE Photonics Technology Letters*, *18*(13), 1472–1474. doi:10.1109/LPT.2006.877577

Sun, X., Chan, C.-K., & Chen, L. K. (2006, February). A survivable WDM-PON architecture with centralized alternate-path protection switching for traffic restoration. *IEEE Photonics Technology Letters*, *18*(4), 631–633. doi:10.1109/LPT.2006.870135

Technologies, W. D. M. (2008, January). *WPON white paper*. Retrieved from http://www.ciphotonics.com/PDFs_Jan08/WPON_White_Paper_v10.pdf

Verbrugge, S., Colle, D., Pickavet, M., Demeester, P., Pasqualini, S., & Iselt, A. (2006, June). Methodology and input availability parameters for calculating OpEx and CapEx costs for realistic network scenarios. *OSA Journal of Optical Networking*, *5*(6), 509–520. doi:10.1364/JON.5.000509

Wang, X., Wang, S., Zhang, A., & Wang, J. (2009, November). A novel highly reliable WDM-PON system. In *Proc. of Communications and Photonics Conference and Exhibition Asia* (ACP) (pp. 1-2).

Wang, Z., Sun, X., Lin, C., Chan, C.-K., & Chen, L.-K. (2005, March). A novel centrally controlled protection scheme for traffic restoration in WDM passive optical networks. *IEEE Photonics Technology Letters*, *17*(3), 717–719. doi:10.1109/LPT.2004.842378

Wang, Z., Zhang, B., Lin, C., & Chan, C.-K. (2005, September). A broadcast and select WDM-PON and its protection. In *Proc. of European Conference on Optical Communication* (ECOC) (vol. 3, pp. 549-550).

ENDNOTES

[1] OCS = Optical Carrier Suppression Unit

[2] NUC = 1 power monitor + 1 optical switch + 1 interleaver filter + 1 modulator and receiver

Chapter 4
Fault Management in Transparent Optical Networks

Carmen Mas Machuca
Technische Universität München, Germany

ABSTRACT

The advantages of transparent optical networks such as high capacity and low cost can be outweighed by their complex fault management and the high impact of the faults occurring within them. Indeed, transparent optical networks reduce unnecessary, complex, and expensive opto-electronic conversion, to the cost of having faults more deleterious and affecting longer distances than in opaque networks. Moreover, transparent optical networks have limited monitoring capabilities, which could hinder efficient and accurate fault detection and localization. Different approaches have been proposed in the literature to perform fault localization, targeting different fault scenarios (e.g. single/multiple faults or looking at the optical/higher layers), and considering different assumptions (e.g. ideal/existence of false or lost alarms). Furthermore, fault management depends on the placement of monitoring equipment, whose optimization has been studied and also presented in this chapter.

INTRODUCTION

The increasing number of users and the creation of new services and bandwidth-hungry applications have precipitated an explosion of bandwidth requirements. This increase in required bandwidth has driven the evolution of backbone, metro and even access networks towards an optical physical layer.

Optical networks can offer the required, high bandwidth to the users and support the constantly increasing traffic. There are two types of optical networks: opaque and transparent. The former perform opto-electronic conversion at their nodes, while having an optical transmission be-

DOI: 10.4018/978-1-61350-426-0.ch004

tween nodes, whereas the latter offer full optical transmission of the signal end-to-end. Optical networks are being migrated from opaque towards the promising transparent implementations that deliver lower cost and high flexibility. In opaque optical networks, the conversion of the received optical signal to the electrical domain at each network before transmitting it optically is very costly and limiting because it depends on the bit-rate, modulation format and so forth. However, this conversion allows a close monitoring of the signal quality (e.g. Bit Error Rate, BER) and the regeneration of the signal overcoming the restrictions of fiber transmission (e.g. attenuation, dispersion and non-linearities) and even possible signal degradations caused by soft failures (e.g. fiber bending). On the other hand, transparent optical networks reduce unnecessary, expensive opto-electronic and electro-optic (O/E/O) conversions, offer high data-rates, provide flexible switching, and support multiple types of clients (e.g. different bit rates, modulation formats and protocols). However, the faults in transparent optical networks are more deleterious and affect longer distances than in opaque networks. Furthermore, the fiber and node transmission limitations (attenuation, dispersion, filter concatenation and so on) become more significant due to the longer distances, their accumulative effects and the possible overlay with faults. The accumulation of impairments and faults as well as their spread through the network, significantly increase the difficulty of the detection, identification and localization of the fault(s).

These facts highlight the importance that fault management has in transparent optical networks: faults are more difficult to detect, their impact is much broader, and due to the high transmission capacity, the information that is lost when a fault occurs is substantial. In the near future, meshed optical transparent networks will commonly be used. One of the main faced problems is the fault identification, localization and management because of the reasons described.

The impact of a network fault is usually measured as a service interruption time. Services may have an associated Service Level Agreement (SLA) that defines the maximum allowed service interruption time. When the service disruption is longer, a penalty may be paid to the user. In the early 1990s, when no transparent networks were deployed, Potter (1991) showed that the cost of faults leading to network unavailability was from $50,000 to $100,000 per hour. Historically, some business users were penalized but today, with the new services having much stricter requirements, there has been a substantial increase in the utilization of penalties, which further increases the cost associated with faults. Hence, operators are investing in increasing their network reliability and reducing the cost associated with faults which are directly related to fault detection and localization accuracy, efficiency and time.

TRANSPARENT OPTICAL NETWORKS

A fault can be defined as the unexpected interruption of component function. Two types of faults can be distinguished: *Soft* faults are the faults causing degradations of the signal quality and may occur due to aging, misalignments or non-ideal environment conditions (temperature, humidity, etc.), whereas *hard* faults are the faults causing a complete signal interruption such as fiber cut. Based on statistics given by Grover (2004), faults at the nodes are rarer than faults at the links. Most of the hard faults of links are due to digs-up, whereas most of the soft faults of links are due to material degradation to undesired environmental conditions.

In addition, not only faults but also attacks should be considered. In this context, an attack is defined as an intentional action against the ideal and secure functioning of the network. Attacks can broadly be classified as eavesdropping or service disruption, based on the effects of the attacks.

Eavesdropping can be addressed with Semantic security, whereas service disruption can be solved with Fault Management (Mas Machuca, Tomkos & Tonguz (2005)). In the remaining chapter, we will consider the attacks aiming the service disruption also as faults.

Faults in Transparent Optical Networks

Faults in transparent networks differ significantly in their cause and as consequence, in their effects. We propose a classification of the faults based on these differences:

- **Hard faults:** Defined as the faults causing a complete signal interruption. They are mostly caused by an external action (erroneous dig-up, rodents, sabotage, etc.). Some examples are:
  - **Fiber cut:** This has been shown to be the most common fault. The causes of a fiber cut can be classified as human error, accidental, or malicious (i.e. attack). According to studies by the FCC (2001) and Crawford (1993), human error is by far the most common cause of cable faults, accounting for over 40% of all cable malfunctions. These faults can be detected by any monitoring equipment since they cause a drop in the optical signal power and the exact location can be found by using Optical Time Domain Reflectometry (OTDR) techniques.
  - **Power cut:** Some network components use power for their functioning. For example, an Erbium Doped Fiber Amplifier (EDFA) requires power for the pumping and opaque cross-connects need power for their switching fabrics. Depending on the equipment, and its sensitivity to sudden loss of power, it can be connected

to an uninterruptible power supply (UPS). Otherwise, the power cut, as a hard fault, can interrupt all the connections going through that particular component.

- **Soft faults:** These are the faults caused by equipment misalignment, aging, and the like, causing degradation of the signal quality. Some examples (see Figure 1) that have been shown in Mas Machuca, Tomkos & Tonguz (2005) and Rejeb, Leeson & Green (2006) are:
  - **Wavelength misalignment:** A transmitter may emit a signal with a wavelength slightly different to the expected one. In this case, the signal will be degraded (its spectrum is not aligned with the remaining equipment in the network such as filters or receivers). The degradation will further increase as the signal propagates through the network.
  - **In-band jamming:** Intra-channel crosstalk, which arises in transmission links due to reflections, causes this type of fault as shown by Ramaswami & Sivarajan (2002). Some examples are multiplex cascading (shown in Figure A (a)) which causes the superposition of the same signal but with different phases or switching (shown in Figure B (a)), where signals using the same wavelength and sharing the same switch may interfere each other.
  - **Out-band jamming:** This fault is related to interchannel crosstalk and non-linearities. An example applied to a demultiplexer has been shown in Figure A (b), and to a switch in Figure B (b). If an attacker inserts power at a wavelength outside the signal window, it may affect the signal when the Raman effect and/or Cross-Gain Modulation, for example in semi-

conductor optical amplifiers (SOAs), occur. The same happens if there is a non-uniform gain in an amplifier, causing pulse broadening, which leaks some power to the neighboring channels. The effects of this fault are limited in opaque networks where the signal is O/E/O converted and some BER measurement and correction can be performed. However, in transparent optical networks, it cannot be detected and eliminated, even when filtering the signal. It can be detected when received and monitored by a BER or Eye monitor.

Monitoring Equipment

Fault management relies on the information received from the monitoring equipment, which generally taps the transmitted signal and evaluates the signal quality. Some monitoring equipment monitors the optical signal, whereas other types perform optoelectronic conversion and monitor the electrical signal (this can be more accurate but

needs to be configured to the bit-rate and other characteristics of the signal).

A summary of the most common monitoring equipment and techniques are given in by Medard et al. (1998) and Mas Machuca et al. (2004a):

*   An Optical Power Meter (OPM) can detect any change in the incoming optical power over a wide band. It may be able to send an alarm when the optical power crosses a defined threshold. A signal with good optical power does not guarantee good quality since it may contain too much noise.
*   The Optical Spectrum Analyzer (OSA) measures the spectrum of the incoming optical signal. There are many implementations of OSAs as presented by Stokes & Derickson (1997). The parameters that can be monitored are: channel power, spectrum shape, channel central wavelength, and optical signal to noise ratio (OSNR). OSNR gives more accurate information on the signal quality than the OPM, since OSNR may degenerate even if the optical power is stable. However, OSA has low response

Figure 1. Soft fault examples

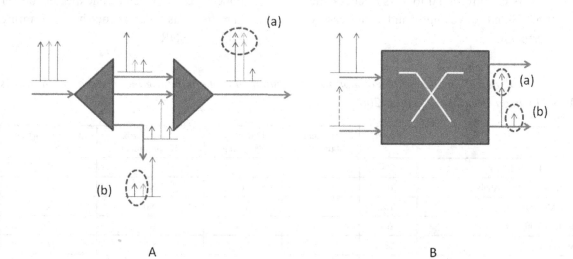

A   B

times and the disadvantages of statistical comparisons.

- Pilot tones are signals traveling with the data through the same nodes and links but used for monitoring purposes. They can be at different carrier frequencies to the transmitted signal, in different time slots or using different codes. They are able to identify general transmission faults such as fiber cut or bending. In some cases, they are combined with Optical Time Domain Reflectometry (OTDR) techniques, which use the echo of the pilot signal to accurately locate the transmission fault location.
- An Optical Supervisory Channel (OSC) is a special channel used to monitor the state of the line amplifiers along the link, especially for amplifiers where direct access is not possible. The OSC can also be used to control the amplifiers, that is, to turn them on or off. The OSC is carried on a wavelength different to the ones used to carry data and at each amplifier, the OSC is received, processed, and resend. Hence, the detection capabilities are very limited and this category has not been included in the comparative table.
- Wavemeters detect any variation in the wavelength and power used. Some operators employ signal lockers to check the transmitted wavelength and if necessary correct it.

- BER monitoring converts the signal to the electrical domain and it calculates the Bit Error Rate which is sensitive to noise and time distortion. This equipment is sensitive to impairments such as crosstalk, chromatic and polarization mode dispersion, and optical non-linearities.
- Eye monitoring is able to monitor the eye diagram, which gives information regarding the signal distortion. After processing the histograms obtained from the eye diagram, statistical characteristics of the optical signal can be obtained. However, to obtain the histogram, the amplitude of the eye should be measured which requires either synchronous or asynchronous sampling of the optical signal.

The comparison of the monitoring equipment based on their fault detection capabilities, has been shown in Table 1.

Monitoring equipment can be classified in different categories depending on the faults they are able to detect:

- **Category I (cat-I):** Monitoring equipment able to detect just unexpected power variations such as PM.
- **Category II (cat-II):** Monitoring components able to detect unexpected power variations as well as out-band jamming such as OSNR.

Table 1. Fault detection capabilities of monitoring equipment that can be used in transparent networks (Mas Machuca, Tomkos & Tonguz, 2005)

	Power	In-band Jamming	Out-band Jamming	Wavelength misalignment	Time distortion
Optical Power Meter	Yes	No	No	No	No
Optical Spectrum Analyzer	Yes	No	Yes	No	No
Eye Monitoring	Yes	Yes	Yes	No	Yes
BER Monitoring	Yes	Yes	Yes	No	Yes
Wavemeter	Yes	No	No	Yes	No
Pilot Tones	Yes	No[2]	No[3]	Yes	No

- **Category III (cat-III):** Monitoring components able to detect unexpected power variations as well as in-band and out-band jamming such as BER.
- **Category IV (cat-IV):** Monitoring components able to detect unexpected power variations, as well as wavelength misalignment.

FAULT MANAGEMENT

The ITU–T has divided the general-management functionality offered by systems into five functional key areas: Fault, Configuration, Accounting, Performance, and Security Management (the so-called FCAPS). This section focuses on the first management area.

Fault management deals with the *prevention of*, *detection of* and *reaction to* faults and attacks. It has a similarity with the health management in preventing illnesses (*prevention*), detecting them based on symptoms such as fever, skin color, etc., (*detection*) and taking the right medication for that particular disease (*reaction*). It may also occur, as in communication networks, that diagnosis is wrong based on lack of symptoms or false symptoms.

Most of the fault detection functions in current fault management schemes are performed locally by an agent, which is responsible for a network component or small set of network components. The result of this detection is sent to the network management system (NMS) via the network management protocol (SNMP or CMIP) as a trap, which request no information from the NMS and therefore can be considered as alarms.

The NMS can be either centralized or distributed. In centralized systems, the NMS is responsible for the whole network and it holds global information such network topology, link states, the wavelength usage on each link, etc. In distributed systems, the control and management is performed by each node of the network, making the management more scalable but adding more traffic to exchange updates between the nodes. Most of the fault localization approaches presented in literature have been designed for centralized systems.

Fault management consists of three different phases as shown in Figure 2:

- **Prevention** deals with the component, system, and network design so that the number of lightpaths interrupted due to faults is minimized. Different techniques can be used:
  - Selection of component architecture and technology which offers required reliability. As an example, Arbués et al. (2007) studied the reliability dependence of different optical add-drop multiplexer (OADM) and optical cross-connect (OXC) architectures. Concerning OADMs, it was shown that parallel architectures are more robust against in-band and out-band jamming faults. However, they offer higher vulnerability against possible power drop situations or misalignment of the spectral channels of the DWDM communication system, especially when AWG technology is used. On the other hand, serial architectures provide a better response against possible power drop malfunctioning (e.g., fiber bending), they present appropriate frequency responses when a misalignment of the communication channels occurs, but they offer a deficient response when an in-band or out-band jamming fault occurs because of their low cross-talk isolation. Concerning OXCs, a broadcast and select architecture seems to be the better choice, since it offers a good compromise among all the failure scenarios (although the wavelength selective architecture provides

Figure 2. Fault/Security management

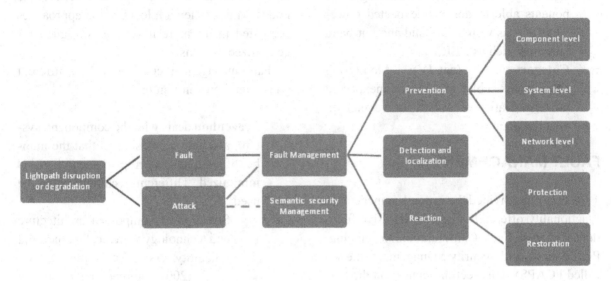

better response against in-band and out-band jamming faults).

- ○ Selection of security requirements such as extra coating for fiber, isolation for humidity and temperature variations for any type of components, connectors, fibers, etc.
- ○ Hardware duplication: Duplication of vulnerable elements, installation of disjoint fibers, etc.
- ○ Routing of lightpaths through the most reliable paths, i.e. through high reliable components.
- When the fault occurs, detection and localization discovers the existence of the fault and performs diagnosis to identify and localize it. The next section is dedicated to this process.
- Finally, reaction manages to restore the lightpaths that have been disrupted by the fault either by protection or restoration mechanisms. Protection mechanisms preplan the recovery path, which is fully signaled before the fault occurs. When it occurs, no additional signaling is required. On the other hand, restoration mechanisms

either preplan the recovery path or compute it dynamically when the fault occurs. Signaling is required to establish the recovery path. Protection recovery times are lower than restoration times. However restoration techniques are more flexible and may use better the network resources.

However, reaction also includes the fault reparation which comprises several processes. First, the distinction of the fault nature, that is, whether the fault is physical or not. In the case that it is a physical fault, the operator should gather the required personnel and equipment, travel to the fault location, repair the fault, and perform the required tests to confirm the fault reparation. The relations between these processes have been depicted in Figure 3. When the fault is not physical and can be repaired remotely, the operator should solve the fault through configuration, and perform the required tests to confirm the fault reparation. In the event of the fault not being successfully repaired, fault detection should be performed again.

Figure 3. Example of fault reparation process modeling

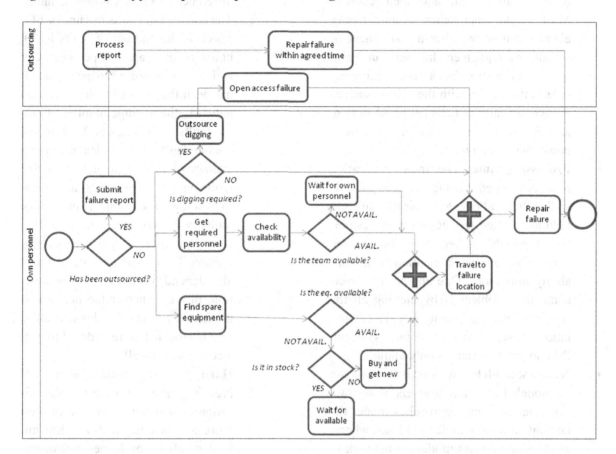

FAULT DETECTION AND LOCALIZATION

Fault detection identifies, based on the alarms received by the network management system, the fault(s) that have occurred as quickly and as accurately as possible. The alarms are issued by some of the network components (so called active components) and the monitoring equipment. The content of the alarms vary depending on the measured value that has exceeded the threshold (e.g. optical power, central wavelength, etc.).

One fault can cause not only one but several alarms generated by different components. When two or more simultaneous faults occur, the received alarms are interleaved, and hence, the faults are much more difficult to identify. It has been shown

by Rao (1993) that the problem of locating multiple faults is NP-hard.

Another severity for efficient fault diagnosis is caused by the possible existence of false and lost alarms. In a real scenario, some expected alarms may not be received (lost) or some unexpected alarms may be received (false) because a wrong configuration of the threshold, or some incorrect trap transmission.

Parameters

Several parameters are used to characterize the suitability of any proposed approach:

- **Alarms:** The fault localization process relies on the alarms received by the network management system and their content. Most of the approaches assume binary alarms, that is, whether a component or monitoring equipment has sent an alarm or not. Few approaches associate more detailed information with the alarms such as the type of fault or detected problem (e.g. an OSA could send alarm of low optical power and wrong wavelength.)

- **Processing time:** The fault localization process is aimed at being as fast as possible so that the total time to repair the fault and hold interrupted services when no restoration is possible is kept low. Some of the approaches presented mention their scalability limitations. One approach to overcome this problem is by moving all the computational complexity in a pre-computation phase, so that when fault(s) occur, the computation time is polynomial.

- **Network model:** Any optical network can be modeled at different levels of abstraction, that is, some approaches model the network as a set of nodes and links, whereas other approaches model the network to the component level (i.e. switches, transmitters, receivers).

- **Target faults:** The faults addressed by the schemes presented differ between them:
  - **Type of faults:** Some approaches target link faults, some others find link and node faults, and few of them target faults at the component level (transmitters, receivers, etc.)
  - **Single vs. Multiple faults:** Some approaches target the identification of single faults while claiming that the number of double faults is much lower than single faults. As an example, the number of single and double faults in the Nobel_EU network with 28 nodes and 41 links (available at SNDlib) has been studied by considering faults occurring with a time

difference of 8 hours as double faults. The considered Faults in Time or FIT, which is the number of faults in 10^9 hours, of links are 570 per kilometer and of nodes are 5376 per node. It has been found that for the European network, the average number of single faults in 20 years is 2719 and of double faults 317, which means an average of 136 single and 16 double faults per year. This is just an indicative study. The number of double faults will increase as the definition considers shorter times between immediate faults. The number of faults also depends on the level of abstraction, that is, whether the network is modeled as a set of nodes and links, or it is modeled more in detail (e.g. at a component level).

  - **Hard vs. Soft faults:** Some approaches have left aside the identification of soft faults because of their more complex diagnosis and less immediate effects of the network operation. However, network operators are interested in the identification of soft faults so that they can prevent more damaging effects by repairing or replacing them.

  - **Network Covered:** Most of the approaches look for the faults that occur in one homogeneous network, that is, the faults that should be identified by the NMS of this network. For example, in transparent optical networks, most of the approaches consider faults that occur at the optical level (e.g. optical fiber, transmitters, switches, etc.). However, a few approaches such as Chao, Yang & Liu (2001), Steinder & Sethi (2002) or Mas Machuca & Thiran (2000) are able to identify faults of elements that belong to heterogeneous networks (e.g. IP over SDH over OTN).

- **Localization efficiency:** Some parameters aim at evaluating the localization efficiency of the proposed approaches. However, these parameters should be only used to compare approaches that consider equivalent assumptions, i.e. approaches covering the same type and number of faults, same covered network, same network model, etc.

  ◦ **Localization degree (LD):** This parameter can be defined as the ratio between the candidates to be faulty and the total number of possible faults (used by Zeng (2006)). For example, if the approach is looking at single link faults, LD could be expressed as: l/L, being l the links candidate to be faulty and L the total number of links in that network. The objective is to achieve an LD as low as possible.

  ◦ **False Positive Rate (FPR)** is defined as the ratio of diagnosed faults, which have not occurred to the number of occurred faults. This ratio increase when alarms may be false and wrongly issued.

  ◦ **False Negative Rate (FNR)** is defined as the ratio of non identified faults to the number of occurred faults. This ratio increases when alarms may be lost or not received by the fault management system.

  ◦ **False Diagnosis Rate (FDR)** is defined as the ratio of the number of wrong diagnosis to the number of occurred faults. This would be the case that a fault occurs, and the diagnosis delivers a wrong fault candidate.

  ◦ **Detection Rate (DR):** It is defined the ration of correctly identified faults to the number of occurred faults. DR can be expressed as DR=1-FDR-FNR. The objective is to find a fault localization technique that achieves a DR-FPR as close as possible to 1. In Lo et al. (2000) this parameter is denoted as the Hit Ratio (see Table 2).

Existing Fault Localization Techniques

Several techniques to perform fault localization have been presented in literature. They have been grouped in three categories depending on whether they are able to localize single or multiple faults and whether they cope with the ideal (none false/ lost alarms) or non ideal scenario.

- Single fault localization, ideal scenario

Probabilistic approaches: Katzela & Schwartz (1995) model the network as a graph, which takes into account the dependencies among the different network components. Nodes are network components, and links are the probability that

Table 2. Fault localization efficiency parameters: FN stands for False Negative, FD stands for False Diagnosis, and FP stands for False Positive

		Actual condition	
		Fault	**Fault free**
Fault Diagnosis Result	No fault found	FN	True Negative
	Wrong fault found	FD	FP
	Right fault found	True Positive	FP

the failure of one particular component propagates to the neighboring one. Once the graph modeling is done, a fault diagnosis algorithm based on the failure probabilities is proposed. The proposed solution to find the most probable explanation of a set of symptoms for a *N*-node dependency graph has complexity of $O(N^3)$.

Wang (1993) models the network as a set of nodes interconnected with links that can fail with a given probability (focused on link hard faults). Depending on the lightpaths that can and cannot be established between each pair of nodes, the proposed solution is able to give the most likely failed link.

Staessens et al. (2010) present a general statement of the fault localization problem based on probabilities. The probabilities are not only the fault probability of a network component, but also the dependency between faults and probability of raised alarms for each fault scenario. Hence, several parameters such as mutual information, self-information and entropy are defined and used to evaluate fault localization.

Finite State Machines (FSM): Wang (1992) treats the system as a discrete event system whose behaviors can be described by concatenation of events. Some of the concatenations correspond to correct network behavior, whereas others correspond to faulty behavior.

Li & Ramaswami (1997) propose FSM to model all possible states of a network link. Each network node runs a FSM for each link connected to it. The FSM for each type of node (bypass and end-nodes) and for different network architectures (broadcast-and-select networks and wavelength-routed networks) are proposed.

Monitoring approaches: These identify single hard faults.

Monitoring cycles (m-cycles), Zeng et al. (2006): This approach discomposes the network into cycles so that each node and link belongs at least to one cycle. Each cycle has one node which has assigned as monitoring node and a supervisory channel is sent along the ring. The key concept

is that a fault is expected to send alarms in the *m*-cycles in which appears, but not others. The alarms received by the network management are associated to a binary vector of as many bits as *m*-cycles.

Monitoring trees/paths/cycles, Harvey et al. (2007 and Ahuja et al. (2009): This approach aimed at extending the monitoring concept to other structures such as trees and paths, while avoiding the use of dedicated lightpath for monitoring. These approaches proactively launch a probe in each tree/path/cycle to identify any single component fault and were extended to localize shared risk link group faults.

Monitoring trail (m-trail): Tapolcai et al. (2009) proposed an approach that overcame the monitoring structure limitations of the previous monitoring approaches by designing *m*-trails. A trail can be any structure of nodes, containing one monitoring node. A supervisory channel is sent through each *m*-trail. The target of this approach is to design the network *m*-trails so that the cost associated to monitoring and fault localization is minimized.

Alarm correlation: Bouloutas et al. (1994) proposed a new system representation using a 4-tuple Phrase Structured Grammar. The alarm correlation is based on the problem-symptom model, where an alarm is associated with a set of possible fault locations (the so called domain of the alarm). Then, the fault lies in the intersection of the set of locations indicated by each alarm and thus, alarms that share a common intersection should be correlated. This is the basis of the proposed Positive Information Algorithm.

Choi et al. (1999) proposed a fault identification scheme based on OSI managed object class dependency graphs, which are small and remain stable for long time. The alarm correlation uses an alarm propagation graph, which suppresses many redundant alarms while keeping the important ones. Based on the alarms retained, an improved version of Bouloutas et al. (1994) was analyzed which uses the alarm causality graph to find the domain of the alarms.

Chao, Yang and Liu (2001) proposed a hierarchical reasoning and alarm correlation method. First, it performs a fault propagation method which finds for each network component fault: (i) the alarms associated with this fault with the probability that they are generated, (ii) the name of the fault, (iii) the time interval when this fault occurs and (iv), the layer where this component belongs. Then, a fault propagation model is used based on a causality graph represented by the belief network, previously presented by Russel and Norvig (1995). The proposed fault reasoning mechanism is based on the hierarchical domain-oriented delegated architecture with each domain having one manager. Domains are arranged in a hierarchical manner. Fault reasoning is based on the fault propagation model at one level and on the information received from lower levels. The mechanism uses the Alarm Correlation Engine (ACEngine) introduced by Chao (1999). The time-complexity of using the hierarchical domain-oriented delegated reasoning mechanism is bounded by $O(sAF)$, with s being the number of segments of the largest domain, A the total number of significant alarms, and F the number of significant faults.

Kim et al. (2008) presented a fault localization method that collects the network alarms from various Network Management systems, that is, it can be applied to heterogeneous networks. The alarms received are analyzed with a rule-based alarm correlation method that uses a consolidated inventory database and some predefined correlation rules. The proposed method extracts the root cause event from the related event group using the alarm correlation rules. In order to cope with faults of elements no able to send alarms, the method has been extended to generate presumptive events.

- Single fault localization, non-ideal scenario

Bouloutas et al. (1992) represented the communication process as a finite state machine, and the possible faults as additions and changes of arcs in that FSM. They used the recorded communication history in order to propose possible hypotheses of multiple faults. They coped with the possibility of missing some events and having some corrupted events. Thus, the problem became one of inferring a finite state structure given unreliable, partially observed information about its event trace history.

Lo et al. (2000) presented two coding based schemes which are based on causality graphs. Causality is a partial order relation between events and is represented by $p \rightarrow s$ (event p causes event s). The causality graph, which is assumed to be correctly pruned, non-cyclic and without many-to-one relations, is reduced to correlation graph. The correlation graph is converted to a set of codes, one for each fault and each alarm. The codes can cope with a tolerance level which specifies the maximal number of symptom loss allowed (non-ideal scenario of having lost alarms). The first proposed scheme has a time complexity of $O(mn^\vartheta)$, being m the number of candidate problems, n the number of observed symptoms and ϑ the maximal number of symptoms that must be chosen from the observed ones.

Steinder & Sethi (2002): This work proposed a technique that allowed lost and false alarms to be incorporated in a fault localization algorithm based on iterative propagation in belief networks presented by Dechter (1996). The complexity of the algorithm when applied to a network of n nodes, is $O(n^5)$.

- Multiple fault localization, non-ideal scenario

Mas Machuca et al. (2005) proposed a multiple fault location algorithm able to cope with false and lost alarms. The algorithm is based on the channels established in a transparent optical network. One channel is defined as a unidirectional connection between two network nodes. The network is modeled down to the component level and hence, any channel is modeled as an ordered set of network components. These network components can be either optical equipment or monitoring equipment. The former takes care of the optical signal transmission, whereas the latter takes care of the optical signal motoring. The monitoring components are able to send alarms when the optical signal is degraded and do not mask any failure since they receive the optical signal through a tapping coupler. On the other hand, the optical components are able to fail and can mask failures (e.g. a filter is able to remove the no desired out-band signal and, therefore, mask any out-band jamming that may have occurred before the filter). Based on their masking properties, the optical components can be classified into different categories. These categories are taken into account when computing the domains of one component fault, which is defined as the set of network elements that will send an alarm when a given optical component fails. Different domains are defined for the different faults covered by the proposed algorithm. The target of the Transparent Fault Localization Algorithm (TFLA) developed in this paper was to minimize the computation time required to give the results to the human manager when receiving alarms. The goal was achieved by building a binary tree with a depth equal to the number of monitoring equipment items. The leaves of the binary tree are filled with the associated component faults, multiple faults and with the existence of lost and/or false alarms. When alarms are received, the tree should simply be traversed and hence the computation required when fault(s) occur is just polynomial. The next section presents this scheme in detail. The complexity of the on-line part of the algorithm when alarms are received is just $O(n_a)$, being n_a the number of monitoring equipment. This low complexity comes with a cost of a higher computation time to build the binary tree which is $O(4^{n_a} n_a^2)$ and an exponential storage requirement of $O(2^{n_a})$.

Rejeb et al. (2006) presented a scheme able to identify faults and attacks in optical networks called Multiple Attack Localization and Identification (MALI) algorithm. This algorithm is distributed and relies on a reliable management system (hence, it assumes ideal scenario with neither false nor lost alarms), and models the network down to the component level. When a fault or an attack occurs, the downstream and upstream nodes detecting the problem send an alarm and perform fault localization based on the correlation of channel state information. The NMS confirms which of the fault candidates delivered by the nodes is in fact faulty.

Pal et al. (2008) proposed a scheme to detect and localize multiple failures in WDM networks. The scheme is based on an algorithm which determines the optimal number of network nodes that should hold monitoring equipment so that all possible node or link faults can be detected. This scheme can consider one false and/or missing alarm.

Mas Machuca et al. (2010) evaluated the performance of the previous TFLA in a high level modeled network. The identification of double faults was not satisfactory and therefore, the addition of a new module was proposed. This method identifies the components that are more likely to issue alarms in case of a fault. The lightpaths using these components (unless they are source or destination), are replaced by an alternative path. The selection of the alternative path is based from a set of alternative paths derived from the network topology. The path that has the fewest common alarming components with the existing connections is chosen. With this new module the localization of double faults could significantly improve without decreasing the accuracy to identify single faults. In the networks considered, Nobel_EU and Nobel_USA from SNDlib, the number of pairs of candidates for the former was reduced by some 65%, whereas for the latter it was reduced by 46%. The average number of single fault candidates was always one.

TRANSPARENT FAULT LOCATION ALGORITHM

This section presents one fault location algorithm that can be applied to transparent optical networks, is able to locate multiple faults and can cope with the existence of fault and lost alarms. This method is based on the Transparent Fault Location Algorithm (TFLA) introduced in the previous section. Since the problem of locating multiple faults has been shown to be NP-complete even in the ideal scenario, this algorithm concentrates all the complexity in a Pre-Computation Phase (PCP) so that when a fault occurs, the Core Phase (CP) of the algorithm exhibits minimum complexity in order to deliver the TFLA result as fast as possible. Another advantage of this approach is that the PCP should be executed only when there are changes in the established channels, which is realistic since connections in transparent networks are not as dynamic as in electrical networks.

The PCP has as input the set of established channels. A channel is defined as an ordered set of network components, which can be done at a very abstract network model level of links and nodes or a more detailed network model level of amplifiers,

transmitters, etc. Each network component may have different faults. For each component and possible fault, the domain is computed and stored in a domain matrix with as many rows as faulty components and as many columns as monitoring equipment monitor the established channels. The domain of a faulty component is defined as the set of alarms that should be received in case this fault occurs to this particular component ("Compute Domains" in Figure 4). Let us present a simple example based on the network shown in Figure 5 (a). This network has been modeled as a set of nodes and links. Three lightpaths have been established: Channel1 from Node 3 (N3) to Node 2 (N2), and Channel2 and Channel3 from Node 1 (N1) to Node 4 (N4). These channels have been modeled as shown in Figure 5 (b). In this example and for reasons of simplicity, only hard faults of nodes and links are considered, and also that a node is not able to send an alarm when failing. Based on these assumptions, the domain of the network components have been shown in Table 3. For the fault of each network component, the alarms that should be sent are represented by a "1".

The next TFLA phase "Group identical domains into C_i" clusters equal domains into

Figure 4. Transparent Fault Location Algorithm

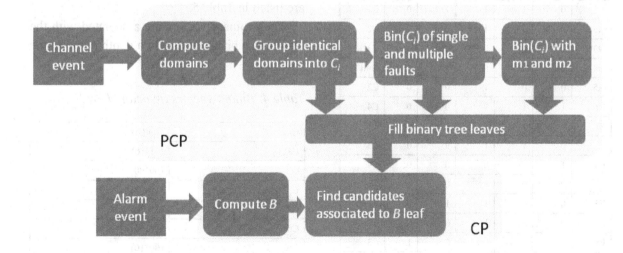

Figure 5. (a) Network and (b) Channel model

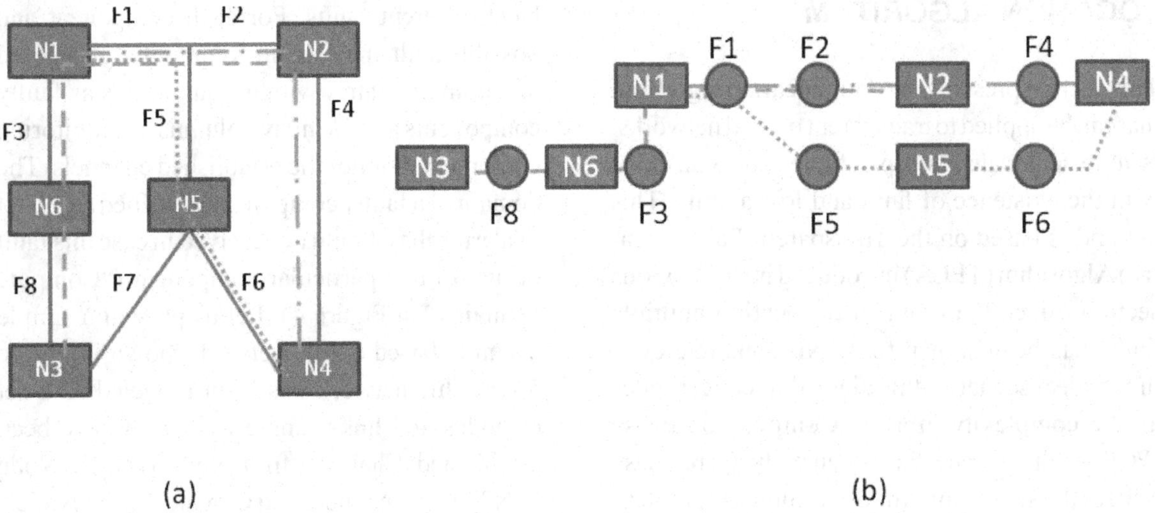

(a)

(b)

equivalent classes C_i, as shown in the last column of Table 3. These classes are converted to binary vectors of size the number of monitoring components in the established channels: $Bin(C_i)$ (shown in Table 4).

At this phase, we have the binary vectors associated to any single fault in the network and the TFLA will be able to identify when no alarm is lost or false. In order to find the binary vector of

multiple faults we can apply the fact that the domain of simultaneous faults is equivalent to the union of each single fault domain. Hence, for double faults, the binary vector of its domain can be obtained by computing the point-wise OR operation of the binary vector of each single fault. When the obtained binary vector is the same as the binary vector associated to a single fault, it can be discarded if we assume the approach that one single fault is more likely than two faults. Hence, only the binary vectors obtained which are different from the ones associated with single faults are kept. The new vectors of the example are listed in Table 5.

These binary vectors are associated with the leaves of a binary tree with a depth equal to the

Table 3. Domain matrix of the example presented

	N1	N2	N4	N5	N6	Ci
N1	0	1	1	1	0	C1
N2	0	0	1	0	0	C2
N3	1	1	0	0	1	C3
N4	0	0	0	0	0	Null
N5	0	0	1	0	0	C2
N6	1	1	0	0	0	C4
F1	0	1	1	1	0	C1
F2	0	1	1	0	0	C5
F3	1	1	0	0	0	C4
F4	0	0	1	0	0	C2
F5	0	0	1	1	0	C6
F6	0	0	1	0	0	C2
F8	1	1	0	0	1	C3

Table 4. Binary vectors associated to C_i

Ci	Bin(Ci)
C1	(01110)
C2	(00100)
C3	(11001)
C4	(11000)
C5	(01100)
C6	(00110)

Table 5. Binary vectors corresponding to double fault scenarios

Ci	Bin(Ci)	Double faults
C7	(11111)	C3U{C1,C6}
C8	(11110)	C4U{C1,C6}
C9	(11101)	C3U{C2,C5}
C10	(11100)	C4U{C2,C5}

number of monitoring equipment m (in our example $m=5$: N1, N2, N4, N5 and N6 since N3 is not receiving and monitoring any signal). The binary tree of the example is shown in Figure 6.

Let us consider the non-ideal scenario of having false or lost alarms. The binary tree can be viewed as a particular block error-correcting code, whose codewords have the property that the logical OR of any two codewords is another codeword. One empty leaf of the tree corresponds to an erroneous word, and we should find the leaf whose codeword has a minimal Hamming distance to the empty one. This module finds the binary vectors that have a minimal distance given the mismatching thresholds m_1 (allowed number of lost alarms) and m_2 (allowed number of false alarms). For example, for $m_1 \neq 0$ and $m_2 = 0$ we accept m_1 lost alarms, i.e., the binary vectors that fall within this margin from the correct codewords are the binary vectors having a '0' when $Bin(C_i)$ has '1' in at most m_1 positions. In our example, and considering $m_1 = 0$ and $m_2 = 1$, it means that we accept one false alarm, i.e., the binary vector *(01111)* could be also associated to the faults of C_1 since $Bin(C_1) = (01110)$. In this way, more leaves of the binary tree are filled and point to non-ideal scenario solutions.

All these modules belong to the PCP, which can be computed as soon as the lightpaths are established. Then, when a fault occurs and the alarms are received, the Core Phase (CP) of the algorithm runs. The received alarms are converted to a binary vector B. The binary tree is traversed to the leaf corresponding to B. The potentially

Figure 6. Binary tree of the presented example with the leaves pointing to the single and double faults in the ideal scenario

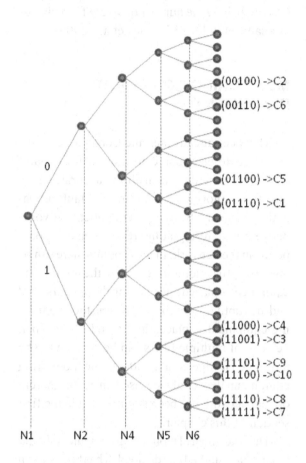

fault candidates are the ones pointed to by this leaf. For example, if the received alarms are from Node 2 and 4, B= (01100) and the candidate is F2. Then, the human manager will have to check the status of this equipment remotely if possible and start organizing the fault reparation (look for personnel, required equipment, licenses, means of transportation, etc.). If the received alarms have been triggered by Nodes 2, 4 and 6, the best candidate to be faulty is F2 accepting one false alarm. The next candidate is the double fault scenario associated to C9 and also accepting one false alarm. The other candidates in the list have a Hamming distance equal or larger than 2.

Regarding the complexity of the algorithm, the computational time of the CP is only $O(m)$, whereas the computational time of the PCP is $O(4^t m)$, being t the number of single fault classes as analyzed in Mas Machuca et al. (2004a).

MONITORING EQUIPMENT PLACEMENT

As it has been already presented, opaque networks perform opto-electronic conversion of the signal at each network node, which allows accurate signal quality monitoring and limiting the fault propagation. Conversely, transparent optical networks perform monitoring at the receiver and at some points in the network. In these points there can be non disruptive monitoring, where the monitoring equipment is placed after tapping the optical signal and disruptive monitoring, where the monitoring equipment is placed at the end of the fiber, i.e., one of multiple fibers can be switched to be monitored (Kilper et al., 2004)). The monitoring equipment that could be used in a transparent optical network has been summarized in the first section of this chapter.

The placement of monitoring equipment should be carefully studied and decided. It has been shown that transparent optical networks have replaced the electrical components by optical ones, which have reduced monitoring capabilities (compared with their electrical counterparts). Hence, there is a trade off between the cost savings of transparent networks versus the extra capital expenditures required for extra monitoring and the operational expenditures associated to a more complex trouble shooting.

Different problems associated to the monitoring equipment placement have been studied and presented in literature:

- One problem studied was to assume a full monitored transparent network and identify the monitoring equipment that could be

removed without reducing the fault localization accuracy. Stanic et al. (2002) proposed a scheme that takes into account the network architecture and the established lightpaths to determine the fault propagation through the network, assuming only power monitoring. This work is based on the domain matrix shown in the previous section and reduces the number of columns so that the redundant monitoring information is eliminated. The redundant monitor deletion problem is claimed to be NP-Complete.

- Another study by Mas Machuca & Tomkos (2004b) focused on the monitoring equipment placement defining as optimal location as the position of the monitoring equipment that minimizes the number of network elements that are candidates to have a failure, so that the effort for the network manager is minimized. The proposed solution was based on the TFLA presented in the previous section, which depends on the established channels. However, network operators may be more interested in the location of new monitoring equipment based on the network topology rather than the channel based approach. In that case, the problem was solved when having ring topologies.

- Mas Machuca & Kiese (2007): This approach overcame the limitations of the previous ones, i.e., it could be applied to any transparent network independently of the lightpaths that are established and of the network topology. In this study, the target was to minimize the cost of the required monitoring equipment and to maximize the number of monitored connections, considering monitoring equipment that could be placed in any node of the network and was able to monitor all the lightpaths going though that node. Three different problems were studied:

- **Optimal Placement of Limited Monitoring Equipment:** Given a network topology with unmonitored nodes and links, find the optical location for a given number of monitoring equipment.
- **Optimal Placement of New Monitoring Equipment:** Given a running network with unmonitored and monitored nodes, and links, find the optimal location of new monitoring equipment so that the fault location is best improved.
- **Optimal Monitoring:** Given a network with unmonitored nodes and links, find the optical location of the minimum monitoring equipment required to so that the longest unmonitored transparent link (LUTL) was shorter than a given threshold. In this case, for the Nobel_EU and Nobel_USA networks 100% of the nodes should be monitored for LUTL lower than 500 km, around 65% of the nodes had to be monitored for LUTL =1250 km, and less than 50% of the nodes for LUTL greater than 2500 km.

- Kiese & Mas Machuca (2007): The novelty of this approach was to consider different types of monitoring equipment and to identify which monitoring equipment should be acquired and where should be placed given a budget, the network topology, the network components and the failure probabilities in order to increase the ability to locate failures as precisely as possible. Monitoring equipment was classified into three different categories depending on the faults they could detect (classification presented in first section): cat-I could only detect power drop, cat-II could also detect out-band jamming, and cat-III could also detect in-band jamming. Node- and link-

failures have been depicted as the predominant source of faults in a transparent optical network and they can be detected by the cheapest available monitoring equipment (i.e. cat-I) and therefore placing them allows the detection of the main cause of failures. A change in both, the cost relationship between the different kinds of ME and the failure probabilities, might lead to a wider variety of ME to be placed in the network. Only large budgets could justify purchasing cat-II and cat-III monitoring equipment.

CONCLUSION

Fault management has been shown to be very important to network operators. The reason is that transparent optical networks offer substantial capacity and carry many services which are sensitive to delays and loss. According to these requirements, operators should operate networks as reliable as possible so that the service interruption times are kept as low as possible and avoid paying very large penalties (previously agreed in SLAs). For that purpose, fault management should be as fast and accurate as possible, especially in transparent optical networks.

This chapter has presented the fault management challenge in transparent optical networks. Fault management deals with the prevention, detection and localization, and reaction to faults. For this purpose, a definition, classification and some examples of faults that may occur in transparent optical network shave been given. The basis of an efficient fault management relies on the existing monitoring equipment, which has also been described in this chapter. Special attention has been given to the fault detection and localization processes by given an overview of the existing approaches and the parameters that should be taken into account when deciding which method is more suitable to a particular scenario. Last but not least, the problem of locating monitoring equipment has also been presented given some existing solutions.

The importance of fault management in future optical networks is based on the impact it has on the operational costs as well as the connection availability. Hence, future optical networks have to guarantee a fast and accurate fault location.

REFERENCES

Ahuja, S. S., Ramasubramanian, S., & Krunz, M. (2008). *SRLG failure localization in all-optical networks using monitoring cycles and paths* (pp. 700–708). IEEE INFOCOM.

Bouloutas, A. T., Calo, S., & Finkel, A. (1994). Alarm correlation and fault identification in communication networks. *IEEE Transactions on Communications, 42,* 523–533. doi:10.1109/TCOMM.1994.577079

Bouloutas, A. T., Hart, G., & Schwartz, M. (1992). Fault identification using a finite state machine model with unreliable partially observed data sequences. *IEEE Transactions on Communications, 42,* 523–533. doi:10.1109/TCOMM.1994.577079

Chao, C. S., Yang, D. L., & Liu, A. C. (1999). Alarm correlation view (ACView). *Proceedings of IASTED International Conference on modelling and Simulation,* Philadelphia, (pp. 291-253).

Chao, C. S., Yang, D. L., & Liu, A. C. (2001). An automated fault diagnosis system using hierarchical reasoning and alarm correlation. *Journal of Network and Systems Management, 9,* 183–202. doi:10.1023/A:1011315125608

Choi, J., Choi, M., & Lee, S.-H. (1999). An alarm correlation and fault identification scheme based on OSI managed object classes based on OSI managed object classes. *IEEE International Conference on Communications, 3,* (pp. 1547-1551).

Crawford, D. (1993). *Fiber optic cable dig-ups: causes and failures Network reliability: A report to the Nation: Compendium of Technical Papers.* Chicago: National Engineering Consortium.

Dechter, R. (1996). Bucket Elimination: A unifying framework for probabilistic inference" In *Proc. Of the Twelfth Conference on Uncertainty in Artificial Intelligence,* Morgan Kaufmann Publishers.

FCC. (2001). FCC reportable network outages. *2001 Annual Report,* (pp. 7-17).

Gil Arbues, P., Mas Machuca, C., & Tzanakaki, A. (2007). Comparative study of existing OADM and OXC architectures and technologies from the failure behavior perspective. *Journal of Optical Networking, 6,* 123–133. doi:10.1364/JON.6.000123

Grover, W. D. (2004). *Mesh-based survivable networks: options and strategies for optical. MPLS, SONET and ATM networking.* Prentice Hall PTR.

Harvey, N. J. A., Patrascu, M., Wen, Y., Yekhanin, S., & Chan, V. W. S. (2007). *Non-adaptive fault diagnosis for all-optical networks via combinatorial group testing on graphs* (pp. 697–705). IEEE INFOCOM.

Katzela, I., & Schwartz, M. (1995). Schemes for fault identification in communication networks. *IEEE/ACM Transactions on Networking, 3,* 753–764. doi:10.1109/90.477721

Kiese, M., & Mas Machuca, C. (2007). Optimal placement of different types of monitoring equipment in transparent optical networks. In *Proceedings of the Workshop on Monitoring, Attack Detection and Mitigation (MONAM 2007)* Toulouse, France.

Kilper, D. C., Bach, R., Blumenthal, D. J., Einstein, D., Landolsi, T., & Ostar, L. (2004). Optical performance monitoring. *Journal of Lightwave Technology, 22,* 294–304. doi:10.1109/JLT.2003.822154

Kim, J., Yang, Y.-M., Park, S., Lee, S., & Chung, B. (2008). Fault localization for heterogeneous networks using alarm correlation on consolidated inventory database. *Lecture Notes in Computer Science, 5297,* 82–91. doi:10.1007/978-3-540-88623-5_9

Li, C.-S., & Ramaswami, R. (1997). Automatic fault detection, isolation, and recovery in transparent all-optical networks. *Journal of Lightwave Technology, 15,* 1784–1793. doi:10.1109/50.633555

Lo, C.-C., Chen, S.-H., & Lin, B.-Y. (2000). Coding-based schemes for fault identification in communication networks. *International Journal of Network Management, 10,* 157–164. doi:10.1002/(SICI)1099-1190(200005/06)10:3<157::AID-NEM360>3.0.CO;2-G

Mas Machuca, C., & Kiese, M. (2007). Optimal placement of monitoring equipment in transparent optical networks. In *Proceedings of the 6th International Workshop on Design and Reliable Communication Networks (DRCN2007),* La Rochelle, France

Mas Machuca, C., Nguyen, H., & Thiran, P. (2004a). Failure location in WDM networks. In Sivalingam, K., & Subramaniam, S. (Eds.), *Optical WDM networks: Past lessons and path ahead.* Kluwer Publishers.

Mas Machuca, C., & Thiran, P. (2000). An efficient algorithm for locating soft and hard failures in WDM networks. *IEEE Journal on Selected Areas in Communications, 18,* 1900–1911. doi:10.1109/49.887911

Mas Machuca, C., & Tomkos, I. (2004b). Optimal monitoring equipment placement for fault and attack location in transparent optical networks. *Lecture Notes in Computer Science, 3042,* 1395–1400. doi:10.1007/978-3-540-24693-0_125

Mas Machuca, C., Tomkos, I., & Tonguz, O. (2005). Failure location algorithm for transparent optical networks. *IEEE Journal on Selected Areas in Communications, 23,* 1508–1519. doi:10.1109/JSAC.2005.852182

Medard, M., Marquis, D., & Chinn, S. (1998). *Attack detection methods for all-optical networks.* Network and Distributed System Security Symposium.

Pal, A., Paul, A., Mukherjee, A., Naskar, M. K., & Nasipuri, M. (2008). Fault detection and localization scheme for multiple failures in optical network. *Lecture Notes in Computer Science, 4904,* 464–470. doi:10.1007/978-3-540-77444-0_48

Potter, D. (1991). The need for network management. *Computer Communications, 14*(2), 121–125. doi:10.1016/0140-3664(91)90042-Y

Ramaswami, R., & Sivarajan, K. N. (2002). *Optical networks: A practical perspective.* Morgan Kaufmann.

Rao, N. S. V. (1993). Computational complexity issues in operative diagnosis of graph-based systems. *IEEE Transactions on Computers, 42,* 447–457. doi:10.1109/12.214691

Rejeb, R., Leeson, M. S., & Green, R. J. (2006). Fault and attack management in all-optical networks. *IEEE Communications Magazine, 44,* 79–86. doi:10.1109/MCOM.2006.248169

Russel, S. J., & Norvig, P. (1995). *Artificial intelligence: A modern approach.* Prentice-Hall Publishing Company.

SNDlib. (n.d.). *Survivable fixed telecommunications network design library.* Retrieved from http://sndlib.zib.de/home.action

Staessens, D., Manousakis, K., Colle, D., Mahlab, U., Pickavet, M., Varvarigos, E., & Demesteer, P. (2010). *Failure localization in transparent optical networks.* 2nd International Workshop on Reliable Networks Design and Modeling, Moscow, Russia.

Stanic, S., Subramaniam, S., Choi, H., Sahin, G., & Choi, H.-A. (2002). On monitoring transparent optical networks. *International Conference on Parallel Processing Workshops* (ICPPW'02), (p. 217).

Steinder, M., & Sethi, A. S. (2002). Increasing robustness of fault localization through analysis of lost, spurious, and positive symptoms. In *INFOCOM 2002. Twenty-First Annual Joint Conference of the IEEE Computer and Communications Societies, 1*, (pp. 322-331).

Stokes, L. F., & Derickson D. (1997). *Lightwave component and system measurements*. Short Course Notes, OFC 97.

Tapolcai, J., Wu, B., & Ho, P.-H. (2009). *On monitoring and failure localization in mesh all-optical networks* (pp. 1008–1016). IEEE Infocom Proceedings.

Wang, C., & Schwartz, M. (1992). *Fault detection with multiple observers* (pp. 2187–2196). IEEE Infocom Proceedings.

Wang, C., & Schwartz, M. (1993). Identification of faulty links in dynamic routed networks. *IEEE Journal on Selected Areas in Communications, 11*, 1449–1460. doi:10.1109/49.257936

Zeng, H., Huang, C., & Vukovic, A. (2006). A novel fault detection and localization scheme for mesh all-optical networks based on monitoring cycles. *Photonic Network Communications, 11*, 277–287. doi:10.1007/s11107-005-7355-3

KEY TERMS AND DEFINITIONS

Alarm: Unrequested information sent by monitoring equipment or network components to the network management system informing about an abnormal event.

Attack: Intentional action against the ideal and secure functioning of the network or one component.

Fault: Unexpected interruption of the expected functioning of the network or one component. It may be cause by aging, misalignment, external conditions, etc.

Monitoring Equipment: The equipment used in the network to monitor the quality of the transmitted signal and/or the status of a network component (e.g. optical fiber).

ENDNOTES

[1] OSAs have low resolution to detect small wavelength misalignments.

[2] Unless the in-band jamming covers the frequencies at which the pilot tones are carried.

[3] Unless the pilot tones traverse the same amplifiers as the signals and therefore, can be affected by competition gain.

Chapter 5
All–Optical Resilient Pulse–Position–Modulation–Based Packet–Switched Routing

Z. Ghassemlooy
Northumbria University, UK

W. P. Ng
Northumbria University, UK

H. Le Minh
Northumbria University, UK

ABSTRACT

In traditional optical networks, configured as static physical pipes, the carrier-grade network resilience is provided by means of protection and restoration capabilities. However, there is a need to develop a new generation of dynamic reconfigurable all optical networks with built in network resilience capabilities. In the next generation, high-speed photonic packet switching networks, ultrafast packet header processing, and packet switching are the vital building blocks. In this chapter, a review of different routing schemes for high-speed photonic packet switching networks and the concept of reducing the size of the look-up routing table are presented. A novel PPM signal format has been introduced in order to reduce the size of the routing table in order reduce packet switching and processing time compared to the conventional routing tables. A failure self detection and a routing table reconfiguration in the optical domain are introduced, and a number of factors such as system performance, reliability, and complexity are also discussed.

INTRODUCTION

In recent years we have seen a remarkable progress in research to increase the long-haul optical fibre back-bone transmission distance as well as

DOI: 10.4018/978-1-61350-426-0.ch005

exploiting its virtually unlimited transmission bandwidth (70 THz ("Annual Report," 2005)) to deliver multiplicity of services to the end users at a global scale. Nowadays, systems capable of delivering Terabits/s over optical fibre communication networks spanning tens of thousands of kilometres are practically feasible (Suzuki, Fujiwara, &

Iwatsuki, 2006; Zhu, Funabashi, Pan, Paraschis, & Yoo, 2006). These outstanding achievements are due to a number of technical breakthroughs in the optical domain such as low attenuation (<0.25 dB/km) optical fibres at S- and L-bands, erbium doped fibre amplifiers and Raman amplification at C, L and S bands, the distributed amplification, effective long-haul dispersion management schemes (Doerr, et al., 2006), tunable lasers, gain equalisation, optical multiplexing/demultiplexing techniques (Bergano, 2005; Ohara, et al., 2004), inline all-optical repeaters and all optical buffering (Dorren, et al., 2003; Hunter, Chia, & Andnovic, 1998; Takahashi, et al., 2004), advanced error detection/correction schemes (Choi, et al., 2005), packet compression/decompression techniques (Takenouchi, Takahata, Nakahara, Takahashi, & Suzuki, 2004), all-optical gates and switches (Guo & Connelly, 2006; Zhang, Wu, Feng, Xu, & Lin, 2007) and high-quality system resilience features (K.K. Lee, Lim, & Ong, 2005).

Nevertheless, the ever-growing demand for the data based services and applications, e.g. e-services and e-applications or broadband access networks are the driving forces for upgrading the existing network and deployment of optical networks at a larger scale at all network levels. With the future data traffic being Internet Protocol (IP) based services (multimedia, e-learning), there is the need for improved interface between the existing transmission protocols (e.g. ISO/OSI-protocol stack) and the physical layer (dominated by the optical transmission). Seamless integration of optical networks with conventional network applications and services has resulted in further developments and a broader deployment of optical networks in all areas of modern telecommunication networks. These networks can be classified and distinguished between Access Networks (including fibre-to-the-home (FTTH) and fibre-to-the-business (FTTB)), the Metropolitan Area Network (MAN), Wide Area Network (WAN), and high-speed indoor multiple access networks. Therefore, the increase in demand for reliable rout-

ing of optical data entirely in the optical domain imposes many challenges such as the ability of the network to efficiently and economically process and switch a huge amount of optical data at every router while maintaining high-quality network management and resilience under faults and malicious attacks (Fu-Tai, et al., 2004).

Traditional optical networks are mainly dominated by the voice and private-line traffics, where all traffics are treated identically with a full protection. The next generation optical networks (opaque, managed reach and transparent (i.e. all-optical)), will support more services and more multimedia type traffics, thus requiring additional link control and management schemes. The quality of service (QoS) of such networks depends on the topology of the layered network (K. K. Lee, Lim, & Ong, 2006). Moreover all-optical networks are susceptible to malicious attacks because the signals remain in the optical domain and are not easy to be monitored closely. In such networks with very high data rates even a single attack of a short duration can result in a large amount of data being lost. Thus, offering the same level of full protection to all traffics in the traditional networks is very costly and wasteful. In such networks, offering the same level of full protection to all traffic is very costly and wasteful. In the next generation high speed IP-centric optical networks there are two main issues of equal importance: (i) the resilience (or reliability) provisioning under faults and malicious attacks, and (ii) the scalability of network management and control systems. In such networks resilience is very important to protect against failure in order to provide high level of network availability (Androulidakis, Doukoglou, Patikis, & Kagklis, 2008; Ramaswam & Sivarajan, 2002). For ultra-high speed optical networks the common level of network availability is 99.999%, therefore a single failure of network will interrupt a huge amount of online services and consequently resulting in a significant loss of revenue. Therefore, the layered transport network is essential in the management of

Table 1. Main characteristic of OCS, OBS and OPS

Switching technologies	Granularity	Utilisation	Complexity	Resilience
Circuit switching	coarse	poor	low	poor
Burst switching	moderate	moderate	moderate	good
Packet switching	fine	high	high	good

the resilient network in the event of a node failure (Xin, Hongxiang, & Yuefeng, 2008).

In optical communication networks there are three fundamental switching techniques categorised as: optical circuit switching (OCS), optical burst switching (OBS) and optical packet switching (OPS). Table 1 shows the summary of the major characteristics of the three different switching technologies based on their scales, utilisations, complexity and resilient properties.

Among these switching technologies, OCS has the coarsest switching granularity owing to fibres and wavelength bands or wavelengths (wavelength routing) being switched in a circuit-switched wavelength division multiplexing (WDM) network. Wavelength routed networks can be realised by utilising commercially available large-scale switching fabrics such as optical cross-connect (OXC) switches (Olkhovets, et al., 2004). This type of network is physically fixed therefore has a poor resilient property as adding an extra stand-by switching fabric is not economically efficient as well as being more complex to implement.

Granularity in the burst switching is moderate since the IP packets are aggregated to bursts with sizes of several tens of kBytes in OBS networks. In this technology, an optical transport network sets up an end-to-end connection and reserves resources for the duration of the burst only, which is not supported in OCS. In contrast, in packet switching networks, routing is implemented on a packet-by-packet basis, with a size of hundreds to a few kBytes, thus providing the finest granularity (Blumenthal, 2001).

In comparison to a fixed-connection OCS, both OBS and OPS are capable of supporting dynamic traffic and therefore, thus improve the utilisation of the network resources. OPS is most suitable for the dynamic and reconfigurable routing schemes as packets can be routed in different paths to reach a destination, thus ensuring a reduced network traffic latency and congestion. Routing-path decisions are managed by a network management mechanism, which constantly updates the routing information at every network routers to deal with network blocking or failure problems. The OPS system therefore offers a superior resilient property.

In a complete resilient OPS network, there is a need for failure detection, failure notification and path re-routing (Gee-Kung, et al., 2009). These network management features should exist in a layer that is parallel to the network without compromising the high speed capability of the optical network. A typical router of this type is depicted in Figure 1. In this model routing process is continuously monitored and appropriate actions are taken at the resilient control layer to ensure that packets are routed accurately.

In resilient photonics networks, the challenge resides on the efficiency of monitoring and network recovery for data transmissions in the all-optical domain. In the following sections all-optical routing techniques will be reviewed and the resilient feature will be discussed.

ALL OPTICAL ROUTING SCHEMES

Optical fibre communication networks were first introduced with a simple topology of point-to-point

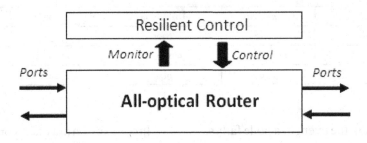

Figure 1. Core all-optical router with resilient control layer

connection, but without any address recognition or packet routing capabilities. In second generation optical networks such as the synchronous optical network (SONET) and the synchronous digital hierarchy (SDH) schemes, additional features such as packet processing, switching and routing in complex network topologies including star, ring and mesh were introduced. However, such networks have a number of limitations. (i) Establishing a new connections is carried out statically and manually, which takes a very long months and may result in lost carrier opportunities, (ii) requiring a large chunk of the bandwidth for protection against failures, (ii) costly to upgrade to meet the traffic growth, and (iv) the main processing and switching at intermediate routers are performed in the electrical domain that requires extensive use of optical-to-electrical-to-optical (O–E–O) conversion modules (Green, 2001). In contrast the emergence of a new generation of a very-high speed optical networking technology, including the dense WDM and optical time division multiplexing transmission technologies, optical multiplexers and optical cross-connect devices, employing O-E-O modules is not possible because of the speed of operation, thus resulting in the data throughput bottleneck. To overcome this problem the third generation all-optical networks were introduced, where data packets processing and routing are entirely carried out in the optical domain. The truly ultrafast photonic networks capable of supporting multi-media broadband

services must be therefore optically transparent regardless of bit rates and protocols.

A photonic packet switching core-network consists of optical fibre systems with optical core routers (core nodes) and edge routers (edge nodes) which are identified by their unique addresses and the fibre links, as shown in Figure 2. In a core network, packets are routed from a source edge router (ingress router) to a target edge router (egress router) via a predefined shortest path through a number of core routers. A packet contains main elements such as the clock signal for synchronisation purpose, a header containing address of the destination node and the payload. The shortest-path details are stored in a look-up routing table at each intermediate router. The entries of the routing table are used to compare with the incoming packet header address in order to determine the next hop where the packet is forwarded to. At an ingress router, data packets from client networks with the same (egress router) destination are temporally multiplexed into optical high-speed packets at high bit rates. Upon entering the core network, synchronisation and address (label) details are optically added to the optical data packet prior to transmission to the core routers. The entire header processing and routing are performed in the optical domain to achieve a maximum data throughput.

Recently there we have seen considerable developments in all-optical packet switching routers for ultrahigh-capacity photonic DWDM and OTDM networks. In such networks, success-

Figure 2. Typical all-optical core network and packet format

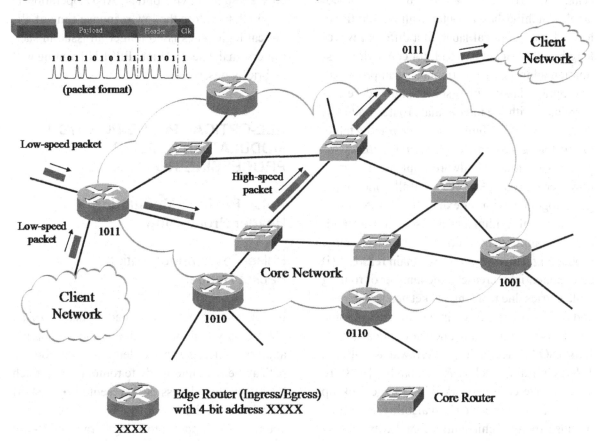

fully carrying out packet routing decision is a major challenging task owing to complex hardware implementation particularly when correlating packets header address bits with every entry of a large size look-up routing table. A number of all-optical packet header processing and routing/switching schemes has been proposed. The simplest is based on the all-optical self-driven packet switching scheme with a predefined look-up routing table at the source router, where the header address is used to directly control the optical switches (i.e. on/off states) in the core router (Ghassemlooy & Ngah, 2005; Yuan, Li, Li, & Wai, 2003) to ensure packet forwarding. Although this scheme is attractive for a small-scale all-optical routing, it is not suitable for large-scale networks consisting of many intermediate routers/

nodes because of a much longer header length required for multiple-hop routing. In addition it lacks routing table reconfiguration capability, since routing decision-making information is embedded in every packet, therefore offering no recovery capability in the case of a node failure.

In (Hauer, et al., 2002), a partial header recognition scheme has been proposed for an optical bypass router in large-scale networks. The router optically processes a small subset of the packet header address that will have a high probability of selecting a specific routing path. The remaining bits of the packet header address, which are not included in the subset, are processed at the main electronic router, thus resulting in a large processing latency. Recently in (Calabretta, Contestabile, Kim, Lee, & Ciaramella, 2006; Ramos, et al., 2005; Yoo, 2003), an optical multi-protocol label-switching-based router with an additional lower bit rate optical label, at a different wavelength to the packet payload, for conveying destination address information has been proposed. The optical label with a short length address is correlated with the intermediate router's address patterns stored in a built-in look-up routing table, thus offering a complete optical transparency for packet switching. Hardware implementation is, however, still complex because of all-optical logic gates together with an exhaustive (brute-force) correlation algorithm that are used for header address recognition. In addition, it introduces a large processing delay owing to two main reasons: (i) an exponential increase in the number of routing table entries due to a long packet header address, and (ii) a slow response time of all-optical logic gates based on the semiconductor optical amplifiers (SOAs) technology (Agrawal & Olsson, 1989; Girardin, Guekos, & Houbavlis, 1998). In a larger-size core network with a large look-up routing table, the router's scalability is limited because of its architectural complexity and its associated cost (optical devices). Although these issues have continuously been addressed with the penetration of the optical technology into the telecommunications market, it is still too costly to put together a large-scale all-optical packet switching network, thus hindering the all-optical transparent network evolution.

As all-optical header processing is an imperative element to achieve a truly all-optical packet-switched network, different approaches have been demonstrated in (Le-Minh, Ghassemlooy, & Ng, 2006a; Le-Minh, Ghassemlooy, Ng, & Chiang, 2009). In (Le-Minh, et al., 2009), an alternative header processing scheme based on the pulse position modulation header processing (PPM-HP) has been introduced, where the incoming packet header address and the routing table entries are both converted into the PPM format. This approach offers a reduced size routing table, thus minimising the packet header processing time. In this scheme,

only a single bitwise optical AND operation is required, therefore the low response-time of all-optical logic gates is no longer an issue in such routers. In the next section, PPM-HP routing will be briefly described.

ALL-OPTICAL PULSE-POSITION-MODULATION HEADER PROCESSING ROUTER

Pulse Position Modulation Header Processing

Pulse-Position-Modulation Packet Header

In optical core network, a packet header of an N-bit binary codeword contains the destination address, which is used to determine the routing path at the each intermediate router (node). Each bit "1" in the address is represented by a short duration optical pulse. In PPM an N-bit binary codeword is mapped into a single frame of 2^N-slot length with a single short pulse located at the m_A^{th} slot. The m_A^{th} pulse position corresponds to the decimal value of the N-bit binary packet header address. For example an N-bit address of $[a_{N-1}a_{N-2} \ldots a_2a_1a_0]$ will have the decimal metric m_A that is computed by:

$$m_A = a_{N-1} \times 2^{N-1} + \ldots + a_{N-2} \times 2^2 + a_1 \times 2^1 + a_0 \times 2^0. \qquad (1)$$

Figure 3(a) illustrates an example of a 4-bit codeword "1001" with $m_A = 9$ and its equivalent PPM address pattern where a single optical pulse is located at the 9th slot.

Pulse-Position Routing Table

In a network with an N-bit length address, a full routing table could have up to 2^N look-up entries. The router makes a routing decision by correlat-

Figure 3. Examples of (a) the "1001" binary address codeword and its equivalence 16-slot PPM frame (e.g. PPM address) and (b) the 3-entry PPRT generated from the conventional look-up routing table, the router has 3 outputs and the address length is 4

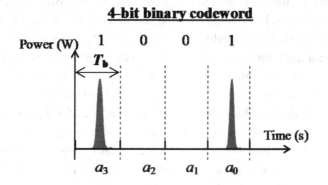

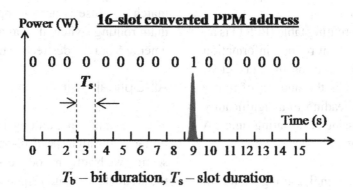

T_b – bit duration, T_s – slot duration

(a)

Sets (P)	Address patterns	Decimal metrics in the set	PPRT entries
P_1	0 0 0 0 0 0 1 0 0 1 1 0 1 0 0 1 1 0 1 0 1 1 0 0 1 1 1 0	{0,2,6,9,10,12,14}	
P_2	0 0 0 1 0 0 1 1 0 1 0 0 1 0 0 0 1 1 0 1 1 1 1 0	{1,3,4,8,13,14}	
P_3	0 0 0 1 0 1 0 1 0 1 1 1 1 0 1 0 1 0 1 1 1 1 1 0 1 1 1 1	{1,5,7,10,11,14,15}	

Packet (PK)

(b)

ing and matching the packet header address with a unique entry in the routing table. In the worst case scenario, i.e. where all possible entries are checked (exhaustive correlation), the router needs to perform 2^N N-bitwise correlations. If N is large, the correlation process requires either (i) a lengthy sequential correlation using a minimum number of optical logic gates (or optical correlators) or (ii) an extensive use of a large number of parallel optical logic gates for parallel matching. In both cases, it requires a compromise between the header recognition speed and the router's complexity and costs. Therefore, the proposed PPM-HP approach offers a routing table with a reduced number of entries while fully preserving the routing information.

The pulse-position routing table (PPRT) is an alternative way to represent routing information with a reduced size by limiting the number of routing entries as well as the number of router output ports M. Thus leading to a significantly reduced packet processing and routing times. A PPRT is constructed as follows:

1. From the conventional routing table, all decimal metrics of address patterns (P) which have the same router output #m are grouped into a set $\mathbf{P}_m$. The number of $\mathbf{P}_m$ elements is $N_m = \text{length}(\mathbf{P}_m)$
2. The mth entry in the PPRT is a 2^N-slot PPM frame, which accommodates only N_m optical pulses. These pulses are located at positions corresponding to the decimal metrics given in $\mathbf{P}_m$.

An example is given in Figure 3(b) for $N = 4$ bits, where the router has three output ports ($M = 3$). Each generated PPRT entry in the table is a 16-slot PPM frame with a number of pulses positioned at dedicated slots corresponding to the decimal weights of address patterns assigned to a particular node output. When a packet header address correlates with entries having decimal metrics of 0, 2, 6, 9, 10, 12 and 14, the packet will be forwarded to the router output port 1. For the set $\mathbf{P}_1 = [0, 2, 6, 9, 10, 12$ and $14]$, the first PPRT entry has a number of short duration pulses located at positions 0, 2, 6, 9, 10, 12 and 14. Similarly for sets $\mathbf{P}_2$ and $\mathbf{P}_3$ the second and third PPRT entries have a number of pulses located at positions 1, 3, 4, 8, 13, 14 and 1, 5, 7, 10, 11, 14, 15, respectively.

Assuming that a packet header address with a decimal metric of 6 is converted to a PPM format with a pulse located at the 6th slot, see Figure 3(b), the correlation (using a single-bitwise Boolean AND operation) of the PPM address with PPRT entries will result in an output signal only for the PPRT entry with a pulse at the 6th position. This matching pulse is used to optically control the main routing switch, thus allowing the packet to emerge from the desired destination output port.

All-Optical Switch

The fundamental building block in all-optical routers is the high speed all-optical switch. In this section we briefly introduce a number of state of the art switches, with response time within a few picoseconds range, such as the terahertz optical asymmetrical demultiplexer (TOAD) (Sokoloff, Prucnal, Glesk, & Kane, 1993), the symmetrical Mach Zehnder (SMZ) (Hirooka, Kumakura, Osawa, & Nakazawa, 2006; Nakamura, Tajima, & Sugimoto, 1994) and the ultrafast nonlinear interferometer UNI (Patel, Rauschenbach, & Hall, 1996).

Figure 4 illustrates the schematic block diagram and the operation of TOAD, SMZ and UNI switches. A TOAD is basically constructed using a short length of mono-mode optical fibre loop with an SOA positioned in a fixed location from the loop centre with an offset time of $T_{sw}/2$. The switching window T_{sw} is defined by the propagation time of signal from the offset SOA to the loop centre, and is normally fixed. The incoming data stream, including the routing information messages, entering the TOAD via the input port, are split by the input coupler into two clockwise (CW) and counter-clockwise (CCW) components. With no control pulse (CP) is injected into the loop, CW and CCW components propagating within the loop will experience the same unsaturated SOA gain G_0. These components are recombined at the input coupler before emerging from the reflected port. The input and reflected pulses are detached using an optical circulator. By injecting a single CP or dual CPs (CP$_1$ and CP$_2$) (Le-Minh, Ghassemlooy, & Ng, 2008) into the fibre loop via a second coupler the nonlinear properties of the SOA can be altered, and as a result, the CW and CCW components propagating through the SOA will experience different levels of gain and phase. Consequently following recombination at the input coupler, the input signal will emerge from the transmitted port (Tx port), thus the switching function.

Unlike TOAD switches, which require a fibre loop to create a switching window for a packet or pulse routing, the SMZ offers a much enhanced integration capability to the optical transmission system as it uses dual sub-millimetre length SOAs for switching. With the SMZ being a balanced optical interferometer, the input data packet is only switched (or extracted) to the output 1 when both SOA1 and SOA2 are excited by CP1 and CP2 (Nakamura, et al., 1994). The switching window is defined by the delay time between CP2 and CP1, which should be longer than the length of the switched data/control packet. UNI is an alternative switching module where the signal is extracted by discrimination of one of the signal projections in the orthogonal polarisation mode. This approach however requires careful preservation of the signal polarisation within a long fibre loop. In general it would be a preferable choice to adopt the SMZ switch in all-optical router due to its compact property compared to the TOAD and UNI.

Single Wavelength PPM-HP Router

The diagram of the proposed single wavelength $1 \times M$ PPM-HP router is depicted in Figure 5(a). The PPM-HP router is composed of a number of main modules including the clock extraction module (CEM), a pulse-position-modulation header extraction module (PPM-HEM), a pulse position routing table (PPRT), AND gates, an optical switch controller (OSWC) and an optical switch (OSW) module. The incoming packet $PK(t)$ is split and applied to the CEM, PPM-HEM and OSW modules with delays of 0, T_{CEM} (required time for the clock extraction for synchronisation) and T_{PPM-HP} (total required time for PPM header processing), respectively. The clock pulse $Clk(t)$ is extracted using the CEM module as reported in (Le-Minh, Ghassemlooy, & Ng, 2006b; Le-Minh, et al., 2009) and is used to initialise the operation of the SMZ-based serial-to-parallel converter (SPC), PPM address conversion module (PPM-ACM) and PPRT modules with delays of 0, T_{ACM} and T_{PPRT}, respectively.

A block diagram of the PPM-HEM is shown Figure 5(b) including 1-N SPC and PPM-ACM modules. The packet header bit stream $[a_{N-1} a_{N-2} \ldots a_0]$, where a_{N-1} is the most significant bit, are extracted by the 1-N SPC module into its parallel format using the control signal $CP_{SPC}(t) = \kappa Clk(t)$, where κ is the input splitting factor. The converted parallel bits are amplified, with gains G_{a0}, G_{a1}, ..., G_{aN-1}, and delayed by T_0, ..., T_{N-1} to ensure that all a_i are simultaneously applied to switches control inputs in the PPM-ACM module before being applied to control inputs of N 1×2 switches for PPM address conversion in the PPM-ACM module. The input to the PPM-ACM module is $CP_{PPM-ACM}(t)$, where $CP_{PPM-ACM}(t) = \kappa CLK(t + T_{ACM})$. Based on values (i.e. 0/1) of address bits a_i ($i = 0, 1, \ldots, N-1$), $CP_{PPM-ACM}(t)$ propagating through N switches will accumulate a total delay of $m_A \times T_s$.

The output of the PPM-HEM module is the converted to the PPM-address format a_{PPM} which is given by:

Figure 4. (a) TOAD, (b) SMZ, and (c) UNI

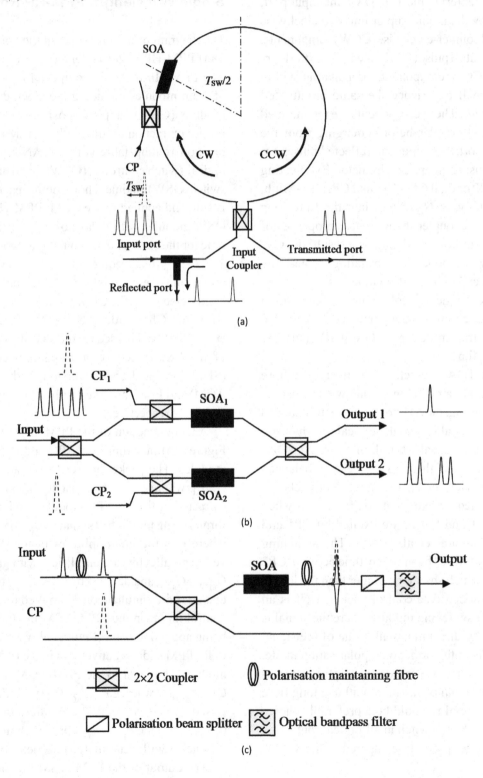

Figure 5. (a) All-optical core-router architecture based on PPM-HP showing packet switching from the input to the router output #2 and, (b) all-optical PPM header extraction module

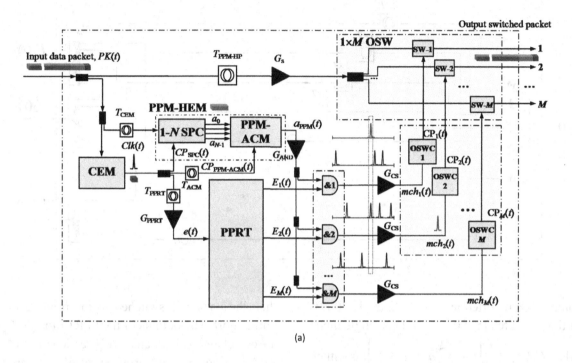

(a)

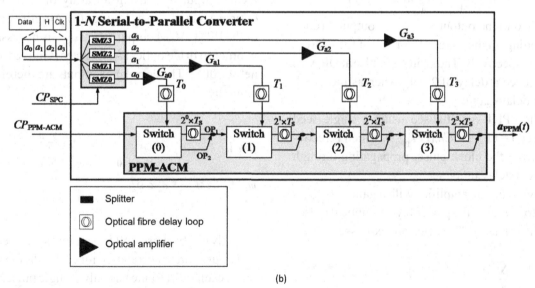

(b)

$$a_{PPM}(t) = CP_{PPM-ACM}\left(t + \sum_{i=0}^{N-1} a_i \times 2^i \times T_s\right), \qquad a_i \in \{0,1\}$$

(2)

Note that the additional delay induced by $CP_{PPM\text{-}ACM}(t)$ is equal to the product of the decimal value of the header address and the PPM slot duration T_s. The delay is generated using a series of 1×2 optical switches with a high on/off extinc-

Figure 6. Pulse position routing table generation

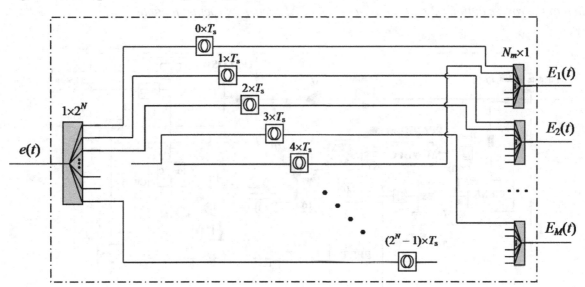

tion ratio as reported in (Le-Minh, et al., 2009). At the nth switch ($n = 0,1,...,N$-1), a single input pulse $CP_{PPM-ACM}\left(t + \sum_{i=0}^{n-1} a_i \times 2^i \times T_s\right)$ is either switched to the output 1 (OP$_1$) or output 2 (OP$_2$) depending on the status "1" or "0" of the address bit a_i, respectively. The switched pulse at OP$_1$ will experience a delay of $2^i \times T_S$ whereas there will be no delays at OP$_2$.

The PPRT is initialised using a single seed pulse $e(t) = (1 - 2\kappa)Clk(t+T_{PPRT})$, which is recovered from the clock pulse, propagating through different delay units, see

Figure 6. The amplifier with a gain of G_{PPRT} is used to maintain the power level of optical pulses in PPRT. The PPRT entries are expressed as:

$$E_m\left(t\right) = \sum_{d_m} e\left(t + d_m \times T_s\right), \qquad \forall\ d_m \in D_m.$$

(3)

Each D$_m$ set contains all decimal values of address patterns assigned to the router output #m ($m = 1,2,...,M$). If the decimal value of an incoming packet header address bits (i.e. target edge-node address) match a value within the D$_m$ set

then the packet is switched to the router's m^{th} output port. The outputs of the PPRT module E_m are correlated with the amplified version of PPM-ACM output a_{PPM} using an array of two-input optical AND gates, see Figure 5(a). Amplification of the PPM-ACM output signal a_{PPM} is necessary in order to drive optical AND gates (Guo & Connelly, 2006). The correlated outputs are therefore given by:

$$mch_m\left(t\right) = a_{PPM}\left(t\right) \times E_m\left(t\right) = \begin{cases} 1 & if \quad d_m = \sum_{i=1}^{N-1} a_i \times 2^i \quad \forall m \\ 0 & if \quad d_m \neq \sum_{i=1}^{N-1} a_i \times 2^i \quad \forall m \end{cases},$$
$$m = 1, 2, ..., M \qquad d_m \in D_m.$$

(4)

Only one bit-wise AND operation is required to carry out address correlation for each PPRT entry. Since each PPM frame has only a single pulse, the matching pulse vector $MCH(t) = [mch_1(t)\ mch_2(t) ... mch_M(t)]$ for M-PPRT entries will have only one non-zero element. Matching pulses $mch_m(t)$ are amplified (with a gain of G_{CS}) and applied to OSWC modules (Le-Minh, et al., 2009) to generate control pulses $CP_m(t)$ having the same optical intensity. This is to ensure a switching window with a flat gain profile during the switching interval. The switched packet $PK_{SW}(t)$ for the mth output is computed as:

$$PK_{SW}(t) = PK(t + T_{PPM-HP}) \times mch_m(t), \qquad (5)$$

where $mch_m(t)$ is the resultant non-zero matching pulse.

PPM-Based Header Processing Gain

Header recognition involves PPM-address conversion (in the PPM-ACM module) and address correlation (a_{PPM} with PPRT entries). Since both tasks could be carried out simultaneously, the header recognition time due to PPM-HP T_{PPM-HP} is determined by the duration of a 2^N-slot PPM-frame as:

$$T_{PPM-HP} = 2^N \times T_s. \qquad (6)$$

Note that M AND gates are used for parallel correlation of a_{PPM} with MPPRT entries. In a conventional header recognition scheme using the exhaustive (brute-force) correlation algorithm employing M AND gates, the required header recognition time T_{EX-HP} is:

$$T_{EX-HP} = 2^N \times N \times T_{AND} \times M^1, \qquad (7)$$

where T_{AND} is the minimum time interval required for AND gates gains to recover between two successive AND operations (i.e. the SOA recovery time (Nakamura, et al., 1994)). Typically T_{AND} (hundreds of picoseconds) is much greater than T_b (few picoseconds) in multi-hundred Gbit/s optical networks. The header recognition time gain R_T is defined as the ratio of the time required for the brute-force approach T_{EX-HP} over the required time for PPM-HP T_{PPM-HP}, and is given by:

$$R_T = \frac{N \times T_{AND}}{M \times T_s}. \qquad (8)$$

Note that M is typically smaller than N and $T_{AND} \gg T_s$.

Multiple-Wavelength PPM-HP Router

The PPM-HP router described in the previous section is used for single-wavelength input data packet(s). In the scenario where multiple packets at different wavelengths arrive at the router, the PPH-HP router should be capable of simultaneously processing and forwarding all the packets to the desirable outputs without any losses and delays (i.e. no congestion). For such systems, the combination of individual PPM-HPs is required. In Figure 7(a) a typical multiple-wavelength PPM-HP router (MW-PPM-HP router) is illustrated. The router comprises of M PPM-HP modules in combination with a common MW-PPRT module, M continuous wave (CW) laser sources and a number of WDM multiplexers and demultiplexers. At the router input, N-bit-address packets with multiple-wavelength (at $\lambda_1, \lambda_2 \dots$ and λ_M) are passed through a WDM demultiplexer before being fed to a bank of PPM-HP modules. Each PPM-HP module k ($1 \leq k \leq M$) will process packets at a given wavelength λ_k. A single pulse e_k extracted from the clock pulse of the incoming packet at λ_k in the kth PPM-HP module is input to the MW-PPRT module to generate M PPM-formatted routing table entries $E_1, E_2, \dots, E_M$ (at the same λ_k), which are then fed back to the kth PPM-HP module for address recognition. The MW-PPRT, shown in Figure 6, could be reused for different wavelengths.

Following address correlation, which is carried out in the kth PPM-HP module, the packet wavelength is converted via the wavelength converter (WC) to the desired wavelength λ_o ($o =$ modulo $(k+m-1,M)$, $\forall k, m, o \in [1, M]$). The wavelength-converted packet is applied to the optical multiplexer to select the dedicated MW-PPM-HP router m output. To avoid PPM-HP modules from switching signals with identical wavelengths to the same router output, permutations in feeding

Figure 7. (a) Block diagram of the proposed WDM 1×M router based on PPM-HPs, a common MW-PPRT and assisted CW laser sources and, (b) the diagram of a 1×M PPM-HP router (operating at λ_k) including wavelength converters showing the input packet at λ_m is converted to λ_2 and switched to the PPM-HP output #2

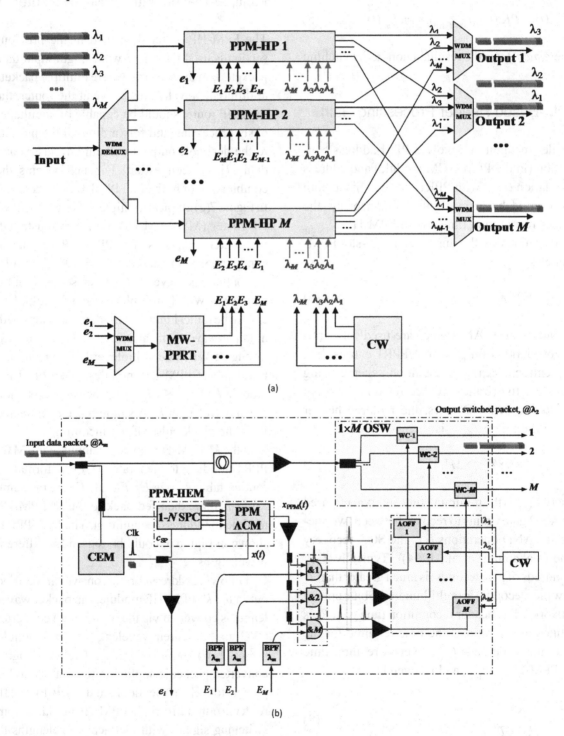

order of MW-PPRT entries to the inputs $e_1, e_2, ...,$ e_M of different PPM-HP modules are required, see Figure 7(a) and (b). In situations where all M-input packets are required to be switched to the router output 1, there are up to M simultaneous signals at M different wavelengths emerging at each router output, thus offering a non-blocking characteristic.

Individual PPH-HP modules in the MW-PPM-HP router are constructed similarly to the router described in Figure 5(b), except for additional external MW-PPRT modules, and the replacement of the switching units by wavelength converters (WCs) and optical switching controls by all-optical flip-flops (AOFFs) (Le-Minh, Ghassemlooy, & Ng, 2007). The structure of MW-PPRT is the same as the single wavelength PPRT.

The matching output from the oth AND gate of the kth PPM-HP is applied to set the oth AOFF to its 'on' state. The wavelength converter based on the SMZ is further explained in (Ueno, et al., 2002). With the oth AOFF being excited by the matching pulse it generates an output signal at a wavelength of λ_o with a constant optical power level and a duration equal to the packet duration T_{pk}. This continuous wave will interact with the input packet due to the cross phase modulation (XPM) effect within WC-o to switch the packet to the oth output port of the kth PPM-HP module. Outputs from individual PPM-HP modules will be wavelength multiplexed to the dedicated router output port.

RESILIENT ALL-OPTICAL ROUTER

Resilience in All-Optical Routing

Network resilience in general refers to the network ability in recovering data and services lost or affected due to failures encountered by the system. It may be considered as a component of QoS. In optical networks additional resource is required for a complete resilience provisioning mechanisms

in order to protect the network against all possible network failures. These network management features including failure detection, failure notification and path re-routing should exist in a layer that is parallel to the network without comprising the high speed capability of the optical network (Mas, Tomkos, & Tonguz, 2005). Resilience schemes can be classified into two categories: (i) *protection*, which uses a pre-assigned alternative path and capacity to ensure resilience, generally offers improved traffic restoration speeds than reactive restoration, and (ii) *restoration*, which refers to the technique that restores the affected traffic by finding an alternative path using all available capacity, is much more flexible at choosing the alternative path, thus resulting in improved sharing of resources. Network topologies adopted have a key role on the resilience provisioning mechanisms. Usually only protection is adopted in the point-to-point / linear and ring topology networks, while both protection and restoration mechanism could be employed in a mesh topology based networks. Restoration is suitable for networks with rapidly changing traffic demands, thus requiring dynamic real-time algorithms. However, the drawbacks of restoration are lack of available spare resource and long restoration time particularly in heavily loaded networks.

Network resiliency schemes can be generally classified into two categories:

1. **Reactive restoration:** where unreserved network resources is allocated to put right the network data flow affected by the fault. However, there are a number of disadvantages including the amount of unreserved resources allocated may not be adequate, therefore leading to data flow rejection, and longer recovery latency time. The latter is more problematic in heavily loaded networks, since more time is needed to find and establish alternative links.

2. **Proactive restoration/protection:** where backup links are indentified to protect traffic

against possible faults at the time of establishing the primary links within a very short time.

In optical networks (opaque, managed reach and transparent (i.e. all-optical)), path layer technologies are used to provide both network control and management to overcome network failures. Wavelength division multiplexing has been proposed for the wavelength path for a network to provide resilience and restoration to the network path (Baroni, Bayvel, Gibbens, & Korotky, 1999; Sato, Okamoto, & Hadama, 1994). Such configurations are complex and highly sensitive to scalability.

The detection and notification of node failure can be carried out in electrical or optical domains utilising path layer technologies. In order to build a resilient optical network where the detection and notification are carried out in the optical domain, new network configurations have been widely proposed (an der Heiden, et al., 2007; Fernandez Vallejo, et al., 2009). In IEEE 802.17 Resilient Packet Ring (RPR), spatial reuse can be maximised through the shortest path routing algorithm within RPR. The RPR is able to accommodate the large bandwidth requirement for downstream communications such as the high definition television (HDTV). Spatial diversity schemes would require redundancy to protect the network against failure and thus giving rise to the challenges in the control or management layers. However, such networks restrict the scalability of the optical network.

Redundancy in optical networks is the most common method of providing resilience against failure; however it is not cost effective to provide redundancy for a node or an optical router. Therefore in the case of a node failure, the re-routing of path becomes more complex. Optical routers should therefore have the capability to re-route packets via different optical paths in the same manner as in electrical routers. Hence a self re-routing node would enable a higher level of scalability as

well as maintaining a higher level of availability. In the following parts, we will discuss such self-routing systems that offer similar performance as redundancy systems.

Figure 8 depicts the detail diagram of the proposed resilient all-optical router where an additional control layer is applied to monitor and control the router performance. The physical layer (optical transponder) is in the optical domain whereas the resilient management is in the electrical domain as the natures of information and transmission rate at each layer are different. At the router input, arriving packets propagate through a 1×2 splitter. The main portion (in term of the signal power) of the packet is delayed via an optical fibre delay loop (OFDL) unit. This delay is to ensure that a routing decision can be made before the packet is switched to the target output port. In a packet, segments are allocated for synchronisation, header extraction and signal processing. The self-extracted clock signal is used for the router synchronisation purpose. The header extraction module extracts and converts packet address bits into a format appropriate for the header recognition module. In the header recognition module packet address correlation is carried out, which is based on the routing information contained in both the packet header address and the routing table. The correlated output signal (pulse) is applied to the control input of the optical switching unit for switching the incoming packet to the correct output port. Inclusion of optional modules such as a reconfigurable look-up routing table, buffering and signal processing will provide additional functions to improve non-blocking characteristic (contention solution) of the router as well as improving the optical signal quality.

As shown in Figure 8, the resilient management layer monitors the data and control signals flowing within the router and responds accordingly. There are two main functions that the resilient management layer will offer:

Figure 8. Self-routing resilient all-optical router

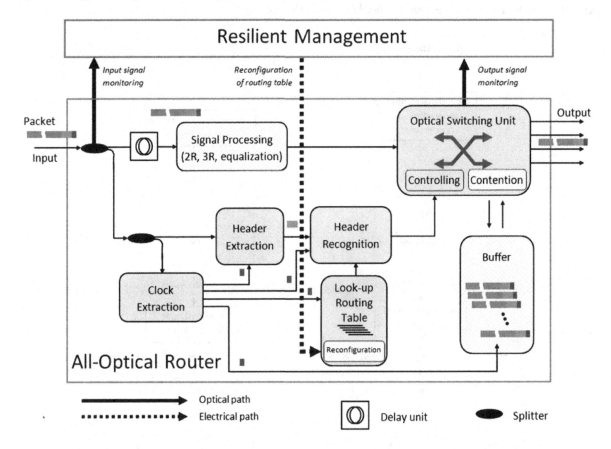

- **Self monitoring router status:** self detection of failures occurred within the router.
- **Updating/refreshing the routing table:** responding to the failure of neighbouring routers by means of self-reconfiguration of the routing table to avoid forwarding packets to failed nodes.

Router Failure Self-Detection

In the self-monitoring mode, the latency of packet propagating through the router is measured to determine the failure of router functionality. The input data packet is detected and the arrival time-stamp is also recorded in the management layer. Since the optical router structure is pre-configured, the packet processing time is approximately predicted. Therefore any late arrival or no show of

packets at the router output will indicate failures within the router, therefore no packet switching. The resilient management layer monitors the incoming data stream, the routing decision signal and the switched data signal. Part of incoming optical signal is converted into an electrical signal whose strength is continuously being measured and compared with a predefined threshold level. Once the measured level is higher than the predefined threshold level, the resilient management layer will record the arrival time of the packet. In the event of malfunctioning comparator units the routing decision signal will not be generated or the switching fabric will fail to function properly, no signal will emerge from the router output. After a certain period of time the management layer will consider that there is a failure within the router.

Figure 9. Flow chart of the proposed resilient all-optical router (Note: timeout is the difference between packet's arrival time and the time the packet leaves the router)

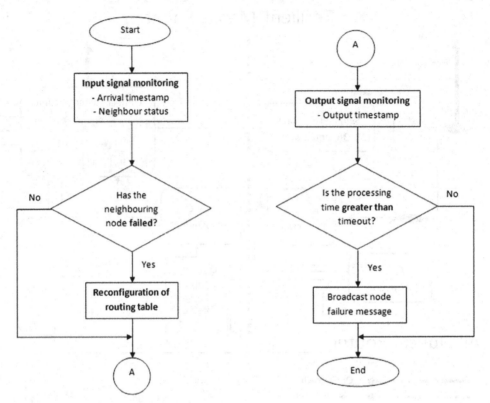

Up to the point where a failure is detected the router could communicate with its neighbouring routers using either an additional electrical channel or an existing optical channel. The routing information message containing information of router unavailability is therefore broadcasted.

In Figure 10(a) and (b), there are two configurations of the control channel in electrical and optical domains, respectively.

In Figure 10(a), messages containing the node failure information could be relayed to adjacent routers via copper wires. The message could be encapsulated in one of the standard protocol formats, e.g. Routing Information Protocol (RIP), and being broadcasted periodically as defined by the protocol. If another router is tapping in to decode the message, it will update its own optical routing table accordingly by disabling the interfaces connection to the path leading to the broken

router and rerouting packets to one of the available adjacent routers. Owing to the routing protocol, other routers within the network will also be able to automatically isolate the broken router until the time when it is back to its normal operation via the next polling of RIP signalling propagation. The advantage of using an additional electrical channel is to elevate the extra complexity added to the optical backbone infrastructure and to minimise the installation cost as the routing information channel does not require a very high bandwidth (e.g. the RIP protocol does not need to refresh neighbour status up to every 30s period) (Hedrick, 1988).

On the other hand migration of the routing information channels into existing optical data channels could be a preferable choice for long haul backbone networks such as transatlantic links. This is due to the additional cost of extra electri-

Figure 10. Resilient router configuration with: (a) electrical control path, and (b) optical control path

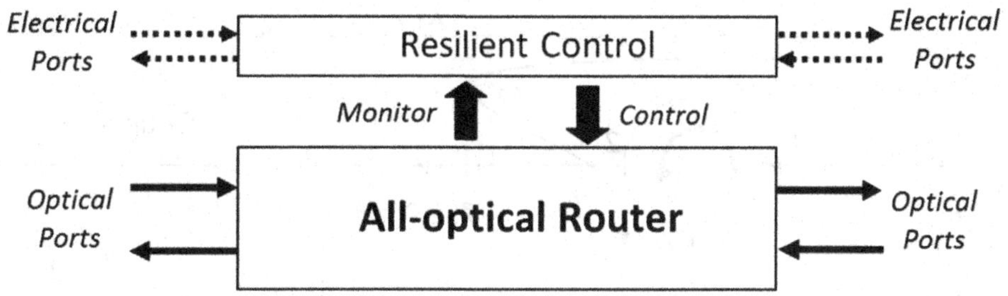

cal cable installation and the signal attenuation management over a long span. In Figure 10(b) the optical interface of resilient router is shown. As both the routing information message and data packets share the same transmission medium, an optical interface with routing information message insertion and extraction capabilities would be a necessary requirement. For this category of optical interface, the optical add-and-drop multiplexer (OADM) in either time or wavelength domain could be utilised.

Optical Interface

The key technologies for providing a better exploitation of the existing transmission infrastructure are the optical time domain division multiplexing (OTDM) (Tucker, Eisenstein, & Korotky, 1988) and WDM. In OTDM schemes slower speed data channels are multiplexed into a higher stream at a much higher bit rate by means of non-overlapping multiple time-slots allocated to the incoming data channels employing a single wavelength. In the WDM technology each channel is assigned a specific wavelength (bandwidth) within a well defined total transmission bandwidth. OTDM systems are more complex and costly to implement than the well established WDM schemes. In such systems optical add-and-drop multiplexers can be used to extract and insert the routing information message from the incoming data stream and to outgoing data flow from the router, respectively. Inclusion of the routing information onto the outgoing data

stream is simple, since an optical coupler could be readily used. However, it is a challenging task to extract the optical message from the data stream at an ultrahigh speed. Here a number of approaches will be discussed.

In OTDM systems the extraction of the OADM module could be realised by adopting one of the all-optical switching technologies such as TOAD, SMZ UNI which have been described in previous section. In those schemes the relevant information is switched out from the data stream entirely in optical domain, therefore it could support data rate up to several hundreds of Gigabit per second. The fibre Bragg grating (FBG) based OADM technique is another attractive method for extracting data in WDM systems (Othonos & Kalli, 1999). As the routing information message is modulated at a different wavelength but transmitted at the same time with the data stream in the same fibre, FBGs together with an optical circulator could be used to filter out the message at the router (or wavelength selection)

FBG is a type of distributed Bragg reflector which is constructed in a short segment of optical fibre that reflects particular wavelengths of light and transmits all others. Fibre core has a periodic variation to the refractive index thus generating a wavelength specific dielectric mirror. In the optical interface, FBG could be used as an inline optical filter at the input and output of the router. The schematic block diagram of FBG based OADM module is depicted in Figure 11. As the optical pulses propagates through fibre

Figure 11. WDM OADM employing FBGs and optical circulators

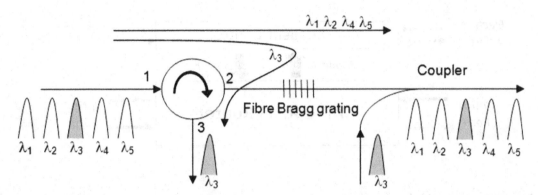

and the circulator λ_3 is reflected by the FBG (by design) and is dropped from the system via port 3 of the optical circulator, while the rest of the wavelengths continue forward alone the fibre. λ_3 containing new information is inserted back (or added) to the data stream via a coupler or a second optical circulator.

Updating Routing Table in Resilient All-Optical Router

The routing table contains a set of addresses for the next hops (i.e. neighbour nodes) that the packet could be forwarded to. In electrical routers, routing table entries are stored in the memory (e.g. RAM/ ROM), which could be readily read whenever the header processing is carried out. In the optical domain, there are however no optical memory or storage devices available in the commercial market with the capacity to hold a large number of data over a very long period. Therefore in most reported all-optical routers optical routing table entries are typically generated on packet arrival. There are two main approaches to generate routing table entries which are widely adopted:

1.   Entries are generated in the electrical domain by using a pulse generator and then imprinted onto the optical wave through an optical modulator, or

2.   Entries are directly generated in the optical domain by using a seed optical pulse, which propagates through a network of optical branches and optical fibre delay lines whose outputs are then recombined as shown in Figure 6.

The first approach is relatively straight forward to implement as a complex entry codeword/pattern could be generated in the electrical domain and then be used to modulate the optical signal. However the pattern generation rate in the electrical domain is typically limited about to about 40 Gbit/s due to the limitation of the electrical device, therefore in general this approach is not suitable for ultrahigh-speed optical systems. Time division multiplexing in the optical domain could be utilised to increase the speed albeit at the cost of a more complex routing entry generator.

For ultrahigh capacity optical switched systems, routing decision should be carried out in the real-time at each intermediate router, e.g. correlation of the packet header address and routing table entries are carried out at the transmission line rate. Therefore, routing table entries should be optically generated immediately following packet detection and clock pulse extraction. The optical entries could be directly generated by a seed optical pulse which is extracted from the incoming packet. Figure 6 is an example showing the conceptual design of all optical routing table entry generator.

Updating the optical routing table therefore could be realised by:

- Modifying the electrical pattern generator output, which is adopted for the approach described in the (i) case. This approach is straightforward to implement as the process is entirely performed in the electrical domain.

- Altering propagation paths for the seed pulse, where pulses propagate through different arms with a different set of delays to generate new patterns for the (ii) case. The paths taken are selected by switches whose operation conditions are strictly controlled by the resilient management layer.

In the next section, the updating process for the PPRT module in the optical domain as in the case (ii) will be presented.

Updating Pulse-Position Routing Table

The conventional PPRT module, which is illustrated in Figure 6, is composed of an optical splitter, couplers and optical fibre delay lines. If 2^N entries are required (in the conventional routing table) all delay paths $(0, T_s, ... 2^{N-1}T_s)$ will be fully utilised to create the PPRT. If the network is fixed and has no resilient capability, the number of delay paths could be reduced. However in a dynamic all-optical router, requiring routing table reconfiguration, full paths facilities should be readily be made available.

In Figure 12 a reconfigurable PPRT is proposed, which is composed of a splitter, optoelectronic switches, optical fibre delay line units and signal combiners. The incoming seed is split via a 1×2^N splitter and propagates through 2^N branches with the appropriate delay times ranging from 0 to $(2^N - 1)T_s$. In each entry, for example i^{th}, the new pulse locations (after refreshing the PPRT)

Figure 12. Reconfigurable pulse-position routing table architecture based on splitter, optical fibre delay line units, switch arrays and a combiner

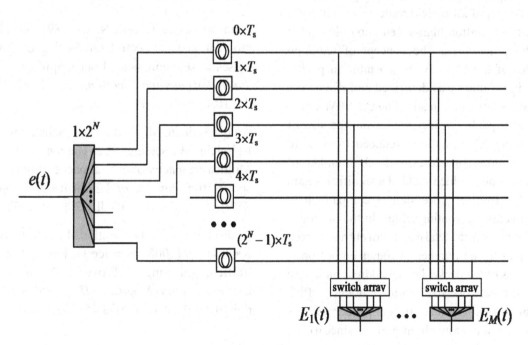

will be determined by the status of each optoelectronics switching array prior to combining them at every entries E_i.

The PPRT refreshing rate mainly depends on changes occurring within the network topology (e.g. addition of new nodes or dropping unavailable nodes), where a very fast response is not a requirement. Therefore a large-scale commercial switching fabric such as micro-electromechanical systems (MEMs) could be readily utilised (Olkhovets, et al., 2004). MEMs consist of an array of mirrors that reflect the optical wave to dedicate output ports, therefore packets with different data rates and wavelengths could be switched at the same time. This design is also transparent to transmission protocols.

CONCLUSION

In this chapter, we have reviewed different routing schemes for high-speed photonic packet switching networks, where all the router elements are designed to operate in the optical domain, in order to achieve a high routing throughput. Header processing and packet switching play a vital role in the design of all-optical routers that will form the next generation high-speed photonic packet switching networks. The concept of reducing the size of the look-up routing table in packet switching routers was described and discussed. We have shown that in adopting the PPM signal format for packet header processing the size of the routing table could be reduced from 2^N to M (N: number of address bits, M: number of router outputs), which yielded a significant gain in packet routing processing in comparison to the conventional routing tables. In the proposed PPRT approach, the total number of entries is fixed and equal to the number of router output ports regardless of the packet header address bit length N and the network size (up to 2^N nodes). PPRT configurations formulate packet header address correlation rather a much simple task, since it only requires a single bit-wise optical AND gate, thus resulting in reduced system complexity as well as processing time. In addition the chapter also discussed the architecture for high-throughput multiple-wavelength PPM-HP routers, based on the multiple-wavelength PPRT.

The network itself could not be fully reliable unless it has a failure protection mechanism to ensure a data flow with a minimum rate of packet loss. In a dynamic optical network, the routers should offer the failure self-detection capability as well as updating neighbouring node status in order to maintain seamless data transmission. Routers could therefore reconfigure their routing table in the optical domain to adapt to a new network topology. In Section 4 a failure self detection and a routing table reconfiguration in the optical domain were introduced and a number of factors such as system performance, reliability and complexity were also discussed. Using a large switching fabric like MEMs could offer reconfiguration of routing table in optical domain regardless of data rates, wavelengths and protocols.

REFERENCES

Agrawal, G. P., & Olsson, N. A. (1989). Self-phase modulation and spectral broadening of optical pulses in semiconductor laser amplifiers. *IEEE Journal of Quantum Electronics, 25,* 2297–2306. doi:10.1109/3.42059

An der Heiden, M., Sortais, M., Scheutzow, M., Reisslein, M., Seeling, P., & Herzog, M. (2007). Multicast capacity of optical packet ring for hotspot traffic. *Journal of Lightwave Technology, 25*(9), 2638. doi:10.1109/JLT.2007.902092

Androulidakis, S., Doukoglou, T., Patikis, G., & Kagklis, D. (2008). Service Differentiation and traffic engineering in IP over WDM networks. *Communications Magazine, IEEE, 46*(5), 52–59. doi:10.1109/MCOM.2008.4511649

Baroni, S., Bayvel, P., Gibbens, R. J., & Korotky, S. K. (1999). Analysis and design of resilient multifiber wavelength-routed optical transport networks. *Journal of Lightwave Technology, 17*(5), 743. doi:10.1109/50.762888

Bergano, N. S. (2005). Wavelength division multiplexing in long-haul transoceanic transmission systems. *Journal of Lightwave Technology, 23*, 4125–4139. doi:10.1109/JLT.2005.858255

Blumenthal, D. J. (2001). Photonic packet switching and optical label swapping. *Optic. Networks Mag.*, 1-12.

Calabretta, N., Contestabile, G., Kim, S. H., Lee, S. B., & Ciaramella, E. (2006). Exploiting time-to-wavelength conversion for all-optical label processing. *IEEE Photonics Technology Letters, 18*, 436–438. doi:10.1109/LPT.2005.863207

Choi, E., Jang, H., Lee, J., Lee, H., Hwang, S., & Oh, Y.-J. (2005). Modeling and verification of FEC performance for optical transmission systems using a proposed uniformly quantized symbol error probability model. *Journal of Lightwave Technology, 23*, 1100–1104. doi:10.1109/JLT.2005.843452

Doerr, C. R., Chandrasekhar, S., Buhl, L. L., Cappuzzo, M. A., Chen, E. Y., & Wong-Foy, A. (2006). Optical dispersion compensator suitable for use with non-wavelength-locked transmitters. *Journal of Lightwave Technology, 24*, 166–170. doi:10.1109/JLT.2005.860475

Dorren, H. J. S., Hill, M. T., Liu, Y., Calabretta, N., Srivatsa, A., & Huijskens, F. M. (2003). Optical packet switching and buffering by using all-optical signal processing methods. *Journal of Lightwave Technology, 12*, 2–12. doi:10.1109/JLT.2002.803062

Fernandez Vallejo, M., Perez-Herrera, R. A., Elosua, C., Diaz, S., Urquhart, P., & Bariain, C. (2009). Resilient amplified double-ring optical networks to multiplex optical fiber sensors. *Journal of Lightwave Technology, 27*(10), 1301. doi:10.1109/JLT.2009.2015774

Fu-Tai, A., Kyeong Soo, K., Gutierrez, D., Yam, S., Hu, E., & Shrikhande, K. (2004). SUCCESS: A next-generation hybrid WDM/TDM optical access network architecture. *Lightwave Journal of Technology, 22*(11), 2557–2569. doi:10.1109/JLT.2004.836768

Gee-Kung, C., Chowdhury, A., Zhensheng, J., Hung-Chang, C., Ming-Fang, H., Jianjun, Y., et al. (2009). Key technologies of WDM-PON for future converged optical broadband access networks [Invited]. *IEEE/OSA Journal of Optical Communications and Networking, 1*(4), C35-C50.

Ghassemlooy, Z., & Ngah, R. (2005). Simulation of 1x2 OTDM router employing symmetric Mach-Zehnder switches. *IEE Proceedings. Circuits, Devices and Systems, 152*, 171–177. doi:10.1049/ip-cds:20041017

Girardin, F., Guekos, G., & Houbavlis, A. (1998). Gain recovery of bulk semiconductor optical amplifiers. *IEEE Photonics Technology Letters, 10*, 784–786. doi:10.1109/68.681483

Green, P. (2001). Progress in optical networking. *IEEE Communications Magazine, 39*, 54–61. doi:10.1109/35.894377

Guo, L. Q., & Connelly, M. J. (2006). All-optical AND gate with improved extinction ratio using signal induced nonlinearities in a bulk semiconductor optical amplifier. *Optics Express, 14*, 2938–2943. doi:10.1364/OE.14.002938

Hauer, M. C., McGeehan, J., Touch, J., Kamath, P., Bannister, J., Lyons, E. R., et al. (2002). Dynamically reconfigurable all-optical correlators to support ultra-fast internet routing. *Proc. OFC 2002, USA*, (pp. 268-270).

Hedrick, C. (1988). *Routing information protocol. STD 34, RFC 1058*. Rutgers University.

Hirooka, T., Kumakura, T., Osawa, K., & Nakazawa, M. (2006). Comparison of 40GHz optical demultiplexers using SMZ switch and EA modulator in 160 Gbit/s-500 km OTDM transmission. *IEICE Electronics Express, 3*, 397–403. doi:10.1587/elex.3.397

Hunter, D. K., Chia, M. C., & Andnovic, I. (1998). Buffering in optical packet switches. *Journal of Lightwave Technology, 16*, 2081–2094. doi:10.1109/50.736577

Le-Minh, H., Ghassemlooy, Z., & Ng, W. P. (2006a). Multiple-hop routing based on the pulse-position-modulation header processing scheme in all-optical ultrafast packet switching network. *IEEE GLOBECOM 2006,* San Francisco, USA, OPN06-03.

Le-Minh, H., Ghassemlooy, Z., & Ng, W. P. (2006b). Ultrafast all-optical self clock extraction based on two inline symmetric Mach-Zehnder switches. *Proc. of ICTON 2006, Nottingham, UK,* (pp. 64-67).

Le-Minh, H., Ghassemlooy, Z., & Ng, W. P. (2007). All-optical flip-flop based on SMZ with a feedback-loop and multiple forward set/reset signals. *SPIE Optical Engineering Letters, 46*, 040501.

Le-Minh, H., Ghassemlooy, Z., & Ng, W. P. (2008). Characterization and performance analysis of a TOAD switch employing a dual control pulse scheme in high-speed OTDM demultiplexer. *IEEE Communications Letters, 12*, 316–318. doi:10.1109/LCOMM.2008.061299

Le-Minh, H., Ghassemlooy, Z., Ng, W. P., & Chiang, M. F. (2009). All-optical packet routing network based on PPM-HP header processing. *IET Communications Proceeding, 3*, 465–476. doi:10.1049/iet-com:20070505

Lee, K. K., Lim, F., & Ong, B. H. (2005). *Building resilient IP networks*. Cisco Press Networking Technology.

Lee, K. K., Lim, F., & Ong, B. H. (2006). *Building resilient IP networks* (1st ed.). Cisco Press.

Mas, C., Tomkos, I., & Tonguz, O. K. (2005). Failure location algorithm for transparent optical networks. *IEEE Journal on Selected Areas in Communications*, 1508. doi:10.1109/JSAC.2005.852182

Nakamura, S., Tajima, K., & Sugimoto, Y. (1994). Experimental investigation on high-speed switching characteristics of a novel symmetric Mach-Zehnder all-optical switch. *Applied Physics Letters, 65*, 283–285. doi:10.1063/1.112347

Ohara, T., Takara, H., Shake, I., Mori, K., Sato, K., & Kawanishi, S. (2004). 160 Gbps OTDM transmission using integrated all-optical MUX/DEMUX with all-channel modulation and demultiplexing. *IEEE Photonics Technology Letters, 16*, 650–652. doi:10.1109/LPT.2003.818953

Olkhovets, A., Phanaphat, P., Nuzman, C., Lichtenwalner, C., Kozhevnikow, M., & Kim, J. (2004). Performance of an optical switch based on 3-D MEMS crossconnect. *IEEE Photonics Technology Letters, 16*, 780–782. doi:10.1109/LPT.2004.823703

Othonos, A., & Kalli, K. (1999). *Fiber Bragg Gratings: Fundamentals and applications in telecommunications and sensing*. Artech House.

Patel, N. S., Rauschenbach, K. A., & Hall, K. L. (1996). 40 Gbps demultiplexing using an ultrafast nonlinear interferometer (UNI). *IEEE Photonics Technology Letters, 8*, 1695–1697. doi:10.1109/68.544722

Ramaswam, R., & Sivarajan, K. N. (2002). *Optical networks: A practical perspective* (2nd ed.). Morgan Kaufmann.

Ramos, F., Kehayas, E., Martinez, J. M., Clavero, R., Marti, J., & Stampoulidis, L. (2005). IST-LASAGNE: Towards all-optical label swapping employing optical logic gates and optical flip-flops. *Journal of Lightwave Technology, 23*, 2993–3011. doi:10.1109/JLT.2005.855714

Report, A. (2005). *Fraunhofer Institut Nachrichtentechnik Heinrich-Hertz-Institut, Berlin, Germany*.

Sato, K. I., Okamoto, S., & Hadama, H. (1994). Network performance and integrity enhancement with optical path layer technologies. *IEEE Journal on Selected Areas in Communications, 12*(1), 159. doi:10.1109/49.265715

Sokoloff, J. P., Prucnal, P. R., Glesk, I., & Kane, M. (1993). A terahertz optical asymmetric demultiplexer (TOAD). *IEEE Photonics Technology Letters, 5*, 787–790. doi:10.1109/68.229807

Suzuki, H., Fujiwara, M., & Iwatsuki, K. (2006). Application of super-DWDM technologies to terrestrial terabit transmission systems. *Journal of Lightwave Technology, 24*, 1998–2005. doi:10.1109/JLT.2006.871115

Takahashi, R., Nakahara, T., Takahata, K., Takenouchi, H., Yasui, T., & Kondo, N. (2004). Photonic random access memory for 40-Gb/s 16-b burst optical packets. *IEEE Photonics Technology Letters, 16*, 1185–1187. doi:10.1109/LPT.2004.824987

Takenouchi, H., Takahata, K., Nakahara, T., Takahashi, R., & Suzuki, H. (2004). 40-Gbit/s 32-bit optical packet compressor/decompressor based on a photonic memory. *Proc. Conference on Lasers and Electro Optics, CThQ, San Francisco, California, U.S.A.*

Tucker, R. S., Eisenstein, G., & Korotky, S. K. (1988). Optical time-division multiplexing for very high bit-rate transmission. *Journal of Lightwave Technology, 6*, 1737–1749. doi:10.1109/50.9991

Ueno, Y., Nakamura, S., Hatakeyama, H., Tamanuki, T., Sasaki, T., & Tajima, K. (2002). 168-Gb/s OTDM wavelength conversion using an SMZ-Type all-optical switch. *Proc. ECOC 2002, Munich, Germany*, (pp. 13-14).

Xin, L., Hongxiang, W., & Yuefeng, J. (2008). Resilient burst ring: Extend IEEE 802.17 to WDM networks. *Communications Magazine, IEEE, 46*(11), 74–81. doi:10.1109/MCOM.2008.4689248

Yoo, S. J. B. (2003). Optical label switching, MPLS, MPLambdaS, and GMPLS. *Optic. Networks Mag., 4*, 17–31.

Yuan, X. C., Li, V. O. K., Li, C. Y., & Wai, P. K. A. (2003). A novel self-routing address scheme for all-optical packet-switched networks with arbitrary topologies. *Journal of Lightwave Technology, 21*, 329–339. doi:10.1109/JLT.2003.808755

Zhang, J., Wu, J., Feng, C., Xu, K., & Lin, J. (2007). All-optical logic OR gate exploiting nonlinear polarization rotation in an SOA and red-shifted sideband filtering. *IEEE Photonics Technology Letters, 19*, 33–35. doi:10.1109/LPT.2006.888991

Zhu, Z., Funabashi, M., Pan, Z., Paraschis, L., & Yoo, S. J. B. (2006). 10000-hop cascaded in-line all-optical 3R regeneration to achieve 1250000-km 10-Gb/s transmission. *IEEE Photonics Technology Letters, 18*, 718–720. doi:10.1109/LPT.2006.871141

Chapter 6
Performance Evaluation of Survivability Approaches in Optical Networks

Abdelhamid Eshoul
University of Ottawa, Canada

Hussein T. Mouftah
University of Ottawa, Canada

ABSTRACT

The chapter outlines the different survivability approaches for mesh networks under static and dynamic traffic environments. It describes the different solution options and their implementations. Also included are detailed performance analyses and evaluations for the difference survivability approaches under both traffic environments. Finally, we present a performance comparison between the different survivability approaches and end the chapter with some concluding remarks.

INTRODUCTION

Network survivability requires reserving enough spare capacity during the connection setup and utilizing the reserved spare capacity upon the occurrence of a network failure. The objectives of any survivability scheme are to allocate network resources efficiently and minimize restoration time during a network failure. Achieving both objectives simultaneously, especially under wavelength continuity constraint, poses a major challenge in survivable *WDM* networks.

Survivability approaches are classified based on the protection ranges as link, segment and path based protection (Eshoul and Mouftah, 2009). In link-based protection, the traffic is rerouted around the end nodes of the failed link; whereas, in path-based protection, a backup path is pre-determined between the source and the destination nodes. On the other hand, Segment-based protection is a trade-off between the link-based and path-based protection schemes. Protection schemes are also classified based on the possibility of resource sharing as dedicated and shared protection. Dedicated protection schemes have fast restoration times

DOI: 10.4018/978-1-61350-426-0.ch006

at the expense of higher resource redundancy. In contrast, shared protection schemes reduce resource redundancy significantly at the expense of increased restoration time. There are two different implementations to shared protection in mesh networks: the diverse routing approach (Dongvun and Subramaniam, 2000) and the p-cycle approach (Grover and Stamatelakis, 1998).

Diverse Routing Approach

The diverse routing approach is a path-based protection scheme, where a working path (primary path) and a backup path are set up during the *RWA* process. The routes of the working and the backup paths must be link-disjoint in order to protect against link cut, or node-disjoint in order to protect against node and link failures. In the shared protection scheme, a number of backup paths can share the same resources as long as their working paths are not under the same risk of failure. The concept of Shared Risk Trunk Group (*SRTG*) is introduced to check when different working paths can share the same backup resource. The *SRTG* of a trunk t consists of the working resources that pass through it. All working paths that are part of *STRG* t fail when t fails. As a result, the working paths of any *SRTG* cannot share any backup resources. Stating the idea differently, any backup resource in a trunk t can only protect one working resource, of the same size or smaller, in any other trunk.

P-Cycle Approach

Unlike the diverse routing approach, the p-cycle approach is a link based approach where one or more pre-configured protection cycles, which may overlap with each other, are formed. The major advantages of p-cycle protection schemes over the diverse routing protection schemes are their ability to achieve both good resource efficiency and fast restoration times simultaneously. The p-cycle approach can achieve fast restoration time

due to its switching over mechanism, during times of failure, which is similar to that of the link-based protection scheme, where switching over to the backup path involves only the end nodes of the failed link. Moreover, p-cycle protection schemes can reach good resource redundancy compatible to that of conventional protection schemes used in mesh networks.

APPLIED TRAFFIC

The traffic applied to wavelength-routed WDM networks is mainly confined to two types: static traffic and dynamic traffic. Under the static traffic environment, the objective is to set up a given set of demands while minimizing network resources. Depending on the complexity of the static traffic problem, Integer Linear Programming (ILP) techniques can effectively be used to optimally solve small to moderate size problems for both diverse routing and p-cycle approaches. Several heuristic algorithms have been proposed for larger problems to solve the survivable Routing and Wavelength Assignment (RWA) problem. In this chapter we mainly use ILP techniques to solve the static survivable RWA.

On the other hand, under the dynamic traffic environment, connection requests arrive and depart the network at random times. So, the objective is to minimize the blocking rate of the arriving requests while at the same time simplifying the complexity and the improving the efficiency of the dynamic RWA algorithm. In order to achieve the objectives, the routing and wavelength assignment decisions must be based on the latest network state information such as traffic congestion and wavelength usages (Haque et al., 2004, Yurong et al., 2004). Moreover, due to the QoS constraints, the RWA scheme cannot disrupt the paths that are active during the setup of newly arriving requests. These constraints dictates the setup of each individual connection separately as it arrives based on the network status at the time. As a result, ILP techniques are unsuitable to solve the dynamic RWA problem, especially for large networks, due to its complexity. Several heuristic have been proposed to solve the dynamic survivability problem.

SOLUTIONS UNDER STATIC TRAFFIC

Due to the space limitations, we only consider Integer Linear Programming (ILP) techniques to solve the static survivability problem. As described above, a set of connection requests is given prior to the RWA process. So, the static RWA problem may be solved using the ILP techniques. However, the complexity of larger problems can be simplified by breaking the problem into smaller sub problems. For example, the diverse routing survivability problem can be divided into the routing sub problem and wavelength assignment sub problem. Similarly, the p-cycle survivability problem can be divided into the primary routing, the p-cycle generation sub problem and the wavelength assignment sub problem.

Diverse Routing Solution under Static Traffic

To minimize the gap between the optimum solution and the generated solution of the survivability problem, we solve the survivable RWA problem jointly. However, to simplify the complexity of the problem, we only include, in the formulation, a subset of all possible routes between each s-d pair to be used as candidates for the working paths. The subset of primary candidate routes for an s-d pair consists of the most likely K routes. Then, for each primary candidate route in the primary set, a set of possible candidate routes for its backup path is generated. Each primary candidate route must be link-disjoint or node-disjoint, as required, with each backup candidate route in its corresponding set. The following two-step algorithm is used to

generate the most likely candidate routes between each s-d pair:

1. **Working path candidates:** For each s-d pair, generate the k-shortest routes to be considered as candidate paths for the primary path(s) of that s-d pair in the formulation of the problem. K is an integer variable ($K = 1, 2...$) that may be *tuned to* simplify the complexity of the problem, minimize the optimality gap and balance the load. The candidate primary routes can be generated using any k-shortest routing algorithm, such as Yen's algorithm (Suurballe and Tarjan, 1984), and do not have to be link or node disjointed.

2. **Backup path candidates:** For each candidate primary route, generated in step one, a subset of all possible corresponding link-disjoint routes is generated. First, the cost of the bidirectional links making the candidate primary route is set to infinity so that these links would not be part of any of the backup routes' candidates. Then, the k-shortest algorithm is invoked to generate a subset of up to D backup candidate routes. Each backup candidate route in D must be link-disjoint with its corresponding primary candidate route.

K and D must always be greater than one, because it is not known prior to the optimization process which of these routes would be part of the optimum final solution when all the demands are considered jointly. Furthermore, some of the primary routes in some trap topologies do not have corresponding link-disjoint routes.

Notations

- N is the number of nodes in the network.
- E is the number of directed links in the network.

- T is the number of trunks in the network. A trunk consists of two unidirectional links.
- Q is the number of s-d pairs, $1 \leq Q \leq N(N-1)$
- W is the number of wavelengths per link (link capacity).
- K is the maximum number of primary paths that an s-d pair can have.
- D is the maximum number of backup paths that a primary path can have.
- C_e is the cost of using a wavelength on link e.
- e is used to index a link $1 \leq e \leq E$;.
- t is used to index a trunk; $1 \leq t \leq T$
- i is used to index s-d pair; $1 \leq i \leq Q$
- k is used to index a primary route; $1 \leq k \leq K$
- d is used to index a backup route; $1 \leq d \leq D$
- L_e is the directed link e.
- T_t is the bidirectional trunk t.
- Λ_i is the number of lightpaths requested by the ith s-d pair; $i \leq Q$.
- p_i is the number of primary paths between the ith s-d pair; $p_i \leq K$.
- b_{ik} is the number of routes to backup the kth primary path of the ith s-d pair; $1 \leq k \leq p_i$ and $b_{ik} \leq D$.
- A_e is the summation of all the primary paths that are assigned a wavelength on link e; A_e is the same for dedicated and shared protection formulation.
- S_e is the summation of the variables representing the backup capacity that assigned a wavelength on L_e. S_e represents different variables for the dedicated and shared protection formulation.
- P_{ik}^{ω} is used as an objective function variable to represent the kth primary route of the ith s-d pair on wavelength ω. In the final solution P_{ik}^{ω} takes the value of 1 if the kth primary route of the ith s-d is assigned wavelength ω; otherwise it takes the value of 0.

- $B_{ik}^{d\omega}$ is used as an objective function variable in the formulation. In the final solution, $B_{ik}^{d\omega}$ takes the value of 1 if the dth backup route is assigned wavelength ω to protect the kth primary route of the ith s-d ; otherwise it takes the value of 0.
- ξ_e^{ω} is used as an objective function variable in the shared protection formulation. ξ_e^{ω} takes the value of 1 unit incurred for assigning wavelength ω on L_e to one or more backup paths; otherwise it takes the value of 0. This cost does not increase with the number of backup paths that are assigned wavelength ω on L_e
- β_{ik}^t is an indicator function that takes the value of 1 if the kth primary route of the ith s-d pair passes through T_t; otherwise it takes the value of 0.
- χ_{ik}^e is an indicator function that takes the value of 1 if the kth primary route of the ith s-d pair passes through L_e; otherwise it takes the value of 0.
- φ_{ik}^{de} is an indicator function that takes the value of 1 if the dth backup route to protect the kth primary route of the ith s-d pair passes through L_e; otherwise it takes the value of 0.

Dedicated Protection Formulation

Objective function:

$$minimize \sum_{e=1}^{E} C_e \left(A_e + S_e \right) \tag{1}$$

Where:

$$A_e = \sum_{i=1}^{Q} \sum_{k=1}^{P_i} \sum_{\omega=1}^{W} P_{ik}^{\omega} \chi_{ik}^e \tag{2}$$

$$S_e = \sum_{i=1}^{Q}\sum_{k=1}^{P_i}\sum_{d=1}^{b_{ik}}\sum_{\omega=1}^{W} B_{ik}^{d\omega}\varphi_{ik}^{de} \qquad (3)$$

$$\sum_{d=1}^{b_{ik}}\sum_{\omega=1}^{W} B_{ik}^{d\omega} \geq \sum_{\omega=1}^{W} P_{ik}^{\omega} \qquad \forall k \in p_i \ and \forall i \in Q$$

$$(7)$$

Constraints

Demand Constraint

The demand constraint is used to ensure that the number of lightpaths requested by each s-d pair is satisfied.

$$\Lambda_i = \sum_{k=1}^{P_i}\sum_{\omega=1}^{W} P_{ik}^{\omega} \qquad \forall i \in Q \qquad (4)$$

Capacity Constraint

The capacity constraint is employed to ensure that a wavelength ω on any link can only be assigned to a primary path or to a backup path.

$$\sum_{i=1}^{Q}\sum_{k=1}^{P_i}\left\{ P_{ik}^{\omega}\chi_{ik}^{e} + \sum_{d=1}^{b_{ik}} B_{ik}^{d\omega}\varphi_{ik}^{de} \right\} \leq 1 \quad \forall e \in E \quad and$$

$$\forall \omega \in W \ (5)$$

Protection Constraint

The protection constraint ensures that if the primary candidate route of an s-d pair is assigned a wavelength, then one of its corresponding back up routes must also be assigned a wavelength to protect it.

$$\sum_{\omega=1}^{W} P_{ik}^{\omega} = \sum_{d=1}^{b_{ik}}\sum_{\omega=1}^{W} B_{ik}^{d\omega} \qquad \forall k \in p_i \ and \forall i \in Q$$

$$(6)$$

The protection constraint can also be written as follows:

Integer Constraint

All variables are binary numbers

Shared Protection Formulation

Objective function:

$$minimize \sum_{e=1}^{E} C_e \left(A_e + S_e \right) \qquad (8)$$

A_e is calculated in a similar way to that of the dedicated protection formulation. However, S_e for shared protection is different because several backup paths can share the same resource on link e.

$$S_e = \sum_{\omega=1}^{W} \xi_e^{\omega} \qquad \forall e \in E \qquad (9)$$

Constraints

Demand Constraint

Same as the dedicated protection formulation.

Capacity Constraint

It is similar to the capacity constraint in dedicated protection. However, for shared protection backup paths, a wavelength may be assigned to more than one backup.

$$\left\{ \sum_{i=1}^{Q}\sum_{k=1}^{P_i} P_{ik}^{\omega}\chi_{ik}^{e} \right\} + \xi_e^{\omega} \leq 1 \quad \forall e \in E \ and \ \forall \omega \in W$$

$$(10)$$

Fixed Cost Constraint

$$\xi_e^\omega \leq \sum_{i=1}^{Q}\sum_{k=1}^{P_i}\sum_{d=1}^{b_{ik}}B_{ik}^{d\omega}\varphi_{ik}^{de} \qquad (11)$$

Protection Constraint

It is the same as the protection constraint employed for dedicated protection.

Sharing Constraint

$$\sum_{i=1}^{Q}\sum_{k=1}^{P_i}\beta_{ik}^t \sum_{d=1}^{b_{ik}}\varphi_{ik}^{de}B_{ik}^{d\omega} \leq 1 \; \forall t\in T, \forall e\in E \text{ and}$$
$$\forall\omega\in W \qquad (12)$$

The same sharing constraint inequalities may be generated more than once if the same group of working paths appears in several SRTGs and their backup routes share several links.

P-Cycle Solution under Static Traffic

For a mesh network to be protected against any single span failure, a subset of p-cycles is pre-configured in the spare capacity. However, the number of possible cycles in a mesh network increases rapidly with the number of nodes and the average nodal-degree of the network. As a result, selecting the most efficient cycles that provide full backup is crucial to make the solution of the survivability problem possible and improves the efficiency of the network. Cycle selection is based on the cycle efficiency, which is the proportion of the number of links that the cycle protects to the number of links making the cycle. The static survivable RWA problem using p-cycles can be solved non-jointly, where either the working paths or the backup p-cycles are established first and then either the corresponding minimum spare capacity or the working paths problem is solved using an

optimization tool such as ILP. Alternatively, for a more optimum solution, the problem can be solved jointly, where the working routes and the backup p-cycles are jointly formulated as an ILP.

Non-Joint Solution

The non-joint approach simplifies the complexity of the problem significantly and as a result, decreases its computational time. All previously proposed non-joint approaches (Grover and Stamatelakis, 1998) solve the routing problem first and then find the minimum backup p-cycles capacity required to protect against single span failures. However, solving the routing problem without taking the required backup capacity into consideration may block the forming of the required p-cycles, especially, under wavelength continuity constraint due to wavelength blocking. For example, if the same wavelength on the two directional links in a single span is assigned to some working paths, these two directional links cannot be protected as on-cycle links.

Minimum Backup Capacity First (MBCF)

Since routing the working paths before finding their adequate backups may block the formation of the required backup p-cycles, the problem can be solved by first reserving the minimum backup capacity required against any single span failure. Then the working paths can be established on the residual capacity. Routing the working paths can be performed either using heuristic methods or it can be formulated as an ILP problem. Additionally, the minimum backup capacity problem, for any network topology, can be solved by formulating it as ILP problem. The input to the problem is a set of all possible candidate cycles in the network and the solution is a subset of the candidate p-cycles, with minimum length, that provide full protection against any single span failure. The subset of p-cycles is selected such that each link in the network is either a link that is a part of at least one cycle or a link that straddles at least one cycle (He et al., 2005). Unless the length constraint is imposed on the selected cycles, the result is always Hamiltonian cycles in networks where Hamiltonian cycles are feasible. To ensure full protection against any single span failure and assuming network links can carry up to W channels (assuming W is an even number), the minimum backup capacity is reserved, on each cycle, according to the following rule:

1.   Channels $1 \rightarrow W/2$ are reserved in the clock-wise direction
2.   Channels $W/2+1 \rightarrow W$ are reserved in the counter-clockwise direction

Constrained K-Shortest Routes Algorithm

Since the backup capacity of any possible working path has been taken care of in advance (see MBCF), the rest of the problem comes down to solving the RWA of the working paths from the residual capacity in the most efficient way. The RWA of the working paths can be solved by formulating it as an ILP problem. The RWA problem can be further simplified by finding the most likely candidate routes to be used in the formulation for the overall problem. However, the candidate routes must be chosen so that a single route does not have links that are part of two complementing backup p-cycles.

The following example explains the problem. Consider the topology shown in Figure 1 having several s-d pairs, among them the s-d pair 2-6. Suppose that the selected backup p-cycles are PC1 [1-2-4-6-5-3-1] and PC2 [1-3-5-6-4-2-1] and that the backup capacities are reserved according to the rule given in subsection 1.4.2.1.1. Additionally, it is required to generate the shortest three routes between each s-d pair to be used in the formulation. Although the route 2-4-5-6 is a possible candidate, it is in fact useless because

there is no single available wavelength on its entire links to be used. Furthermore, no backup p-cycles of the same wavelength are available to back it up. Consequently, the routing algorithm, used for calculating the K-shortest routes, must be modified accordingly to avoid selecting useless routes. The idea is to add the necessary constraints to the K-shortest routes algorithm so that the choices of the shortest routes are limited to those routes that do not span over two complementing cycles.

The routing algorithm should be prohibited from selecting two links that belong to two complementary backup p-cycles, in the same route. This added constraint can be easily satisfied when at the first inclusion of a link that is part of a backup p-cycle in a route, all the links that belong to the complementary p-cycles are opened before including any subsequent links in the same route. As a result, any primary candidate route can only be one of the following:

1.   The primary route has one or more of its links on one of the backup p-cycles. If this primary route is selected to carry traffic, it can only be protected by the complementary p-cycle. As a result, if a working lightpath is to use this route, it must be assigned one of the wavelengths that are assigned to the complementary p-cycle.
2.   The primary route has its entire links straddling the backup p-cycles. Therefore, if this

Figure 1. Six-node topology

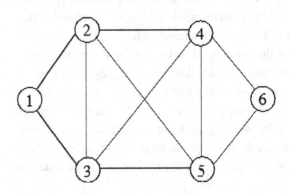

primary route is selected to carry traffic, it can be protected by any p-cycle where the failed link is straddling. Consequently, this primary route can be assigned any wavelength.

Joint Solution

To reduce the gap between the optimum and the generated solutions, the problem can be solved jointly, where the candidate working routes and the candidate backup p-cycles are jointly formulated as an ILP problem. However, the number of p-cycle candidates grows exponentially with the average nodal degree and the number of nodes in the network. Furthermore, the number of variables required to represent the flow in each link due to each path increases with the number of links and wavelengths. As a result, the complexity of the problem grows rapidly, which makes the computational time unacceptable. To simplify the complexity of the problem, the routing problem is first solved using path flow technique, where a number of routes are generated each s-d pair and used as candidate routes in the optimization model. The candidate working routes together with the candidate cycles are then used to formulate the overall problem as a joint ILP. To reduce the number of candidate cycles in the formulation, we use a subset of high merit cycles selected based on the ranking policy presented by Eshoul and Mouftah (2009b).

Notations

We use the notations, used in Section, to represent the number of nodes, number of links, the number of channels, the number of s-d pairs, the number of primary candidate routes, the cost of using a wavelength on a link, the number of lightpaths requested by an s-d pair, the index variable for a link, the index variable for an s-d pair and the index variable for a working route.

- Z is the number possible unidirectional cycles.
- M is the number selected unidirectional cycles to be used in the model.
- Λ_i^e is the number of channels assigned to the ith s-d pair on L_e.
- P_{ik} is an objective function variable used to represent the number of channels assigned to the kth primary route of the ith s-d pair (used for FWC).
- P_{ik}^ω is a binary objective function variable. It takes a value of 1 if the kth primary route of the ith s-d pair is assigned wavelength ω in the final solution; otherwise it takes the value of 0 (used for the NWC).
- X_j is an objective function variable that represents the number of channels required on p-cycle j (used for FWC).
- $X_{j\omega}$ is a binary objective function variable. It takes a value of 1 if the jth cycle is assigned wavelength ω; otherwise it takes the value of 0 (used for NWC).
- ω_{ik} is the lower index for the range of wavelengths which the paths between the ith s-d pair are allowed to use on the kth primary route.
- W_{ik} is the upper index for the range of wavelengths which the paths between the ith s-d pair are allowed to use on the kth primary route.
- φ_{je} is an indicator function which takes the value of 1 if the jth cycle protects L_e; otherwise it takes the value of 0.
- Φ_i^e is an indicator function which takes the value of 1 if the ith s-d pair uses the L_e on its working path; otherwise it takes the value of 0.
- Φ_{ik}^e is an indicator function which takes the value of 1 if the kth primary route of the ith s-d pair passes through L_e; otherwise it takes the value of 0.

- Ψ_{je} is an indicator function which takes the value of 1 if the jth cycle pass through L_e; otherwise it takes the value of 0.
- j is used to index a cycle, $1 < j < Z$.
- A_e is the summation of all the primary paths that are assigned a wavelength (active lightpaths) on L_e.
- S_e is the summation of the variables representing the backup p-cycles that are assigned a wavelength on L_e

Cycle Ranking

The idea of cycles ranking is to sort the set Z of all unidirectional cycles in the network in a descending order according to some efficiency metric. Then, a subset M of the best cycles is selected to be used in the optimization model $M \subseteq Z$. The cardinality of M is tradeoffs between the optimality and the complexity of the problem. Several ranking metrics have been proposed in the literature, however, in this chapter, we use the Route Sensitive Efficiency (RSE), proposed by Eshoul and Mouftah (2009b). RSE is computed based on a route sensitive score RSS. The RSS of a cycle due to a link which the cycle is protecting reflects the relative cost of the candidate primary routes traversing the link. In order to calculate an effective sensitive score, two scoring functions are employed in the calculation of the RSS. The first scoring function reflects how good the candidate route is, relative to the shortest route in its s-d pair routing table (Low Ratio); while the second scoring function reflects how good the candidate route is relative to the longest route in its s-d pair routing table (High Ratio). The Low Ratio and the High Ratio of the kth primary candidate route of the ith s-d pair are denoted by R_{ik} and R^{ik} respectively and are calculated as shown below:

$$R_{ik} = \frac{\sum_{e=1}^{E} \Phi_{i1}^e C_e}{\sum_{e=1}^{E} \Phi_{ik}^e C_e} \tag{13}$$

$$R^{ik} = \frac{\sum_{e=1}^{E} \Phi_{iK}^e C_e}{\sum_{e=1}^{E} \Phi_{ik}^e C_e} \tag{14}$$

The idea is to assign a score to a cycle for each link it protects that is dependent on the relative cost of each primary route candidate traversing the link as well as the amount of traffic requested by the corresponding s-d pair. As a result, RSS is not only sensitive to the traffic demand distribution in the network, but it is also sensitive to the priority and cost of each primary candidate route in the routing table. The RSS of a cycle due to a link l depends on the following:

1. Whether link 1 is protected by the cycle.
2. The number of the candidate routes traversing link 1.
3. R_{ik} and R^{ik} of each candidate route traversing link 1
4. The number of cycles protecting the link 1.
5. The number of channels requested by the s-d pairs of the candidate routes traversing link 1

The following rule is used to compute the sensitive routing efficiency SRE of cycle j:

$$SRE = \frac{RSS}{cost\ of\ the\ cycle} \tag{15}$$

where:

$$RSS = \sum_{e=1}^{E} \sum_{i=1}^{Q} \sum_{k=1}^{K} \frac{\Phi_{ik}^e \varphi_{je} \Lambda_i R_{ik} R^{ik}}{\sum_{z=1}^{Z} \varphi_{je}} \tag{16}$$

Problem Formulation

Non Joint Formulation

It is assumed that the minimum backup capacity for any single span failure has been reserved and configured as described in Subsection 1.4.2.1.1. Additionally, the K-shortest candidate routes for each s-d pair have been generated

Objective function

$$minimize \sum_{e=1}^{E} C_e A_e \qquad (17)$$

Where:

$$A_e = \sum_{i=1}^{Q} \sum_{k=1}^{K} \sum_{\omega=\omega_{ik}}^{W_{ik}} P_{ik}^{\omega} \Phi_{ik}^{e} \quad \forall e \in E \qquad (18)$$

Constraints

Demand Constraint

$$\Lambda_i = \sum_{k=1}^{K} \sum_{\omega=\omega_{ik}}^{W_{ik}} P_{ik}^{\omega} \quad \forall i \in Q \qquad (19)$$

Capacity Constraint

A wavelength on a link can only be assigned to one working path

$$\sum_{i=1}^{Q} \sum_{k=1}^{K} P_{ik}^{\omega} \Phi_{ik}^{e} \leq 1 \ \forall e \in E \ \text{and} \ \forall (\omega_{ik} \leq \omega \leq W_{ik}) \qquad (20)$$

Integer Constraint

All variables take binary values

Joint Formulation

All Z possible cycles are generated and sorted based on the RSE scoring criterion. A subset, M, of the best merited cycles are then selected to be used in the formulation. Additionally, the K-shortest candidate primary routes are generated for each s-d pair. In the absence of wavelength converters, all the links of a working route must be assigned the same wavelength. Furthermore, since p-cycle protection is link- based protection; both a backup p-cycle and all the working paths it protects must be assigned the same wavelength. As a result, two bidirectional cycles on a shared link must not be assigned the same wavelength if any of them is to protect traffic traversing the shared link. Moreover, wavelength blocking blocks the establishment of backup capacity of some working paths and must be avoided. It occurs where the same wavelength is used by working paths in both directions of the same link. Consider the example shown in Figure 1, where two bidirectional p-cycles C1, C2 and two working paths P1, P2 are shown. P-cycle C2 cannot protect P2 if both C1 and C2 are assigned the same wavelength. Similarly, P1 and P2 cannot be protected if both are assigned the same wavelength.

Objective function

$$minimize \sum_{e=1}^{E} C_e \left(A_e + S_e \right) \qquad (21)$$

Where:

$$A_e = \sum_{i=1}^{Q} \sum_{k=1}^{K} \sum_{\omega=1}^{W} P_{ik}^{\omega} \Phi_{ik}^{e} \quad \forall e \in E \qquad (22)$$

$$S_e = \sum_{j=1}^{M} \sum_{\omega=1}^{W} \psi_{jeX_{j\omega}} \qquad (23)$$

Table 1. Topology specifications

Network	Number of nodes	Average Nodal degree	# of unidirectional cycles
ARPANET	20	3.2	2722
COST239	11	4.73	7062
NSFNET	14	3	278
TESTNET	15	3.47	1040

Constraints

Demand Constraints

$$\Lambda_i = \sum_{k=1}^{K}\sum_{\omega=1}^{W} P_{ik}^{\omega} \quad \forall i \in Q \qquad (24)$$

Capacity Constraints

A wavelength on a link can only be assigned to a working path or to a backup p-cycle

$$\left\{ \sum_{j=1}^{M} \psi_{jeX_{j\omega}} + \sum_{i=1}^{Q}\sum_{k=1}^{K} P_{ik}^{\omega}\Phi_{ik}^{e} \right\} \leq 1 \quad \forall e \in E \text{ and }$$

$$\forall \omega \in W \qquad (25)$$

Protection Constraints

$$\sum_{j=1}^{M} \psi_{jeX_{j\omega}} \geq \sum_{i=1}^{Q}\sum_{k=1}^{K} P_{ik}^{\omega}\Phi_{ik}^{e} \quad \forall e \in E \text{ and } \forall \omega \in W$$

$$(26)$$

Integer Constraints

All variables take binary values.

Performance Evaluation of Static Survivability Approaches

In this subsection, the performances of the static traffic survivability schemes, presented above, are evaluated and compared. The evaluations are carried out under different static traffic loads on several network topologies.

Table 1 summarizes the specifications of each network used in the evaluation. Random sets of static traffic demands are generated, for each topology. Each set is generated such that the number of lightpaths is randomly selected for each s-d pair between 0 and a maximum number shown under the demand column. For each s-d pair, the shortest three routes are used as primary candidate routes for the diverse routing and the p-cycle schemes. For the diverse routing approach, the shortest three link-disjoint routes are also generated for each primary candidate route to be used as the candidate routes for backup. The CPLEX 7.0 ILP solver was used to solve the formulated problems on a 2.4 GHz Pentium IV machine.

Table 2 shows the final solution of the dedicated protection formulation on the COST239 and ARPANET topologies. The final solution representing the objective function is given in Channel-Link. Also included are link capacity, number of variables and the computational time. It can be deduced from the results that the computational time increases with the number of

Table 2. Dedicated protection

Topology	Demand	Link capacity (channle)	No. Of variables	Objective function (channel-link)	Computational time (sec)
COST239	3	16	15168	608	61
COST239	3	14	13272	609	182
COST239	7	30	33840	1417	4640
ARPANET	1	40	95080	1369	43610

variables used in the formulation. However, the computational time increases at a much faster rate than the number of variables. The number of variables required to formulate the dedicated protection problem is a function of the number of s-d pairs, the number of channels per link and the number of primary and backup candidate routes.

Due to the increased complexity of the constraint equation in the shared protection scheme, we use smaller problems on the COST239 topology for illustration purposes. Table 3 shows the results of two different static loads. It is observed that the computational time of the shared protection problem is much longer than that of the dedicated protection problem. For example, the solver required 559800 seconds to generate a feasible solution for a shared protection problem with 8000 variables, compared to only 61 seconds to generate the optimum solution for a dedicated protection problem with 15168 variables. The exponential increase in the computational time for the shared protection scheme is mainly due to

the fixed cost constraint rather than the increase in the number of variables. The fixed cost constraint is used in the shared protection formulation to ensure that a backup channel incurs a fixed cost of one unit, regardless of the number of backup paths sharing it.

For the non-joint scheme of the p-cycle approach, the minimum backup capacity required against any single span failure is setup, for each topology, using the MBCF scheme described above. The results are shown in Table 4. The major advantages of the non-joint scheme are its simplified complexity and fast computational time. However, the non-joint scheme makes an acceptable compromise between the optimality and the computational time of the solution. Table 4 shows the minimum working capacity required to establish each set of demand. The total cost is the sum of the minimum working capacity and the backup capacity. If the backup cycles used are Hamiltonian cycles, then the total capacity required is equal to the working cost + (N×W). The capacity redundancy is calculated and given in

Table 3. Shared protection

Demand	Capacity	No. Of variables	Cost (channel-link)	Computational time (sec)
3	8	8000	366	559800
7	18	21240	869	468830

Table 4. P-cycle non-joint solution

Topology	Demand	Capacity	Working cost	Computational time (sec)	Redundancy (%)
COST239	3	10	250	0.03	44
COST239	7	20	603	0.2	36.5
ARPANET	1	30	561	3.6	107
ARPANET	3	84	1711	65	98.2
NSF	3	32	646	3.3	69.3
NSF	7	74	1381	8.8	75
TestNET	3	38	739	1.25	77
TestNET	7	98	1857	7.08	39.6

Table 5. Problem complexity

Number of variables		
P-cycle		Diverse Routing
Joint	non-Joint	
$W(M + QK)$	QKW	$W[QK(1 + D) + E]$

the last column. The capacity redundancy is the proportion of the backup capacity to the working capacity. The COST239 and the TESTNET showed good capacity redundancy, especially at high load while the ARPANET had very poor capacity redundancy, especially at low load.

Comparison between Static Traffic Algorithms

Intuitively, the computational time of any computing problem is a measure of the amount of input data that describes the problem. Consequently, the time complexity of a computing problem is expressed as a function of the amount of data required to describe the problem. Accordingly, the complexity of the static traffic algorithms is expressed as a function of the number of variables required to formulate the problems as ILP. Table 5 shows a comparison between the complexities of the diverse routing and the p-cycle algorithms to solve the survivable RWA under static traffic. Table 5 shows that the p-cycles approach requires more variables than the diverse routing approach only if M>(K×Q×D+E). Other important variables

in the comparison include the solution time and the optimality of the solution in each case.

Where:

- D = number of candidate backup routes per primary route
- E = number of directional links in the network
- K = number of candidate primary routes per s-d pair
- M = number of high merit candidate cycles used
- Q = number of s-d pairs
- W = number of channels on a link

Table 6 shows a performance comparison between the diverse routing and the p-cycle approaches when used to set up two different load matrixes on the COST239 network. The comparison includes the required number of variables, the computational time and the optimality of the solution in each case. From the comparison, it can be seen that the diverse routing approach takes significantly longer to solve than the p-cycle approach. The considerable solution time in the case of the diverse routing approach is mainly due to

Table 6. Comparison between the final solutions of each approach

Demand	Approach		W	NV	TC	T (sec)	GAP (%)
	p-cycle	Joint	8	1952	338	56.53	0
		Non-joint	10	1520	360	0.03	0
	Diverse routing		8	11088	366	559800	7.12
	p-cycle	Joint	18	5274	779	265	0.13
		Non-joint	20	3660	823	0.2	0
	Diverse routing		18	21240	869	468830	11

the fixed cost constraint, which is only required in the diverse routing formulation. The choice between the joint and the non-joint formulations in the case of the p-cycle approach is tradeoffs between the optimality and the time of the solution. For larger problems, the non-joint approach is recommended due to its fast solution time despite the small degradation in the network efficiency.

Where:

- NV = Number of variables
- T = Solution time (sec)
- TC = Total cost (link-channel)
- W = Number of channels

SOLUTION OPTIONS UNDER DYNAMIC TRAFFIC

Survivable RWA under dynamic traffic involves the setup and tear down of lightpaths dynamically at random times with the objective of reducing blocking rates. Due to the unawareness of future demands, and the inability to reroute existing paths, the RWA of each path is performed separately at the time of the arrival. To reduce the blocking probability, the selection of the working and backup paths must be based on some link state information. Therefore, it is desirable to have full knowledge about the routing and wavelength assignment of existing paths. However, due to the significant control overhead, complete information may not be feasible in all network topologies. Another concern which must be taken into consideration during the RWA process is the complexity and the scalability of the algorithm. In fact, the practical application of any dynamic RWA algorithm depends a great deal on its complexity and scalability.

Diverse Routing Approach

Most researchers solved the RWA problem for shared protection by decomposing it into two smaller sub-problems, the routing sub-problem and the wavelength assignment sub-problem. However, the routing sub-problem is not straightforward because the link state during the calculation of the backup path differs from that used to calculate its corresponding working path. Consequently, known algorithms to find the two disjoint paths such as Dijkstra and Suurballe cannot be applied. So, in order to solve the routing problem for shared protection, a straightforward way is to use a two-step approach to generate the two disjoint paths. In the first step, the working path is calculated using any shortest path algorithm based on network status at the time of the arrival. The network status includes the reserved capacity for both working and backup paths at the time of the request arrival. The spare link state of the network is then calculated based on the location of the working path. From the spare link-state, the protection path is derived by removing the spans traversed by the working path. Following this, the determination of resource sharing status will update the costs of all other links. However, in some trap network topologies, the two-step approach cannot find the two disjoint routes. Furthermore, in some topologies, the two-step approach, cannot find the most optimum disjoint pair.

To alleviate these problems, Ho and Mouftah introduced the iterative two step approach (Ho and Mouftah, 2004). For the purpose of this chapter, we describe our algorithm, Survivable Algorithm for Dynamic Routing and Wavelength Assignment (SAD-RWA), presented by Eshoul and Mouftah (2006). The algorithm is based on the iterative two-step algorithm (ITSA) with no wavelength conversion. To ensure the optimum choice of the working and backup light paths for the online connection as well as wavelength continuity constraint, the algorithm inspects the k-shortest routes for each incoming demand on all wavelength planes. To make intelligent decisions during the RWA process, we introduce the following variables:

$S_{l\omega}$ represents the status of wavelength ω on link l, which can be WORKING, BACKUP or

FREE. $\chi^t_{l\omega}$ is an indicator function which takes the value 1 if the wavelength ω is used on link l to protect a working path passing through trunk t. Otherwise, it takes 0. Where $0<\omega<W-1$, $0<1<E-1, 0<t<T$, W is the number of wavelengths per link, E is the number of unidirectional links in the ne $C^p_{l\omega}$) is set according to the rule:

$$C^p_{l\omega} = \begin{cases} \text{actual link cost if } S_{l\omega} = FREE \\ \infty \text{ otherwise} \end{cases}$$

1. For each primary candidate route in step 1, provided that its cost is less than the total cost of the pair (primary and backup) before it, generate a set of d backup routes on each wavelength ω. The cost of ω is set according to the following:

2. Find the sub set $P_t \in T$ making the candidate primary route. P_t is the set of primary candidate trunks. The idea is to calculate the cost of assigning wavelength ω to a backup path on link l, based on the entries $S_{l\omega}$ and $\chi^t_{l\omega}$

3. To disjoin the working and backup routes, the costs of all bidirectional links that are part of P_t are set to ∞, $\forall\,(\omega \in W)$.

   ◦ The cos $S_{l\omega}$ is set to *WORKING* if ω is assigned to a working path on l.

   ◦ The cost of non-sharable wavelength ω on link l is set to ∞ if $S_{l\omega}$ is set to *BACKUP* and any of the entries $\chi^t_{l\omega} = 1$, $\forall(t\in P)$, ω is in the non-sharable state, because of the SRTG constraint.

   ◦ The cost of ω on link l is set to the working cost if $S_{l\omega}$ is set *FREE*.

   ◦ The cost of ω on l is set to a very small value ($\ll 1$), if $S_{l\omega}$ is set to *BACKUP* and $\chi^t_{l\omega} = 0 \forall (t \in P)$

   ◦ Putting it all together, to calculate the backup routes for the any candidate

primary route (P), the cost of wavelength ω on link $l (C^b_{l\omega})$ is set according to the rule:

$$C^b_{l\omega} = \begin{cases} \text{actual link cost if } S_{l\omega} = FREE \text{ and } l \notin P_t \\ \ll 1, \text{if } S_{l\omega} = BACKUP \text{ and } \chi^t_{l\omega} = 0 \forall (t \in P) \\ \infty \text{ otherwise} \end{cases}$$

4. Select the costly pair (primary and backup) of each ω.
5. Repeat steps 3 and 4 for each ω and select the least costly combination.

A flow chart, describing the steps of the SAD-RWA algorithm, is shown in Figure 2.

P-Cycle Approach

We use the two configuration methods for dynamic p-cycle protection described by Eshoul and Mouftah (2009a):

- Offline Configuration Policy (*OCP*), where the minimum backup capacity, required for full protection against any single span failure, is reserved offline. The optimum number and locations of backup p-cycles is calculated using *ILP* techniques. The online computation complexity is then reduced to finding a working route and a suitable wavelength for each arriving demand. The required backup capacity for any possible working path regardless of its location and wavelength has been configured offline.

- Dynamic Configuration Policy (*DCP*), where the locations and the number of p-cycles are re-optimized online. The DCP can be implemented in three different modes:

   ◦ Re-optimization per demand, where the backup capacity is released prior to the establishment of each incom-

Figure 2. Flowchart for the SAD-RWA algorithm

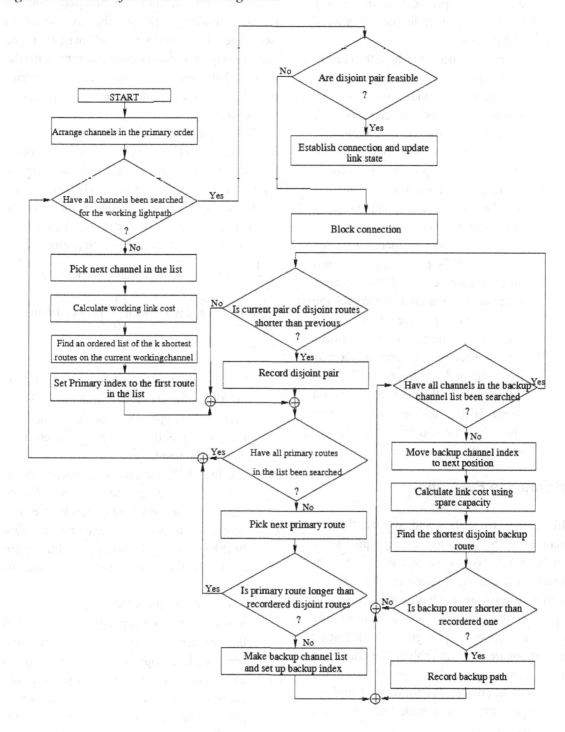

ing demand. The request is accepted if a working path is found and sufficient backup p-cycles capacity can be configured; otherwise the request is blocked. Depending on the dynamic nature of the traffic, the re-optimization per demand mode may pose a considerable computational complexity.

○ No re-optimization, where the current configured p-cycles are left intact and the incoming demand may only be routed on the residual capacity. If a working path for the new request is found, the algorithm configures the current backup p-cycles to provide the necessary backup for the incoming demand. More backup p-cycles are configured as required.

○ Re-optimization on need strategy, offers tradeoffs between the computational complexity of the re-optimization per demand mode and the blocking performance of the no re-optimization mode. The re-optimization on need tries to establish a new lightpath and configure its backup p-cycles using the no re-optimization mode. If it is not successful, it then changes to the re-optimization per demand mode.

Performance Evaluation

In this section, we analyze and compare the performances of the two survivability approaches using the SAD-RWA, OCP and the DCP algorithms described above. To carry out the analyses, a simulation software tool has been developed to compare their blocking performances on a number of different network topologies. The topology specifications are shown in Table 1. The simulation tool is used to carry out all the tests that form the remainder of the chapter. The following simplifying assumptions have been made throughout the simulation experiments:

The network consists of N switching nodes interconnected by *WDM* bidirectional fiber-optic links to construct an irregular mesh topology. The physical topology has a $N \times N$ distance matrix D,

where the elements in the matrix represent the physical distance between the corresponding two nodes. For example, the element D_{ij} is the matrix entry in row i and column j, represents the distance between nodes i and j. Each link carries up to W wavelengths and there is no wavelength conversion.

Traffic is dynamic in nature, where connection requests arrive at each network node randomly and independently according to Poisson's process with an average arrival rate λ. The average arrival rate can vary from one node to another. The destination of a connection demand is uniformly distributed to all other nodes with probability of $\frac{1}{N-1}$. The connection holding time is exponentially distributed with mean $\frac{1}{\mu}$. Blocked demands are dropped and do not return.

- For the diverse routing app *SAD-RWA* algorithm, the shortest 3 routes (K=3) on each wavelength plane are examined for the primary and backup paths for each incoming demand.

- For the *OCP*, the minimum backup p-cycles are reserved offline to ensure full protection against any single span failure. The shortest path algorithm (SP) and the first fit (FF) wavelength assignment scheme are used, with the OCP, to establish the incoming demands.

- For the *DCP*, upon the arrival of each demand, the backup capacity is released and the new demand is routed from the residual capacity. If the lightpath is successfully established, new set of backup p-cycles are configured to protect the working lightpaths. The optimum path and p-cycles (*OPP*) routing scheme, proposed by Eshoul and Mouftah (2009a), is used to establish the incoming lightpaths. Additionally, the backup p-cycles are selected using the heuristic algorithm described by Zhong and Zhang (2005).

- Centralized control entity is used to calculate the working and backup capacity and store the network state information. However, the algorithms can also be implemented using distributed control.

The aim of this experiment is to compare the blocking performances of the SAD-RWA algorithm with known algorithms for the survivable RWA with full wavelength conversion capability (FWC). The experiment is carried out on the NSF topology with a link capacity of 12 channels. Figure 3 shows that at low load (< 82Erlang), the network has similar blocking performances with and without wavelength conversion. The main reason for the similarity in the blocking performances at low load is that the network has more resources than it is required to accommodate the demands. Therefore, the network can still achieve good blocking performance with no wavelength conversion (NWC). At high loads, blocking performances with FWC and with NWC

have deteriorated. The excessive load becomes the dominant factor of the blocking, which explains the reason for the similarity in the blocking performance at higher loads.

At moderate loads, the advantage of the FWC over the NWC becomes apparent due to the wavelength continuity constraint. However, one might argue that under NWC, demands traverse more hops to satisfy the wavelength continuity constraint; as a result, consume more resources causing the blocking probability to increase. On the hand, requests with high hop count tend to have higher blocking probability. Consequently, under NWC, the majority of successful demands are of low hop count which bears the questions of fairness and accuracy of the conclusion. Therefore, to draw a more accurate conclusion about the blocking performance, other experiments need to be carried out.

Figure 4 shows a comparison of the average hop count, per demand, between FWC and NWC. The total cost to establish a lightpath is the sum of its working and backup path hop counts. From the graph, it can be seen that at very low load, the

Figure 3. Blocking performances in the NSF with link capacity of 12 channels

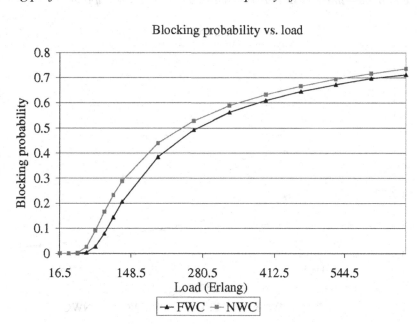

NWC consumes more resources than does the FWC, especially for the working paths, because of the wavelength continuity constraint. This finding is intuitive because in order to satisfy the wavelength continuity constraint, the routing scheme often selects routes with high hop counts. At low loads, it is possible to select routes with high hop count, with a high probability of success, due to the availability of resources. The average hop count, for the FWC working paths, increases gradually, as expected, with the load before sharply declining at excessive load as shown in Figure. The increase in the average hop count of the working paths, for the FWC, is partly due to the increase in load and partly due to the selection metric of the routing scheme. During the RWA process, the routing scheme selects the least costly combination of working and backup path. As a result, the routing algorithm may select a working route with more hop counts, even when paths with fewer hop count are available, because of the sharable resources (free resources) along its backup path. As a result, the gradual increase in the cost of the

working paths is followed by a gradual decrease in the average total cost of the demand.

In contrast, using NWC causes a sharper decrease in the average hop count of the working paths with the load, because of the wavelength continuity constraint. As the load increases only those demands with low hop count have any real chance of succeeding, because of the remote probability to find the same channel available on all hops. Accordingly, the graph shows that the average hop count of the working path is well below the average shortest route of the network. Additionally, those demands that do succeed, for the most part, use shared resources for their backups, resulting in a sharp reduction in the average total cost shown in the graph.

Finally, the graph clearly shows a large difference, in the average consumed resources per demand, between the FWC and NWC under the same load. The FWC, for the most part, consumes more resources per demand than the NWC. Consequently, demands with high hop count have better chances of succeeding with FWC. Therefore, FWC has better resources utilization than the

Figure 4. Average path cost in the NSF network with 12 channel link capacity

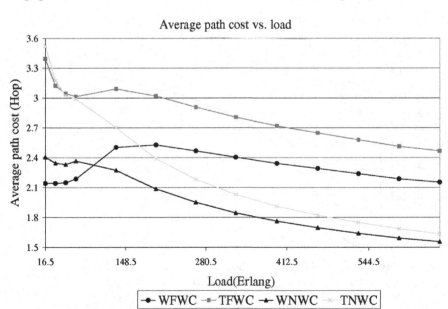

NWC. This finding suggests that although the average percentage of blocked demands is close in both FWC and NWC, as shown in Figure, the blocking comparison alone is not enough to draw an accurate conclusion.

Where:

- WFWC is the working path cost under Full Wavelength Conversion
- TFWC is the total path cost under Full Wavelength Conversion
- WNWC is the working path cost under No Wavelength Conversion
- TNWC is the Total path cost under Full Wavelength Conversion

Therefore, it can be claimed with high degree of certainty that the increase of the blocking probability in the NWC is mainly due to the wavelength continuity constraint resulting on network resources being underutilized. This finding also raises the question of fairness, since only those demands with low hop count have any real chance of being admitted into the network, especially at higher load. One way to address unfairness problem is to use other reservation algorithms (Pramod et al., 2005) which dedicate a number of wavelengths on each link to those demands with high hop counts. Although these reservation schemes can improve the blocking probability of s-d pairs with high hop count, they increase the overall blocking probability of the network.

Performance Comparison

Table 7 gives a comparison between the computational complexities of the SAD-RWA and the OCP algorithms. It states the huge advantage which the OCP algorithm has over the SAD-RWA algorithm in terms of the computational complexity.

Where:

- N = number of nodes in the network
- W = number of channels on a link

Blocking Performance

Simulation experiments are used to analyze and compare the blocking performances of both survivability approaches. The simulation experiments have been carried out on the well known NSF and COST239 European networks, each with a link capacity of 12 channels. The specifications of both topologies are given in Table 1.

Figure 5 shows a blocking performance comparison between the two survivability approaches on the COST239 topology. The Figure shows that while, at low loads, all algorithms have low blocking probability; the DCP has the best blocking performance at higher loads. The DCP derives its improved blocking performance from its ability to release the backup capacity and provide more resources for the working lightpths. As a result, the DCP stands better chances of finding shorter routes for its working lightpaths. Furthermore, the DCP adapts to changes in traffic by dynamically configuring new p-cycles resulting on fewer resource for its backup requirements. The Figure also shows that the SAD-RWA and the OCP algorithms have similar blocking performance at higher loads.

Figure 6 shows blocking performance comparison between both survivability approaches on the NSF topology. The Figure shows that all algorithms have low blocking probability at low load, similar to that of the COST239 topology. However, as the load increased slightly, the SAD-RWA algorithm seems to have better blocking performance than the OCP and the DCP algorithms. As the load increased farther, both the SAD-RWA and DCP algorithms have similar blocking performances.

Table 7. Computational complexity

Computational complexity	
SAD-RWA	**OCP**
$0(WN^2 \log N + W^2 N \log N)$	$0(N \log N)$

Figure 5. Blocking probabilities versus load in the COST239 network

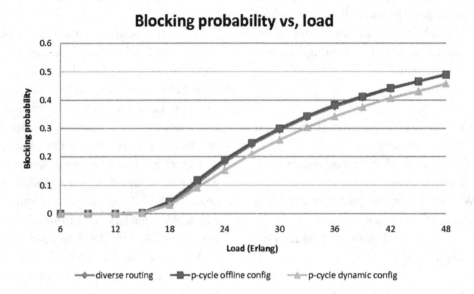

Since the blocking performance comparison on the two topologies suggest different findings and can be confusing, more analyses are carried out to investigate the impact of other important parameters on the blocking performance. For example, the number of nodes in the network and the average nodal degree in the network can have a significant impact on the average amount of resources (wavelength-link) required per working path, which in turn has a direct impact on the blocking probability. Additionally, the wavelength continuity constraint can also have a significant impact on the blocking performance.

Figure 6. Blocking probabilities versus load in the NSF network

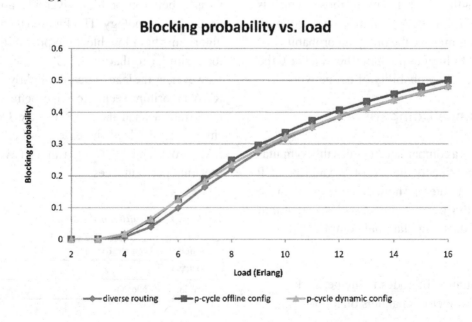

Figure 7. Average hop count of working paths against the load in the COST239 network

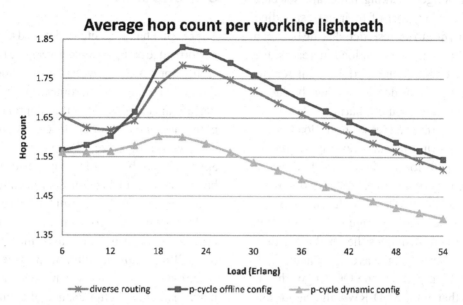

Figures 7 and 8 show the average channel-link per working path against the load on the COST239 and NSF topologies respectively. For the OCP, the NSF topology consumed 33% of its capacity for backup, whereas the COST239 topology consumed only 21% of its capacity for backup due to its higher nodal degree. Moreover, the COST239 topology consumes fewer resources per demand than does the NSF topology which explains the improved blocking performance of the COST239 topology over the NSF topology. At low loads, the SAD-RWA algorithm consumes more resources per working path than do the OCP and DCP due to its selection policies for the working paths. For each demand, the SAD-RWA selects the least costly combination of working and backup paths.

Figure 8. Average hop count of working paths against the load in the NSF network

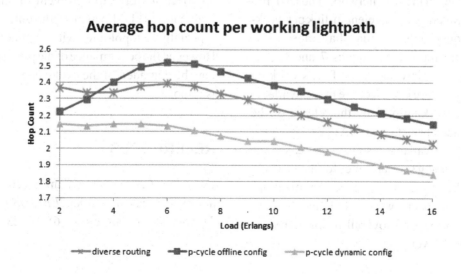

As a result, longer working paths are selected because of the low cost of their corresponding backup paths, due to the resource sharing between several backup paths. As the load increases, the average hop count of the working path for the SAD-RWA algorithm decreased slightly due to the increase in the resource sharing.

However, farther increase in the load is accompanied by an increase in the average hop count per working path for all algorithms, because of the unavailability of resources along shorter paths. The OCP and the SAD-RWA algorithms show sharper increase in the average wavelength-link per working path than does the DCP algorithm. In fact, at higher loads, the average wavelength-link per working path for the OCP algorithm is higher than that of the SAD-RWA algorithm. The OCP algorithm has higher average wavelength-link per working path than the other two algorithms, because of the resources that it reserves, permanently, for its backup p-cycles which force it to select longer working paths. Another important reason is that the SAD-RWA algorithm is more likely to block demands with high hop count, because of the increased difficulty to satisfy the wavelength continuity constraint for the working and the backup paths. This finding also explains why the SAD-RWA algorithm has slightly better blocking probability (at low load) than the other two algorithms in the NSF network. The DCP has the smallest average of wavelength-link per working path (equal to the calculated average at low load) on both topologies Figures 7 and 8. The DCP achieves its low hop count for its working paths because more resources are made available to route the working path by the release of the backup capacity prior to the RWA of each arriving request. These available resources increased the likelihood of using shorter routes for the working paths. Furthermore, the dynamic optimizations of the backup capacity with each lightpath establishment improved the overall resource redundancy in the network.

CONCLUSION

The chapter has presented a detailed analyses and performance comparison between the two survivability approaches (diverse routing and p-cycle) for WDM wavelength routed mesh networks. Performances of survivability approaches under both static and dynamic traffic environments are compared and analyzed. The p-cycle survivability approach has better blocking performance and better network utilization than does the diverse routing approach. The computational complexity of the SAD-RWA algorithm, used for the diverse routing approach, is significantly more than that of the OCP algorithm used for the p-cycle approach. Additionally, s-d pairs which are located several hops apart stand better chance of connecting using the OCP or DCP algorithms than using the SAD-RWA algorithm.

The low computational complexity of the OCP algorithm for the p-cycle approach makes it more suitable and scales better for the dynamic traffic than the diverse routing approach, particularly at higher loads. Furthermore, the blocking performance of the OCP algorithm can be improved by employing any load balancing routing scheme instead of the SP routing scheme at no added complexity. The p-cycle approach presents itself as a better alternative than the diverse routing approach to solve the problem of survivability in dynamic WDM wavelength-routed networks, especially in topologies with high nodal degree. However, the performance of the p-cycle approach may be degraded, if the cycle length constrained is imposed.

REFERENCES

Dongvun, Z., & Subramaniam, S. (2000). Survivability in optical networks. *IEEE/OSA. Journal of Lightwave Technology, 14*(6), 16–23.

Eshoul, A., & Mouftah, H. T. (2006). Survivable algorithm for routing and wavelength assignment under dynamic traffic and no wavelength conversion in mesh networks (SAD-RWA). *GESTS International Transactions on Computer Science and Engineering, 28*(2).

Eshoul, A., & Mouftah, H. T. (2009a). *Performance evaluation of dynamic p-cycle protection methods in WDM optical networks*. Paper presented at International Conference on Transparent Optical Networks (ICTON), Miguel, Azores, Portugal.

Eshoul, A., & Mouftah, H. T. (2009b). Survivability approaches using p-cycles in WDM mesh networks under static traffic. *IEEE/ACM Transactions on Networking, 17*(2), 671–683. doi:10.1109/TNET.2008.2001467

Grover, W. D., & Stamatelakis, D. (1998). Cycle-oriented distributed pre-configuration: Ring-like speed with mesh-like capacity for self-planning network restoration. *IEEE International Conference on Communications (ICC'1998), vol. 1*, (pp. 537-543).

Haque, A., Ho, P., Boutaba, R., & Ho, H. J. (2004). Group shared protection (GSP): A scalable solution for spare capacity reconfiguration in mesh WDM networks. *IEEE Global Telecommunications Conference (GLOBECOM), vol. 3*, (pp. 2029-2035).

He, W., Fang, J., & Somani, A. K. (2005). *A p-cycle based survivable design for dynamic traffic in WDM networks*. Paper presented at IEEE Global Telecommunications Conference (GLOBECOM), St. Louis, Missouri, USA.

Ho, P.-H., & Mouftah, H. T. (2004). Shared protection in mesh WDM networks. *IEEE Communications Magazine, 42*(1), 70–76. doi:10.1109/MCOM.2004.1262164

Pramod, S. R., Siddiqui, S., & Mouftah, H. T. (2005). *Novel distributed protocol for dynamic routing and load balancing for optical networks*. Paper presented at OFC/NFOEC Optical Fiber Communication Conference, Anaheim, California.

Suurballe, J. W., & Tarjan, R. E. (1984). A quick method for finding shortest pairs of disjoint paths. *Networks, 14*(2), 325–336. doi:10.1002/net.3230140209

Yurong, H., Wushao, W., Heritage, J. P., & Mukherjee, B. (2004). A generalized protection framework using a new link-State availability model for reliable optical networks. *IEEE/OSA. Journal of Lightwave Technology, 22*(11), 2536–2547. doi:10.1109/JLT.2004.836764

Zhong, W., & Zhang, Z. (2005). P-cycle-based dynamic protection provisioning in optical networks. *IEICE Transactions on Communications, E88*(B(5)), 1921-1926.

Chapter 7
Maximizing Primary Capacities in Survivable Networks

Arun K. Somani
Iowa State University, USA

David W. Lastine
Iowa State University, USA

ABSTRACT

Achieving low blocking probability and connection restorability in the presence of a link failure is a major goal of network designers. Typically fault tolerant schemes try to maintain low blocking probability by maximizing the amount of primary capacity in the network.

In this chapter, we assume the total capacity on each link is fixed, and then it is allocated into primary or backup capacity. The distribution of primary capacity affects blocking probability for dynamic traffic. This can be seen by simulating dynamic traffic with different ways to distribute capacities in a network. A Hamiltonian p-cycle is a capacity optimal way of allocating primary and backup capacity. However, different Hamiltonian p-cycle may deliver different blocking probability for dynamic traffic. In general, more evenly distributing the backup and primary capacity lowers the blocking probability.

This chapter provides upper bounds on how much primary capacity a network can provide if it uses a link based protection strategy to guarantee survivability for one or more link failures. Using integer linear programs we show that requiring preconfiguring carries a cost in terms of capacity if the solution is structured as a set of cycles.

DOI: 10.4018/978-1-61350-426-0.ch007

INTRODUCTION: SURVIVABILITY AND CAPACITY PLANNING

The problem of providing dependable connections in networks is receiving more attention due to growing transport bandwidths and the consequent huge losses associated with failures of components in such high speed networks. While the current deployed networks operate at 10 Gbps speeds, networks operating at 40 Gbps are also gaining in popularity. The high vulnerability of the networks can be seen in that failures have been observed to happen as often as once every four days (Frederick, Datta, & Somani, 2006). Several researchers have studied dependability problem in various context.

Utilizing Primary Capacity Effectively

Today the usable bandwidth even in a single fiber is more than a user requires. To maximize the primary capacity usable in a network, the bandwidth must be shared. Several techniques exist to allow bandwidth sharing. Many of these techniques can be used simultaneously.

One way to share bandwidth is to use different wavelengths for different connections. Two connections using different wavelengths can use the same fiber at the same time as long as the frequencies are sufficiently different that receiver nodes can distinguish between them. Depending on the minimum spacing between distinguishable frequencies this method is referred by names such as wavelength division multiplexing (WDM) or dense wavelength division multiplexing (DWDM). Past a certain intensity of the electric field, the response of the fiber becomes non-linear. This results in phenomena such as four wave mixing, which prevent some combinations of wavelengths from being usable. In this chapter we assume wavelengths have been spaced appropriately to avoid this.

Optical transmitters and receivers components may be designed for a fixed frequency or be tunable over a range of frequencies. Signals in the optical domain can be routed using an Optical Cross Connect (OCX). Complexity of hardware to route optical signals varies in the degree of granularity for routing wavelengths. Some hardware can route individual wavelengths while other hardware routes groups of wavelengths referred to as a waveband. Support for splitting a signal to create a multicast tree exists. Optical or electronic amplifiers maybe required to maintain signal viability.

A second way bandwidth can be split on a fiber is by the different users of a fiber taking turns transmitting. This technique is known as time-division multiplexing (TDM). Of course TDM and WDM can be done at the same time, in which case it's the users of a frequency that take turns using the frequency.

Reserving frequencies and time slots for point to point connections can result in bandwidth being wasted when a network has bursty traffic since one connection may have extra capacity at one instant while another connection needs more capacity. For bursty traffic, use of a light trail can maximize the amount of capacity available to serve traffic. A light trail is a unidirectional optical bus that allows for all nodes on the bus to timeshare the bus. Since a light trail has multiple possible source nodes, it has a greater chance of using available capacity than a dedicated light trail between two nodes.

Another way of sharing bandwidth is to use code division multiplexing. This method allows multiple users to use a wavelength at the same time by using different codes. Codes spread out data bits into many chirps. For some or all codes, receivers can detect a specific code by looking at the correlation of chirps over time. Increasing the number of users that can share a wavelength results in codes which need more time to transmit.

Restoration vs. Protection

The survivability techniques can be broadly classified into two (Zhou & Subramanian, 2000), (Mohan, Murthy, Somani, 2000): dynamic restoration and preplanned protection.

In case of dynamic restoration, resources to be used to reestablish an affected connection upon a failure are identified after a failure occurs. It is possible that resources may not be available to reestablish a failed connection. In this case, either the connection is dropped or reestablished after dropping a low priority connection if such a discipline is allowed. In either situation, this approach maximize the primary capacity, since initially all resources are dedicated for primary connection establishment only. Such a scheme inherently will cause more delay in restoration and can only be used if the network operation is delay-tolerant.

In case of preplanned protection resources that must be used upon a failure are identified at the time of establishing a connection. If such resources are not available, connection request may be denied (or blocked). These resources remain idle until a failure occurs. Preplanned protection is better in terms of the ability to provide service guarantees.

There are two variation of preplanned protection scheme. In the first variation, the resources are not only planned, but configured to carry out the same connection. This is called 1:1 protection or dedicated protection and the second path is called a secondary or backup path. As soon as a failure is detected, the affected connection is carried on the secondary or backup path. This minimizes the delay and such a scheme is deployed to support almost loss free service. In fact, to avoid any loss it is possible that the two parallel connections may carry all the data all the time and the destination may simply pick one in case both connections are operational. In this case it is also possible for the destination to verify the integrity of data as it is receiving two copies of data. This scheme obviously utilizes maximum resources for backup and thus does not meet the goal of maximizing primary capacity. In the second variation, for each connection the resources to be used to carry the connection in case of a failure are identified and

reserved, but not configured to carry out the connection at the time of establishing a connection. If a connection fails due to a resource failure on the connection path, the identified resources are configured to carry the connection. The advantage of this scheme is that same backup resources can be used to restore two resource disjoint paths as a single failure will only affect one of the two paths. This is called a backup multiplexing scheme (Mohan, Murthy, Somani, 2000). It is easy to visualize that delay in restoration of a connection is moderate in this scheme, but the goal of maximizing primary capacity is also met mostly.

A variation of protection scheme to maximize primary capacity utilization is called partial protection that provides support for most but not all connections and to some extent and not full connection capacity based on specified parameters has been developed in (Sivakumar, Fang, Sivalingham, & Somani, 2008). In this scheme a performance metric called quality of protection (QoP) is used to maximize the protection level while minimizing resource utilization for secondary or backup paths.

Link-Based vs. Path-Based Protection

Protection and restoration techniques can be further classified as link-based or path-based.

In the path based approach, a connection is provided two link diverse paths from source to the destination before being accepted into the network. The original service route of a connection in a survivable network is called the primary path. When a failure occurs on the primary path, the connections on the path get re-routed over backup paths. The path based approach plans a new route for a connection and has many varieties. The backup route may share no resources with the initial route (dedicated backup), it could have minimal overlap with the primary (primary-backup multiplexing), or several backup routes could have resources reserved with one being

Figure 1. Path and link based protection

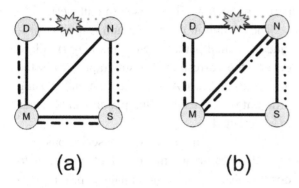

(a)   (b)

chosen based on which network component failed (shared backup). Often path based recovery schemes reserve network bandwidth for their backup paths, though there are exceptions such as sub-graph routing(Frederick, Datta, & Somani, 2006). Typically path based strategies have been more resource efficient.

Recently link-based protection mechanisms have been receiving more attention. In link-based protection, the connections are rerouted around the failed link. Typically, link based approaches lead to long backup routes. There have been many different approaches for link protection.

Figure 1 shows examples of path and link based protection. Part (a) of the Figure shows how path protection works. The path S-N-D fails due to the failure of link N-D, so backup path S-M-D is used. Part (b) of the Figure shows how link protection works. When the path S-N-D is disrupted by the failure of link N-D, the traffic is routed around the failed link using path N-M-D.

L+1 Fault Tolerant Routing

The *L+1* routing strategy is a passive redundancy scheme to tolerate failures. It is a path-based survivability strategy to tolerate link failures as the connection path is modified upon a failure. The secondary or back up connection resources are identified, but not reserved permanently. All

connections may use a different set of routes depending on the actual location of the fault.

A connection is reconfigured upon a failure using the identified resources. The guaranteed provided is that upon failure, each connection will be routed. When a fault occurs, the network restores itself as if that link did not exist or eliminates the defective link from routing consideration. Thus, it is possible that sometimes some of existing connection may have to be rerouted in order to accommodate failure. Thus, the end users of a connection may experience nominal interruption in service when the network experiences a link failure even when the failed link was not on the path used by the connection.

The network control remembers how each connection must be rerouted in case a failure occurs. In fact the routing strategy directly provides the reconfiguration strategy when the failure occurs. The way the routing strategy works is straightforward. The network control and management maintains the list of all routes and paths they use in the network topology. The original graph is called the *base graph*. In addition, it also maintains L additional copies of the graph topology, each representing a particular fault combination. It is assumed that each fault affect some network resources and L such fault patterns are of interest. The easiest way to explain this is to consider link failure in a graph with L links. One may consider double link or some other complex link-combination as one of the failure. For every fault location $i=1$, $2, \ldots, L$, a new topology G_i is maintained. G_i is a copy of original topology G excluding the fault resources affected by fault pattern i. In essence, each G_i is a different subgraph of G.

When a request arrives, first a path is found in the original graph topology using unused resources. Then it is checked if the same resources or some other set of resources available in each of the L topologies to route the request. If such is the case, then the request is accepted, and the corresponding resources are reserved for each of the topology that will be used to route the request in that topology. Figure 2 shows such a scenario. Graph *G* is the original graph with 4 nodes (A, B, C, and D) and five links (1, 2, 3, 4, and 5). We assume that the system is required to tolerate a single link failure. Thus the five link failures are represented by five subgraphs, G_1 to G_5. We assume that the request are bidirectional and use the capacity of the whole link.

When the first request (A-C) arrives, the path used is link A-C. This is available in all subgraphs except G_5 which will have to use path A-B-C. Thus the request is accepted and the corresponding resources (shown in blue dotted lines) are reserved in base graph and all subgraphs. Next the next request (A-D) arrives, the base graph G will use path A-D and so will all subgraphs except G_4 which will use path A-B-C-D as link A-D does not exist in G_4. Now G_4 is fully saturated and therefore no more requests can be accepted unless the new request is accepted with a condition that

the request may not survive certain failure (provide partial protection with respect to fault set).

The routing in the base graph is used to provide the actual path. When a link (say *i*) failure occurs, the corresponding subgraph (G_i) is the new configuration. Any connection that is using different resources than that planned in G_i needs to be rerouted.

We make some interesting observations. The first observation we make is that it is easy to observe that this scheme enhances primary capacity utilization. For example, in this network, one request A-C is routed; its backup path will be either A-B-C or A-D-C. After this dedicated backup path no other request can be accepted whereas with *L+1* scheme, that is not the case. In fact many other possible requests will be accepted. The second observation we make is that in this example no other requests need to be rerouted except the one that is directly affected. That will not always be the case in a complex

Figure 2. L+1 routing strategy

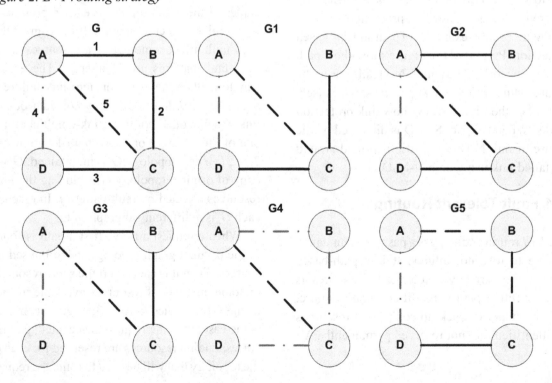

scenario as connections use different resources in different subgraphs. Thus the path to be used in a reconfiguration graph may be being used by some other connection. Such a set of routes are easy to identify (by comparison of two graphs) and then those connections can be rerouted. In fact reconfiguration problem is an interesting problem in itself and efficient solutions have been developed for this purpose (Jose, Somani 2003). The third observation we make is that the scheme inherently uses backup multiplexing. For example links A-B and B-C are used as backup paths for both requests in the example describe above. The fourth observation is that the concept of partial protection with respect to fault sets can be easily accommodated.

It has been shown in (Frederick, Datta, & Somani, 2006), that the blocking probability, network utilization, and effective network utilization improves in this scheme with respect to backup multiplexing. The scheme therefore maximizes primary capacity.

P-Cycle Protection and Its Variations

Several other approaches to provide protection have been developed by various researchers. The ring cover approach presented in (Gardner,

Heydari, Shah, Sudborough, Tollis, & Xia, 1994) tries to find the minimum cost for equipment that will enable the network to survive an arbitrary link fault. The goal is to find minimum cost ring covers for the network N, where a ring cover is a set C of rings such that every link in N is covered by at least one ring in C. If a network N is augmented with enough equipment to support a given ring cover C, it can respond to a link failure immediately by routing the disrupted traffic through surviving links in the ring that covers the failed link.

Generalized loop back recovery in mesh based optical networks was studied in (Médard, Barry, Finn, & He, 2002). The work in presents algorithms for loop-back recovery in the presence of both ring and node failures, and describe a network management protocol that enables distributed coordination in the presence of a failure.

Another link based approach is based on p-cycles (Grover & Stamatelakis, 1998). While the ring cover approach allows cycles to protect only on cycle links, the p-cycle approach allows links straddling a cycle also to be protected at no extra capacity. When present, a Hamiltonian cycle can serve as a p-cycle that protects the maximum possible primary capacity under our model.

Figure 3 depicts three ways to find an alternate path in p-cycle. The original proposal for p-cycle

Figure 3. P-Cycle protection and alternate path

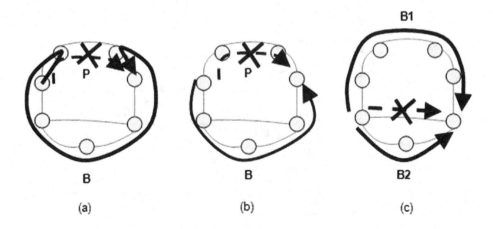

(a)   (b)   (c)

protection is to use a link based schemes as shown in Figure 3 (a). The primary path is marked P and the corresponding back path is marked B when link marked X fails. An alternate approach will be to use a path-based protection as shown in Figure 3 (b) which will be slow to recover, but will not use the same link twice, once in each direction. Figure 3 (c) depicts how to protect a straddling link.

One issue with p-cycle is that while being efficient in terms of total network primary capacity, each p-cycle uses at most half of the bandwidth on all on-cycle links to carry primary capacity. As these links only have half their capacity without carrying any traffic, one might imagine they could easily saturate under heavy network load and raise blocking probability.

An alternate way is to improve blocking probability to reconsider distribution of backup capacity on all links more evenly, in particular for a denser graph. This scheme is named as distributed p-Cycles link protection (DPLP)(Lastine & Somani, 2008).

DISTRIBUTED P-CYCLE LINK PROTECTION

The distributed p-cycle link protection scheme is similar to p-cycle protection scheme except that in DPLP each link is protected by multiple cycles which are not preconfigured and each link provides only a small fraction of bandwidth for protection. To achieve this goal, a set of cycles is found and the amount of backup capacity that needs to be reserved on each link is determined. The rest of the capacity is used to establish primary paths for incoming requests. All requests are routed using primary capacity only. Since the design of cycles is such that the entire primary capacity upon a link failure is protected by the DPLP cycles, the network can provide protection for any link failure.

Figure 4 shows a COST 239 network with capacity that can used to route request assuming that each link has a capacity of 10. Figure 5 shows a set of three DPLP cycles to protect link 3-4. It should be noticed that each link will be protected by a different set of p-cycles. The total capacity available on these DPLP cycles is sufficient to

Figure 4. COST 239 network with 11 nodes, 26 links

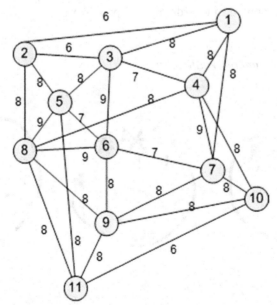

Figure 5. Distributed p-cyles in COST 239 network

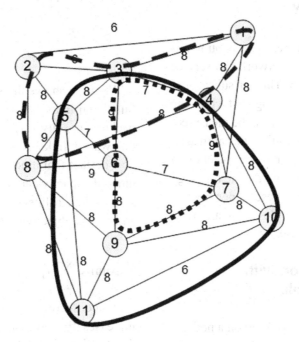

protect the entire primary capacity of link 3-4. Notice that link 3-4 is a straddling link on one of the cycles. Thus the protected on that cycle is twice the backup capacity available on that cycle.

One important question to answer is about the minimal backup capacity needed to protect a given network? This question is answered below along with an integer linear programming (ILP) formulation to solve for the ideal capacity reservation on a set of cycles.

NETWORK MODEL

Our network model is based on an undirected graph denoted by $G=(U,E)$ where U is asset of nodes with $|U|=K$ being the total number of nodes, and E is the set of edges, with $|E|=L$ being the total number of links. A network may have any arbitrary set of link failures given by F which is a subset of the set E and the fault set is bounded by $|F| \leq m$. All links have the same capacity C. Links are taken to be bi-directional where traf-

fic in both directions count against the capacity. Equivalently one could assume that the link level control will only allocate connections on a link that are full duplex.

Figure 6. Link i fails and its traffic uses other links

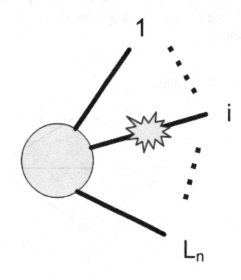

UPPER BOUND ON PRIMARY CAPACITY

This section establishes upper bounds on maximum primary capacity any network can service for the fault scenario described in each subsection. It does so to provide a standard against which the capacity usage of an algorithm can be measured. All such bounds establish conditions that are necessary, but not sufficient to guarantee protection in any given network. In other words, the topology of a network may prevent these upper bounds from being achievable. Examples to illustrate this point will be presented below.

Link-Based Protection with all Traffic Recoverable

We first define a link to be *incident* on a node, if the node is one of the two end nodes of the link.

Lemma: Links incident on a node must allocate at least C units to backup among them.

On any given node, n, with L_n incident links, the primary capacity on a given link i must be equal to or less than the sum of the backups on the other links. Otherwise there is some link that if it fails, there is not enough backup capacity in the other links to cover it.

Algebraically these requirements can be stated as where $P_{i,n}$ is the primary capacity on link i on node n and $B_{i,n}$ is the backup capacity on link i on node n:

$$\forall i \in \{1, ..., L_n\}$$
$$P_{i,n} + B_{i,n} \le C$$
$$P_{i,n} \le \sum_{i \ne j} B_{j,n}$$

Figure 6 shows the scenario where link i has failed and links $1...., i-1, i+1, ...L_n$ must use their backup capacity to pick up the primary traffic that was on link i.

The fact that a node reserves at minimum C units of capacity can be seen by turning the constraints into a linear programming problem by adding the objective function.

$$\text{Maximize } \sum_{i=1}^{L_n} P_{i,n}$$

Since decreasing the value of any given B_i increases the corresponding P_i, the objective function is maximized when the second constraint becomes the strict equality of $P_{i,n} = \sum_{i \ne j} B_{j,n}$

From this it follows that the maximized objective function is

$$\sum_{i=1}^{L_n} P_{i,n} = \sum_{i=1}^{L_n} \sum_{i \ne j} B_{j,n} = (L_n - 1) \sum_{1}^{L_n} B_{i,n}$$
$$= (L_n - 1) B_n$$

Noting that capacity is conserved for all links incident on a node we see

Figure 7. Topologies

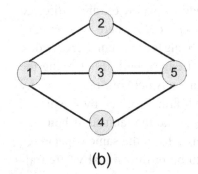

(a)   (b)

$$L_n C = P_n + B_n = (L_n - 1)B_n + B_n = L_n B_n$$

$\therefore B_n = C$ as claimed.

Given this lemma the K nodes in a network would collectively see KC units of capacity being reserved at minimum. Since each link is incident on exactly two nodes this figure double counts the capacity needed, thus the minimum required backup capacity is $KC/2$. Or equivalently the maximum possible primary is $EC-KC/2$.

Achievability of Bounds

The bound are lower bounds that may or may not be achievable, depending on the network topology. In this section, we give examples to illustrate this point.

The bounds can be achieved by the network in Figure 7(a). The network configuration that can achieve this bound allocates primary capacity $C/2$ to links 1-2, 2-3, 3-4, 4-1 and primary capacity C to link 1-3. This gives a total primary capacity of $4C/2+C=3C$. Using the bounding expression the value calculated is $5C-4C/2=3C$ which is exactly the total primary capacity achievable by the network configuration.

A network where the bound cannot be achieved is shown in Figure 7(b). In this topology, the links are effectively grouped into pairs where each link in the pair protects the other link in the same pair. For instance if link 1-2 fails, the only link that could pick up its traffic is link 2-5. The reverse is also true, if link 2-5 fails only link 1-2 can back it up. Link pairs (1-3, 3-5) and (1-4, 4-5) behave in a similar fashion. From this it can be clearly seen that the way to maximize the total primary capacity on the network is to set all link pairs to $C/2$ primary capacity. This gives a total of $6C/2=3C$ primary capacity. However the bound computed using the formula is $6C-5C/2=3.5C$. This is $C/2$ greater than is possible given the topology of the network.

Link-Based Protection with Partial Traffic Recoverable

The above bound can be generalized to the case of only guaranteeing protection to a fraction f of the traffic on a given link. In this case, the constraints become

$$\forall i \in \{1, ..., L_n\}$$
$$P_{n,i} + B_{i,n} \leq C$$
$$f * P_{i,n} \leq \sum_{i \neq j} B_{j,n}$$

Once the protection factor has propagated though the equations the end result is a network's primary capacity that is upper bounded by:

$$EC - \frac{K}{2} \frac{L_n Cf}{(L_n - 1 + f)}$$

Link-Based Protection with Multiple Failures

The bound can be generalized further to the case of allowing m link failures. To see this, let the set of links on a node be divided into two sets, one set D represents an arbitrary set of m links that might be destroyed. The other set R represents the rest of the links and contains all the links not in D.

From link capacity being conserved the following hold:

$$\sum_{i \in D} P_{i,n} + \sum_{i \in D} B_{i,n} = mC$$
$$\sum_{i \in R} P_{i,n} + \sum_{i \in R} B_{i,n} = (L_n - m)C$$

Since the links in D must be protected a third constraint is:

$$\sum_{i \in D} P_{i,n} \leq \sum_{i \in R} B_{i,n}$$

To find the maximum possible useful network load:

$$\text{Maximize} \sum_{i=1}^{L_n} P_{i,n}$$

This sum is maximized when the third constraint becomes a strict equality. It can then be substituted into the first constraints to get:

$$\sum_{i \in R} B_{i,n} + \sum_{i \in D} B_{i,n} = mC$$

One can see an upper bound on the total usable capacity as:

$$EC - \frac{mKC}{2}$$

An Observation about P-Cycles

When a network contains a Hamiltonian cycle, it can be used as a single p-cycle to protect the network. In this case the p-cycle uses capacity $C/2$ on each of its on cycle links. The p-cycle provides the maximum possible primary capacity any link based scheme can achieve when protecting against a single link fault.

DISTRIBUTED P-CYCLE LINK PROTECTION ALLOCATION

In this section, we examine allocating primary and backup capacity for link based protection using three integer linear programs (ILP). The three ILPs vary in the amount of structure imposed on the solution.

The solution to the first ILP is a way to divide a networks capacity into primary and backup capacity along with the backup routes to use when a link fails. The ILP is very simple since it has no constraints to control backup path length or guarantee that the backup routes can be preconfigured to allow for fast recovery. The second ILP

has additional constraints that limit the length of the backup path. The third ILP limits the backup path length and selects p-cycles so the backup capacity can be preconfigured.

Minimal Structure

To attack the problem of maximizing the primary capacity of a network with L links, the ILP divides the task into L sub-problems where each sub-problem is that of finding the backup route(s) to use to protect one of the L links. It is assumed that the flow of information across a link may be split into several paths when it is re-routed because of a link failure.

Each sub-problem is formulated as a flow problem on a network where the capacities of the links are the backup capacities, except a unique link is omitted in each sub-problem. In each sub-problem the nodes that were connected by the omitted link act as a source or sink. One of the nodes acts as a source with a supply equal to the primary capacity of the omitted link, and the other acts as a sink with demand equal to the primary capacity of the omitted link.

The ILP maximizes the sums of the primaries, while requiring the total space allocated between primary and backup on a link to be less than the total capacity of a link. All links were assumed to have the same capacity but this assumption could easily be removed. Link capacity is assumed to be only finitely dividable, if this assumption were to be dropped, the problem would reduce to an LP.

The ILP has three sets of variables, one set P_e represents primary capacity allocated on the links, one set B_e represents backup capacity allocated on the links, and the third set $F_{e,g}$ represents the flows. Links are numbered in $[1, L]$ and the index 'e' on a variable is in reference to the number given a link. Flow sub-problems are also numbered $[1, L]$ where the number given a flow sub-problem corresponds to the link that has been omitted. The index 'g' is used to indicate the number of a flow sub-problem. The index '*n*' indicates the node number numbered [1, *K*].

The mechanism of removing a link from a flow sub-problem is by multiplying every instance of it by zero, effectively causing the variable not to exist. The sign of the flow variable determines the direction of the flow in a link. The values of $d_{n,e}$ is zero if link *e* is not connected to node *n*, and when link *e* is incident on node *n*, $d_{n,e}$ takes on the value of one for the lower node number and negative one for the higher node number.

Objective: Maximize $\sum_{e=1}^{L} P_e$

Subject to:

$$\sum_{e=1}^{L} \left(\left(1 - \delta_{e,g}\right) * d_{n,e} * F_{e,g} - \delta_{e,g} * d_{n,e} * P_e \right) = 0$$

(1)

The first equation in the ILP says that flow is conserved except for the source or sink. That is the flow into a node is equal to the flow out of the node. The kronic delta in the equation causes a node to be a source or sink with supply or demand equal to the capacity of the link omitted from the network.

$F_{e,g} \leq B_e$ 　　　　　　　　　　　　　 (2)

$-F_{e,g} \leq B_e$ 　　　　　　　　　　　　(3)

The second and third inequalities ensure that the backup capacity on a link is large enough for the flow that wants to cross the link. This is required since the flow represents a backup path to be used if the link fails.

$B_e + P_e \leq C$ 　　　　　　　　　　　　(4)

The combined primary and backup capacity allocated on a link cannot be greater than the total capacity on the link.

$B_e, P_e \in [0, C]$
$F_{e,g} \in [-C, C]$
$d_{n,e} \in \{-1, 0, 1\}$

Solution Structured into Dynamic Cycles of Fixed Maximum Length

To randomize the distribution of backup capacity and to minimize the amount of resources used by the network when a link fails, it may be desirable to have the backup paths be as short as possible. To accomplish this, the ILP sub problems can have the additional constraint added that each unit of backup flow is constrained to use one of a number of pre-computed cycles that have a fixed maximum length. These cycles become the DPLP cycles. To prevent the ILP from selecting all possible cycles in a given sub-problem, a penalty is added for each instance of a cycle selected. The reward for increasing primary capacity is set much higher than the penalty for adding more backup cycles to ensure increasing primary capacity always more than offsets the penalty for adding an extra cycle.

Objective: Maximize $\beta \sum_{e=1}^{L} P_e - \sum_{g=1}^{L} \sum_{c=1}^{cycles} Y_{c,g}$

Subject to:

$$\sum_{e=1}^{L} \left(\left(1 - \delta_{e,g}\right) * d_{n,e} * F_{e,g} - \delta_{e,g} * d_{n,e} * P_e \right) = 0$$

(5)

$F_{e,g} \leq B_e$ 　　　　　　　　　　　　(6)

$-F_{e,g} \leq B_e$ 　　　　　　　　　　　(7)

$B_e + P_e \leq C$ 　　　　　　　　　　　(8)

Equations (5)-(8) perform the same function as equations 1-4 in the basic ILP.

$f_{e,g} \, {}^3 - F_{e,g}$ 　　　　　　　　　　(9)

$f_{e,g} \, {}^3 F_{e,g}$ 　　　　　　　　　　(10)

$$f_{e,g} \le (1 - \delta_{e,g}) \sum_{c=1}^{cycles} \gamma_{c,g} * \psi_{e,c} * Y_{c,g} \qquad (11)$$

$$B_e \ge (1 - \delta_{e,g}) \sum_{c=1}^{cycles} \gamma_{c,g} * \psi_{e,c} * Y_{c,g} \qquad (12)$$

Equations (9)-(12) ensure the flows that represent the backup paths can only travel along paths where capacity has been reserved by the selected DPLP cycles.

$$B_e, P_e \in [0, C]$$
$$F_{e,g} \in [-C, C]$$
$$f_{e,g} \in [0, C]$$
$$d_{n,e} \in \{-1, 0, 1\}$$
$$\gamma_{c,g} \in \{1, 0\}$$
$$\psi_{e,c} \in \{1, 0\}$$
$$Y_{c,g} \in [0, C]$$

$\gamma_{c,g}$ takes the value of 1 if both ends of the omitted link are in cycle c, and zero otherwise.

$\psi_{e,c}$ takes the value of 1 if link e is in cycle c, and 0 otherwise.

$\delta_{e,g} = 1$ iff $e = g$, otherwise $\delta_{e,g} = 0$

β is a constant selected to be large enough such that the objective function has its maximum value when the maximum primary capacity has been allocated.

The value of $Y_{c,g}$ is amount of backup flow following this particular cycle. The maximum value could be made less than the total capacity to encourage spreading of the flow.

It should be noted that when two different choices of maximum cycle length result in the same total primary capacity used, and there are multiple equivalent solutions, changing the maximum cycle length may result in a different solution being found by the ILP.

Solution Structured into Preconfigured Cycles of Fixed Maximum Length

If cycles are not only precomputed but also preconfigured, switching to backup can be done faster. This section modifies the ILP from the previous section to select cycles to be preconfigured.

Objective: Maximize $\beta \sum_{e=1}^{L} P_e - \sum_{c=1}^{cycles} Y_c$

Subject to:

$$\sum_{e=1}^{L} \left((1 - \delta_{e,g}) * d_{n,e} * F_{e,g} - \delta_{e,g} * d_{n,e} * P_e \right) = 0 \qquad (13)$$

$$F_{e,g} \le B_e \qquad (14)$$

$$-F_{e,g} \le B_e \qquad (15)$$

$$B_e + P_e = C \qquad (16)$$

Equations (13)-(16) perform the same function as equations 1-4 in the basic ILP.

$$f_{e,g}{}^3 - F_{e,g} \qquad (17)$$

$$f_{e,g}{}^3 - F_{e,g} \qquad (18)$$

$$f_{e,g} \le (1 - \delta_{e,g}) \sum_{c=1}^{cycles} \gamma_{c,g} * \psi_{e,c} * Y_{c,g} \qquad (19)$$

$$B_e \ge \sum_{c=1}^{cycles} \psi_{e,c} * Y_c \qquad (20)$$

$$Y_c{}^3 Y_{c,g} \qquad (21)$$

Equations (17)-(19) ensure the flows that represent the backup paths can only travel along paths where capacity has been reserved by the selected DPLP cycles.

Equation (20) ensures that the backup capacity on a link is sufficient for all p-cycles that will be using the link. Equations (21) ensure that the cycles used for individual failures correspond to cycles that will be preconfigured.

$$B_e, P_e \in [0, C]$$
$$F_{e,g} \in [-C, C]$$
$$f_{e,g} \in [0, C]$$
$$d_{n,e} \in \{-1, 0, 1\}$$
$$\gamma_{c,g} \in \{1, 0\}$$
$$\psi_{e,c} \in \{1, 0\}$$
$$Y_{c,g} \in [0, C]$$
$$Y_c \in [0, C]$$

$\gamma_{c,g}$ takes the value of 1 if both ends of the omitted link are in cycle c, and zero otherwise.

$\psi_{e,c}$ takes the value of 1 if link e is in cycle c, and 0 otherwise.

$\delta_{e,g} = 1$ iff $e=g$, otherwise $\delta_{e,g} = 0$

β is a constant selected to be large enough such that the objective function has its maximum value when the maximum primary capacity has been allocated.

Behavior of Primary Capacity with Cycle Length

We examine the behavior of the available primary capacity as a function of the maximum allowed cycle length. For several networks, we solved our more structured ILPs several times, each time the set of candidate cycles provided was all possible simple cycles up to a selected size. The value of C was set to eight.

Figure 8, 9, and 10 show the primary capacity found for different networks that are shown in Figure 11. On the x-axis is the maximum allowed cycle length and on the y-axis the total primary capacity achieved. For all three networks the use of small cycles with maximum lengths between three and seven results in low available primary capacity as well as unallocated capacity. Requiring that cycles used to protect a link be preconfigured results in less available primary capacity. The impact in the Italian network is small for cycle medium to long cycle lengths. When cycle length is restricted to be around half the total number of nodes in the network, Arpanet and NSFNet can provide significantly less primary capacity when cycles are preconfigured.

Figure 8. Italian primary capacity

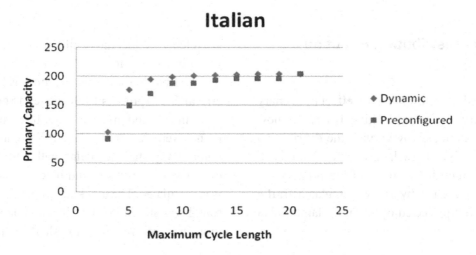

Figure 9. Arpanet primary capacity

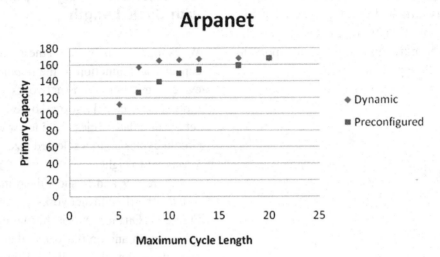

Figure 10. NSFNet primary capacity

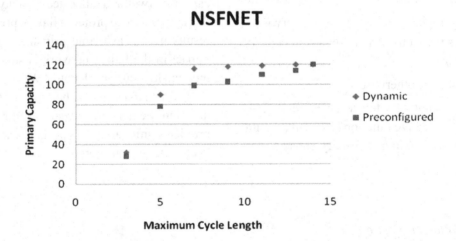

Capacity Distribution Effect on Blocking

In this section we examine the effect of primary capacity distribution on blocking. To examine how blocking is effected by varying the distribution of primary capacity each network was simulated using different distributions of the primary capacity. For each network we simulated traffic on a network protected by a single Hamiltonian

p-cycle as well as the results from one the ILPs given in the chapter.

The first step of the simulation was for a given arrival rate to process 100K request to get the network into a random state. After this a number of rounds were simulated, where 10K events occurred without any statistics being gathered, then 90K events were simulated and blocking probability was calculated. The blocking probability for each round was stored in a text file, and data analysis tools in Excel were used to calculate the average

Figure 11. Networks

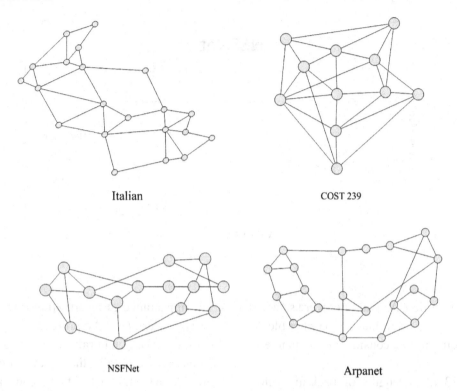

Italian   COST 239

NSFNet   Arpanet

blocking probability and the 95% confidence interval. The number of rounds simulated was one hundred.

Routing was done by finding the shortest route in the network between source and destination only considering links that had enough bandwidth free to accommodate the requested bandwidth.

Figure 12 shows the blocking produced by using a single Hamiltionan (labeled as Ham-

ming) and the more evenly distributed capacities produced by the most structured ILP (labeled as Multicycle) in two different networks. The x access is the arrival rate of traffic, and the y access is the percent of connections blocked. Increasing arrival rate logically leads to increased request blocking. Increasing arrival rate also increases the difference in how blocking occurs between the use of a single Hamiltionan p-cycle and more even

Figure 12. Blocking for Arpanet and Italian networks

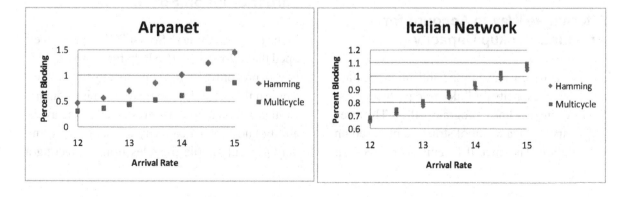

Figure 13. ILPs select from cycles of length at most nine

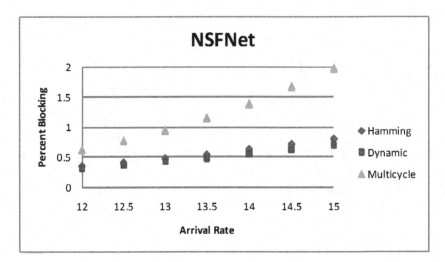

capacity distribution. Arrival rates were selected so blocking would be around 1%. Higher blocking rates would not be considered acceptable in a network.

Figure 13 plots lengths of backup cycles, preconfiguration, and blocking for dynamic traffic. The results show that there is a tradeoff between these parameters. The primary capacity distribution created by the ILP that does not preconfiguration (labeled as Dynamic) results in lower blocking than a single p-cycle. However, recovery time due to a link failure would be slower due to the fact the cycles will not be created till a failure has occurred. The solution found by the ILP using p-cycles at length at most nine does not suffer from this problem. However, restricting the cycle length to nine resulted in less primary capacity in the network resulting in higher blocking.

Alternative Way to Account for Provided Backup Capacity

In this section, we examined allocating primary and backup capacity for link based protection using three integer linear programs (ILP). The three ILPs vary in the amount of structure imposed on the solution. The three ILPs built on each other.

The appearance of information flow as a set of variables results in the protective capabilities of p-cycles arising naturally from the solutions. If a link on a cycle fails, the remainder of the cycle provides a backup path. If both ends of the failed link are in a cycle, but the link itself is not in the cycle, two backup paths exist. Many ILPs in the literature do not model information flow; instead the number of backup paths provided to a given link from a cycle is encoded as a variable coefficient. Using this method equivalent ILPs to the more structured versions can be found in (Zhang, Xu, & He, 2008)and (Shen & Grover, 2004). In these works, the idea of dividing capacity into backup capacity and capacity ready for dynamic traffic is referred to as Protected Working Capacity Envelope (PWCE).

Larger Solution Space

In the paper (Lastine & Somani, 2008) we developed the following ILP to help find a way to plan how to preconfigure backup capacity to tolerate a single link failure. The motivation for introducing another ILP is that constrains on ether the availability and type of capacity available on a link may prevent simple p-cycles from protecting a

Figure 14. Need for non-simple cycle/line

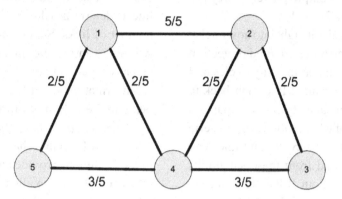

network efficiently or at all when a more complex preconfigurion scheme can provide efficient protection. The advantages of non-simple p-cycles are explored in (Gruber, 2003). The use of p-cycles with preconfigured paths that protecting straddling links only is explored in (Schupke, 2006).

The ILP in this section can be used in finding a solution that can be a combination simple p-cycles, non-simple p-cycles, simple paths, and non-simple paths. We refer to simple and non-simple paths as lines to emphasize the fact that we ignore the cycle like properties a non-simple path possesses on part of its length.

A situation where a combination of a non-simple p-cycle and a line are beneficial is depicted in Figure 14. Each link is labeled with the fraction of the bandwidth that we wish to provide as primary capacity. This can be accommodated if there is a line 5-1-4-2-3 that uses one fifth a links capacity, and a non-simple p-cycle that uses two fifths of the capacity on links which have otherwise unallocated capacity.

The above paragraph tells us all we need to know to see the solution is valid despite the fact the fact the non-simple p-cycle is not explicitly listed. To see why this is so, consider a graph that has only the edges to be used in the non-simple p-cycle and the nodes they are connected to. If the edges in this graph can be connected to form a non-simple p-cycle, the edge between nodes 1 and 2 would be a straddling link and receive four

fifths of a links bandwidth worth of protection. This is exactly the amount of protection still needed after the line has been taken into account. For all other edges the amount of protection needed beyond the line is the same as the capacity reserved on the link for the non-simple cycle. This means, if the link fails the rest of the cycle provides a backup path with exactly this much capacity. Therefore if a non-simple p-cycle exists in the graph, it provides the needed protection. But how do we know if a cycle exists?

A cycle exists if we can find a path in the graph of only selected edges that uses each of the edges only once and we can return to the node we started on. Such a path is an Euler tour and it exists as long as all nodes have an even number of edges. If we are interested in an Euler tour that ends on a different node than it starts on, two nodes have an odd degree. To find an Euler tour one picks a node with odd degree if present or arbitrarily otherwise and travel from node to node using only edges that has not been crossed before. The correctness of this algorithm can be seen by noting that except for the start/end node(s) reaching a node implies the ability to leave the node since the number of available edges at the node before it is reach is even so there must be an available edge to leave again once the node has been reached. An Euler tour need not be unique. In the example under consideration there are two Euler tours that could serve as the non-simple p-cycle. They are 5-1-4-2-3-4-5 and 5-4-2-3-4-1-5.

This example provides insight into how an ILP can be designed to plan preconfigured protection without the need to enumerate many cycles and lines. It is sufficient for an ILP to plan backup capacity and how that capacity can be separated into subgraphs each of which satisfies the criteria for having an Euler tour. P-cycles and lines can then be efficiently found from these subgraphs calculated by the ILP. In our ILP there is one set of variables for finding subgraphs corresponding to cycles (variables are capital letters) and another almost identical set for subgraphs corresponding to lines(mostly in lower case letters). Subgraphs corresponding to cycles, provide protection twice the bandwidth of the cycle to straddling links, and equal to the bandwidth to on cycle links. Subgraphs corresponding to lines provide protection equal to the bandwidth to straddling links.

There is the possibility a non-simple structure maybe able to provide more protection than this in rare cases. For instance a non-simple line will contain a loop; a straddling link from the node visited multiple times to a node in the loop does in fact have two possible backup paths in the non-simple line. In order to keep the ILPs complexity reasonable, we ignore the extra protection that may on rare occasions exist.

Notation in this section follows (Lastine & Somani, 2008) and may not agree with notation used elsewhere in this chapter. The variable indexes are l which ranges over the links, c which ranges over the cycle subgraphs of the same size, L which ranges over the lines, s which ranges over cycle/line graphs capacity, and n which ranges over the nodes.

P is the primary traffic on a link. B is the backup traffic on a link. BB and BBB have no physical meaning; these exist to cause solutions that spread capacity evenly to be favored. E/e indicates if an edge is used in a line or cycle. H/h relates to how many links a cycle/line is using at a node. R/r, F/f, T/t are used to set up a flow problem that

ensures selected links form a connected cycle/line. O determines the two line nodes with an odd number of links. S/s indicates if a link connects two nodes in a cycle/line. This link could ether be an on cycle link or a straddling link.

Information about the network topology appears in the ILP in the form of some constants. C, D, and *capacity* have the same meaning as in the other ILPs in this chapter. The subgraph for a given cycle/line provides a fixed capacity w. To allow fine the ILP to allocate backup capacity in a fine grained way, the values of w are set using its index s as 2^{s-1}.

Minimize: $\sum_{l=1}^{edges} B_l + BB_l + BBB_l$

The number of edges at a node is twice the number of times it will appear in the cycle.

$$\sum_{l=1}^{edges} C_{n,l} S_{l,c,s} = 2H_{n,c,s} \qquad (22)$$

If a link is protected, then a node it is incident on is in the cycle.

$$C_{n,l} S_{l,c,s} \leq H_{n,c,s} \qquad (23)$$

Only if the cycle is connected can the flow problem in the next six equations be satisfied. All nodes will sink some flow. For convince the amount is the number of times the node appears on the cycle. Flow conservation is, of course, observed at all nodes.

$$\sum_{l=1}^{edges} D_{n,l} F_{l,c,s} = T_{n,c,s} - H_{n,c,s} \qquad (24)$$

The selected source node will be able to satisfy all the sinks.

$$T_{n,c,s} \leq 2 * edges * R_{n,c,s} \qquad (25)$$

The amount of source flow will balance the demand of the sinks.

$$\sum_{n=1}^{nodes} H_{n,c,s} = \sum_{n=1}^{nodes} T_{n,c,s} \tag{26}$$

There can be only one source node.

$$\sum_{n=1}^{nodes} R_{n,c,s} \leq 1 \tag{27}$$

Flow can use only selected edges. The use of two inequalities is the result of the direction of flow being encoded in the sign of the amount of flow.

$$C_{n,l}F_{l,c,s} \leq 2 * edges * E_{l,c,s} \tag{28}$$

$$-C_{n,l}F_{l,c,s} \leq 2 * edges * E_{l,c,s} \tag{29}$$

A line has two nodes of odd degree. This is written as an inequality since if the line is not needed the selected number of nodes to have odd degree is zero.

$$\sum_{n=1}^{nodes} O_{n,L,s} \leq 2 \tag{30}$$

The node degree is the number of times the line visits it.

$$\sum_{n=1}^{edges} C_{n,l}e_{l,L,s} = 2h_{l,L,s} + O_{n,L,s} \tag{31}$$

Potentially protected links are incident on nodes in the line.

$$C_{n,l}s_{l,L,s} \leq h_{n,L,s} \tag{32}$$

Only if the line is connected can the flow problem in the next six equations be satisfied. All nodes will sink some flow. For convince the

amount is the number of times the node appears on the line. Flow conservation is, of course, observed at all nodes.

$$\sum_{l=1}^{edges} D_{l,n}f_{l,L,s} = t_{n,L,s} - h_{n,L,s} - O_{n,L,s} \tag{33}$$

There can be only one source node.

$$\sum_{n=1}^{nodes} r_{n,L,s} \leq 1 \tag{34}$$

The amount of source flow will balance the demand of the sinks.

$$\sum_{n=1}^{nodes} h_{n,L,s} + O_{n,L,s} = \sum_{n=1}^{nodes} t_{n,L,s} \tag{35}$$

There can be only one source node.

$$t_{n,L,s} \leq 2 * edges * r_{n,L,s} \tag{36}$$

Flow can use only selected edges.

$$C_{n,l}f_{l,L,s} \leq edges * e_{l,L,s} \tag{37}$$

$$-C_{n,l}f_{l,L,s} \leq edges * e_{l,L,s} \tag{38}$$

The bound on primary protected of a link is calculated by assuming all links are straddling links and subtracting of a correction factor for all links that are on cycle/line links. This is a bound since there may not be enough total capacity on a link to reach the bound and provide backup capacity for other links.

$$P_l \leq \sum_{s=1}^{\log(capacity)} \sum_{c=1}^{cycles} (2W_s S_{l,c,s} - w_s E_{l,c,s})$$
$$+ \sum_{s=1}^{\log(capacity)} \sum_{L=1}^{lines} (w_s s_{l,L,s} - w_s e_{l,L,s}) \tag{39}$$

Every cycle and line that uses a link independently reserves backup capacity. This is necessary to allow for all p-cycles and lines to be preconfigured.

$$B_l = \sum_{s=1}^{\log(capacity)} \sum_{c=1}^{cycles} w_s E_{l,c,s}$$
$$+ \sum_{s=1}^{\log(capacity)} \sum_{L=1}^{lines} (w_s E_{l,L,s}) \tag{40}$$

The total capacity allocated for use as primary capacity and backup capacity cannot be more than the total capacity that exists on a link.

$$P_l + B_l = capacity_l \tag{41}$$

Links with high backup capacity disproportionately raises the value of the objective function. The benefit of this is that solutions where some links have a small amount primary capacity are seen as unfavorable. The use of multiple variables that penalize the objective function allows the creation of threshold values where the penalty to the objective function increases as a link becomes increasingly dedicated to providing backup capacity. The constant values shown are based off the assumption that the total capacity is eight. They will need to be adjusted for other values of the total capacity.

$$BB_l + 3 \geq B_l \tag{42}$$

$$BBB_l + 5 \geq B_l \tag{43}$$

FUTURE RESEARCH DIRECTIONS

Considerable effort in optical network research has been invested in formulating routing strategies as optimization problems. Given traffic demands or traffic estimates, commercial packages for solving linear integer programs found the routes for the traffic. Since the time between a network operator receiving a request for a connection, and actually creating connection can take months so using several days to calculate the routes has been acceptable.

Research in this field is moving towards more dynamic networks that can respond quickly to changes in traffic demands. Connection routing and backup planning are moving towards being entirely dynamic or hybrid schemes where some of the protection and connection routing is done as an optimization problem and the rest is dynamic.

Research needs to be done into more heterogeneous networks that support a wider range of traffic demands along with routing and protection options. A network may supports point to point optical connections, network coding, light trails, and optical multicast using trees. In such a network, when a request arrives which option should be used to route the request? Does it make sense to use the same protection mechanism for the different connection methods? The ability to support a connection type may vary per node.

CONCLUSION

Link based protection allows for planning how to restore from a failure and request routing to be done separately. For a network, there may be multiple capacity optimal ways to configure the backup capacity. Both the amount and placement of backup capacity affect the blocking probability for a connection. Limiting cycle length has an impact on how efficiently backup capacity can be used.

Only using small cycles of lengths of only three or four hops may prevent some links from even being used. Somewhat larger cycles may allow for the protection of all links but require more bandwidth than a network protected with large cycles. For a number of networks, it is possible to protect the maximum possible primary capacity using cycles that are smaller than a Hamiltonian cycle.

REFERENCES

Frederick, M. T., Datta, P., & Somani, A. K. (2006). Sub-graph routing: A generalized fault-tolerant strategy for link failures in WDM optical networks. *Computer Networks, 50*(2), 181–199. doi:10.1016/j.comnet.2005.05.025

Gradner, L. M., Heydari, M., Shah, J., Sudborough, I. H., Tollis, I. G., & Xia, C. (1994). Techniques for finding ring covers in survivable networks. *Global Telecommunications Conference* (pp. 1862-1866). San Francisco, CA: IEEE.

Grover, W. D., & Stamatelakis, D. (1998). Cycle-oriented distributed preconfiguration: Ring like speed and mesh-like capacity for self-planning network restoration. *IEEE International Conference on Communication ICC 98* (pp. 537-543). Atlanta, GA: IEEE.

Gruber, C. G. (2003). Resilient networks with non-simple p-cycles. *10th International Conference on Telecommunications, 2003* (pp. 1027-1032). IEEE.

Jose, N., & Somani, A. K. (2003). Reconfiguring connections in optical networks. *Proc. of DRCN, October 2003*.

Lastine, D., & Somani, A. K. (2008). Supplementing non-simple P-cycles with preconfigured lines. *IEEE International Conference on Communications* (pp. 5443-5447). Beijing, China: IEEE.

Médard, M., Barry, R. A., Finn, S. G., & He, W. (2002). Generalized loop-back recovery in optical mesh networks. *IEEE/ACM Transactions on Networking (TON), 10*(1), 153–164. doi:10.1109/90.986592

Mohan, G., Murthy, C. S., & Somani, A. K. (2001). Efficient algorithms for routing dependable connections in WDM optical networks. *IEEE/ACM Transactions on Networking, 9*(5), 553–566. doi:10.1109/90.958325

Schupke, D. (2006). Analysis of p-cycle capacity in WDM networks. *Photonic Network Communications, 12*(1), 41–51.

Shen, G., & Grover, W. D. (2004). *A framework for dynamic survivable service provisioning based on p-cycles and the protected working capacity envelope (PWCE) concept*. Edmonton, Canada: TRLabs.

Sivakumar, M., Fang, J., Sivalingham, K., & Somani, A. K. (2008). Design and analysis of partial protection mechanisms in groomed optical WDM mesh networks. *Journal of Optical Networking, 7*(6), 617–634. doi:10.1364/JON.7.000617

Zhang, Z., Xu, A., & He, Y. (2008). Dynamically survivable WDM network design with shared-cycles-based PWCE. *International Conference on Advanced Infocomm Technology* (pp. 1-5). Shenzhen, China: ACM.

Zhou, D., & Subramanian, S. (2000). Survivability in optical networks. *IEEE Network, 14*(6), 16–23. doi:10.1109/65.885666

ADDITIONAL READING

Kodian, A., Sack, A., & Grover, W. D. (2004). p-cycle network design with hop limits and circumference limits. *First International Conference on Broadband Networks* (pp. 244-253). IEEE.

Sack, A., & Grover, W. D. (2004). Hamiltonian p-cycles for fiber-level protection in semi-homogenous homogenous and optical networks. *IEEE Network*, 49–56. doi:10.1109/MNET.2004.1276611

Schupke, D. A., Gruber, C. G., & Autenrieth, A. (2002). Optimal Configuration of p-Cycles in WDM Networks. *IEEE International Conference on Communications. 5*, pp. 2761-2765. IEEE.

Shen, G., & Grover, W. D. (2003). Extending the p-cycle concept to path segment protection for span and node failure recovery. *IEEE Journal on Selected Areas in Communications*, *21*(8), 1306–1319. doi:10.1109/JSAC.2003.816598

Stamatelakis, D., & Grover, W. D. (2000). *Network Restorability Design Using Pre-configured Trees, Cycles, and Mixtures of Pattern Types*. Edmonton, Alberta, Canada: TRLabs.

Stamatelakis, D., & Grover, W. D. (2000). Theoretical underpinnings for the efficiency of restorable networks using preconfigured cycles ("p-cycles"). *IEEE Transactions on Communications*, *48*(8), 1262–1265. doi:10.1109/26.864163

Wu, B., Yeung, K. L., & Ho, P.-H. (2010). ILP Formulations for p-Cycle Design Without Candidate Cycle Enumeration. *IEEE/ACM Transactions on Networking*, 284–295.

Zhang, Z., Lixin, Z., & Xu, A. (2009). Shared-p-cycles method for design of survivable WDM networks. *Frontiers of Electrical and Electronic Engineering in China*, *4*(4), 362–370. doi:10.1007/s11460-009-0057-3

Chapter 8
Dynamic Traffic Grooming under a Differentiated Resilience Scheme for WDM Mesh Networks

Taisir E.H. El-Gorashi
University of Leeds, UK

Jaafar M. H. Elmirghani
University of Leeds, UK

ABSTRACT

Due to its huge bandwidth, optical fibre is currently widely deployed to provide a variety of telecommunications services and applications. Wavelength-division multiplexing (WDM) has emerged as the technology of choice to harness the huge bandwidth available in an optical fibre. Traffic grooming supports efficient utilization of network resources by allowing sub-wavelength granularity connections to be groomed onto a single lightpath. Fault-tolerance for WDM networks is a major architectural and design issue as a single link failure can cause loss of an enormous amount of information. However, providing 100% guaranteed resilience to all types of traffic supported by existing and future networks may be unnecessary and wasteful in terms of resource utilization and cost efficiency. This chapter investigates the problem of dynamic traffic grooming for WDM networks under a differentiated resilience scheme. We propose two differentiated resilience schemes at different grooming levels— Differentiated Resilience at Lightpath (DRAL) level scheme, and Differentiated Resilience at Connection (DRAC) level scheme. These schemes explore different ways of provisioning backup paths and tradeoff between bandwidth efficiency and the number of required grooming ports. Both schemes support three resilience classes: dedicated protection, shared protection, and restoration. Simulation is carried out to evaluate and compare the two differentiated resilience schemes. Simulation results show that the DRAL scheme is not very sensitive to the changes in the number of grooming ports, while the DRAC scheme utilizes grooming ports more aggressively as it trades grooming ports for bandwidth efficiency in routing and grooming.

DOI: 10.4018/978-1-61350-426-0.ch008

INTRODUCTION

The introduction of optical fibres as a transmission medium began a revolution in telecommunications. Optical fibre offers a number of advantages and capabilities that can meet the requirements of modern telecommunications networks (Shepherd, 2004). The huge bandwidth is considered the main advantage of optical fibre; however the challenge is to develop the necessary technologies to exploit the huge bandwidth promised by optical fibre to keep network capacity at pace with the enormous bandwidth demand of current and future applications. WDM has emerged as the technology of choice to harness the huge bandwidth available in an optical fibre. Currently WDM is widely deployed in networking infrastructures and is expected to play a significant role in supporting the requirements of the next generation networks in terms of capacity, latency and reliability.

In wavelength-routed WDM networks, the provisioning of network resources to satisfy connection requests is known as the routing and wavelength assignment (RWA) problem (Chlamtac, Faragó, Zhang, 1996) where lightpaths are created to span multiple fiber links. In all-optical networks, a lightpath remains in the optical domain by optically bypassing intermediate nodes. Most of the previously studied RWA algorithms assume that connections require a full wavelength. However, while the wavelength transmission rate has reached OC-192 (10Gbps) and is expected to reach OC-768 (40Gbps) and beyond in the future, networks are still required to support traffic connections at rates lower than the full wavelength capacity (as low as OC-3 (155Mbps)). In addition, for networks of practical size, the number of available wavelengths is still a few orders of magnitude lower than the number of source-destination connections. The bandwidth gap between the low-rate connections and the high-rate wavelengths is addressed by allowing sub-wavelength granularity connections to be groomed onto a single lightpath which results in efficient utilization of network resources.

Fault-tolerance for WDM networks is a major architectural and design issue as a single link failure can cause loss of an enormous amount of information. Network resiliency is defined as the network ability to reconfigure and re-establish communication incase of failure. Different resilience approaches for WDM networks have been extensively studied in the literature (Aneja, Jaekel, & Bandyopadhyay, 2007), (Autenrieth, & Kirstädter,2002), (Bouillet, 2002), (Van Caenegem, 1998). Resilience procedures can be classified depending on various criteria. The backup path establishment method was considered for many years as the main classification criterion as it implies the general quality parameters. Based on this criterion recovery methods can be classified to: protection and restoration methods. Under protection the protection path or the set of possible protection paths are calculated while the working path is established. Resources can be allocated before a failure occurs or they can be allocated after the failure. Under the second option backup resources can be effectively shared among different working paths. In this case a recovery time is considerably longer as signaling is required to allocate the backup recourses. On the other hand, under restoration procedures the backup path is computed on demand to re-route the traffic affected by the failure. Restoration disadvantages include long switching time, temporary instabilities, and the risk of loop creation. However, it is more efficient in terms of bandwidth utilization compared to protection. In practice, protection can be combined with restoration where fast but not efficient protection techniques are used to recover the traffic, and then using restoration techniques a better route is computed and the traffic is switched. According to the allocation of backup resources, protection methods can be classified into two major classes: protection with dedicated or shared backup resources. In the case of dedicated backup resources, backup resources can be used exclusively to establish a backup path related to a particular working path. On the other hand, backup resources are shared among several disjoint working paths. Obviously, the dedication of backup resources is considered to enable fast recovery compared to sharing backup resources where signaling is required to associate backup paths with a traffic coming from faulty working paths. However, the dedication of backup resources is very expensive due to exclusiveness usage of backup resources (Vasseur, 2004).

A further classification of the recovery mechanisms can be done depending on the extent of the network connection involved in the recovery process. Two main options exist: local and global recovery. The global method recovers the whole working path. The ingress and the egress nodes are responsible for the path recovery. In local recovery, only the faulty network elements are bypassed. The nodes adjacent to the failed element are responsible. Local recovery has the advantage that no end-to end failure notification and signaling is required: A node detecting a physical failure may immediately trigger the appropriate recovery actions. Another option in-between global and local recovery is the segment repair, where some of the nodes in the working path but not necessarily the ingress/egress nor nodes which detect a failure are involved in the recover process.

Providing 100% guaranteed resilience to all types of traffic supported by existing and future networks may be unnecessary and wasteful in terms of resource utilization, resulting in cost inefficiencies. A more efficient resilience scheme capable of supporting a variety of applications is required to maximize network utilization by providing different levels of network resilience according to the Service Level Specifications (SLS). The work in (Cholda, 2007) presents a comprehensive survey of research efforts related to resilience differentiation in the Internet and telecommunications networks.

Previous Work

In WDM mesh networks, the traffic grooming problem has mainly addressed static traffic demands where the traffic demand matrix is known a priori (Zhu,& Mukherjee, 2002), (Zhu 2003). Online approaches for traffic grooming in WDM mesh networks have been reported in (Zhu & Mukherjee, 2002b) (Zhu, 2002). The work in (Zhu & Mukherjee, 2002b) proposed different grooming policies and route-computation algorithms for different network states. The work in (Zhu, 2002) developed an algorithm for dynamically grooming low-speed connections to meet different traffic-engineering objectives based on the generic graph model proposed in (Zhu 2003). Survivable traffic grooming, in which sub-wavelength granularity connections need to be protected against failures, is a less explored topic. A number of research papers in the literature have considered survivable traffic grooming. The work in (Xiang, Wang & Li, 2003), (Thiagarajan & Somani 2001) proposed shared path protection algorithms considering traffic grooming. In (Yao & Ramamurthy, 2004) two grooming algorithms were proposed to provision bandwidth based on per-connection requirements. In (Ou, 2003), three traffic grooming approaches are proposed for shared protection in the context of dynamic provisioning - protection at lightpath (PAL) level, mixed protection at connection (MPAC) level, and separate protection at connection (SPAC) level. These three schemes explore different ways of backup sharing, and tradeoff between wavelengths and grooming ports. In (Ou, 2004) traffic grooming under dedicated protection is investigated. In (Yao & Ramamurthy, 2004b), rerouting is employed to improve the network throughput under a dynamic traffic model. Two rerouting schemes were proposed, rerouting at lightpath (RRAL) level and rerouting at connection (RRAC) level. Simulation results show that rerouting significantly reduces the connection blocking probability.

In this study, connections have different resilience requirements. We assume three classes of resilience based on the bandwidth assigned to backup paths of these connections. Class 1 is fully protected by 1+1 or 1:1 dedicated protection. Class 2 is recovered by shared protection. Class 3 assures restoration using the spare capacity left after recovery of Class 1 and Class 2. If necessary, connections of the Class 3 can be rerouted (i.e. they can be torn down and setup on a different lightpath) to establish Class 1 and Class 2 connections. Two differentiated resilience schemes supporting traffic grooming are proposed: (i) Differentiated Resilience at Lightpath (DRAL) level and (ii) Differentiated Resilience at Connection (DRAC) level. These schemes explore different ways of provisioning backup paths and tradeoff between bandwidth efficiency and the number of required grooming ports. The proposed differentiated resilience scheme is verified through simulation conducted on the Italian network as an example of a real world network. The connection requests blocking probability, wavelength efficiency and grooming port efficiency are used as the comparison metric.

PROPOSED SCHEMES

Problem Statement

A connection request is represented as C(s, d, B, CR), where s is the source node, d is the destination node, B is the traffic bandwidth requirement, and CR represents the class of resilience required by the connection. Two types of resource constraints are considered when provisioning a connection request — wavelengths and grooming ports. Typically, as the number of wavelengths in the network increases, the number of grooming ports a node requires decreases, and vice versa. Given the current network state, including the network topology, existing lightpath/connection information, wavelength and grooming-port utilization,

new arriving connections should be routed according to their bandwidth and resilience requirements while minimizing the total cost of the working and backup paths. In the rest of this section, we present the proposed two differentiated resilience schemes: DRAL and DRAC and illustrate them via examples. The initial network configuration used in the illustrative examples is shown in Figure 1. Edges in the figure correspond to bidirectional fibers and each fiber has 2 wavelengths in each direction. The wavelength capacity is OC-192. Every node has 4 grooming ports where T and R represent the number of available grooming-add and grooming-drop ports, respectively.

Differentiated Resilience at Lightpath (DRAL) Level

The DRAL scheme provides differentiated end-to-end resilience with respect to lightpaths. For Class 1 and Class 2 where protection is required, a connection is routed through a sequence of protected lightpaths (P-lightpaths). In addition to the normal working path traversing a sequence of lightpaths where grooming ports are used at the source and destination nodes, each P–lightpath has a link-disjoint path serving as a backup path. However, the two classes differ in the way the backup path is provisioned. During normal operation of Class 1 (dedicated protection), the backup path of this class is set as a sequence of lightpaths using additional grooming ports at the source and destination nodes. On the other hand, the wavelengths used by the backup path of Class 2 (shared protection) are only reserved but they are not set up, and therefore, no additional grooming ports are required. In case of working path failure, the backup path is set up as a lightpath by utilizing the grooming ports previously used by the working path. Two P-lightpaths of Class 2 can share capacity on wavelengths along common backup links if their working paths are link-disjoint. For Class 3, where protection is not required, the working path is provisioned as a sequence of

Figure 1. Initial network configuration

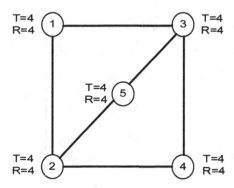

lightpaths. In case of working path failure, the spare capacity in the network is used to restore Class 3 connections in a sequence of lightpaths utilizing the grooming ports previously used by the working path. Rerouting of Class 3 connections is only executed when the normal routing algorithms fails to establish a path for an arriving connection of Class 1 or Class 2. The basic idea of rerouting under the DRAL scheme is to reroute some of the existing lightpaths of Class 3 so that otherwise blocked connection requests of Class 1 and Class 2 can be established. Obviously this implies that Class 3 connections should be routed on separate lightpaths from Class 1 and Class 2 as under the DRAL scheme the whole lightpath will be rerouted. Under both the DRAL and the DRAC schemes (discussed in the next subsection), to reduce the complexity of the rerouting algorithms and the amount of traffic affected by the rerouting operation, the rerouting operation is restricted to one lightpath or connection for a connection request. The DRAL scheme is illustrated through the following example.

Assume the arrival of the first connection request C_1 (3, 4, OC-48c, Class 3) to the empty network of Figure 1. One way of provisioning C_1 under the DRAL scheme (shown in Figure 2(a)) is to route it via the lightpath L_1. The lightpath consumes a grooming-add port at the source node

(node 3) and a grooming-drop port at the destination node (node 4). The remaining capacity on L_1 is OC-144.

Suppose that C_1 remains in the network when C_2 (3, 2, OC-12c, Class 2) arrives. Based on the current network state, one way of provisioning C_2 under the DRAL scheme (shown in Figure 2(b)) is to route it via the P-lightpath L_2 which has lightpath L_{2w} (3,2) as a working path and path (3,4,2) as a backup path. L_{2w} consumes a grooming-add port at the source node (node 3) and a grooming-drop port at the destination node (node 2). The remaining capacity on L_{2w} is OC-180. Two wavelengths need to be reserved along links (3,4) and (4,2).

Suppose that C_1 and C_2 remain in the network when C_3 (3, 4, OC-48c, Class 1) arrives. According to the current network status, to establish C_3, C_1 should be rerouted to L_3 as shown in Figure 2(c). C_3 is routed via the P-lightpath L_4, which consists of two link-disjoint lightpaths L_{4w} (working) and L_{4b} (backup). Both lightpaths consume a grooming-add port at the source node (node3) and a grooming-drop port at the destination node (node 4). The remaining capacity on P-lightpath L_4 is OC-144.

Differentiated Resilience at Connection (DRAC) Level

The DRAC scheme provides differentiated end-to-end resilience with respect to connections. For Class 1 and Class 2 where protection is required a connection is routed via link-disjoint working and backup paths, each of which traverses a sequence of lightpaths. Under the DRAC scheme the rerouting algorithm is executed to reroute some of the existing connections of Class 3 so that otherwise blocked connection requests of Class 1 and Class 2 can be established. The constraint that Class 3 connections should be routed on separate lightpaths from those of Class 1 and Class 2 does not apply to grooming at the connection level. As mentioned before, the rerouting operation

Figure 2. Example illustrating provisioning connections under the DRAL scheme

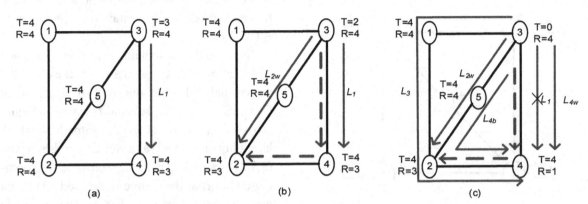

(a)   (b)   (c)

is restricted to one connection for a connection request. The DRAC scheme is illustrated through the following example:

Assume the arrival of the first connection request C_1 (3, 4, OC-48c, Class 3) to the empty network of Figure 1. One way of provisioning C_1 under the DRAC scheme is shown in Figure 3(a). Connection C_1 is routed via the lightpath L_1. The lightpath consumes a grooming-add port at the source node (node 3) and a grooming-drop port at the destination node (node 4).The remaining capacity on L_1 is OC-144.

Suppose that C_1 remains in the network when C_2 (3, 2, OC-96c, Class 2) arrives. Based on the current network state, one way of provisioning C_2 under the DRAC scheme (shown in Figure 3(b)) is to route it via two link-disjoint paths- lightpath L_3 considered as the working path, and the two-lightpath sequence (L_1, L_2) considered as the backup path. The free capacity of lightpaths L_2 and L_3 is OC-96. The free capacity of lightpath L_1 is OC-48. Each lightpath of L_2 and L_3 uses a grooming-add port at its source node and a grooming-drop port at its destination node.

Suppose that C_1 and C_2 remain in the network when C_3 (3, 4, OC-192c, Class 1) arrives. As shown in Figure 3(c) connection C_3 is routed via two link-disjoint paths- lightpath L_4 and L_5. No remaining capacity on L_4 and L_5.

While all the previous connections remain in the network suppose that C_4(1, 2, OC-192c, Class 2) arrives. According to the current network state, C_4 has to be blocked as it is not possible to provision it both a working and a backup path. However, C_4 can be provisioned by applying rerouting at the connection level. C_1 is rerouted so enough capacity becomes available at L_1. As shown in Figure 3(d), C_1 is rerouted via the two-lightpath sequence (L_3, L_8) to allow C_4 to be routed via two link-disjoint paths- the lightpath L_6 considered as the working path, and two-lightpath sequence (L_7, L_1) considered as the backup path.

Comparison between the DRAL and the DRAC Schemes

In this section we compare the characteristics of the DRAL and the DRAC schemes in terms of grooming ports consumption, backup sharing, rerouting and complexity.

Grooming ports consumption: Under both the DRAL and the DRAC schemes a connection with dedicated protection requirements (Class 1) is provisioned a sequence of lightpaths as its working and backup paths, however the two schemes differ in the utilization of these lightpaths. Under the DRAL scheme the two lightpaths of a P-lightpath are considered as an integrated unit and cannot be utilized individually. This constrain does not apply

Figure 3. Example illustrating provisioning connections under the DRAC scheme

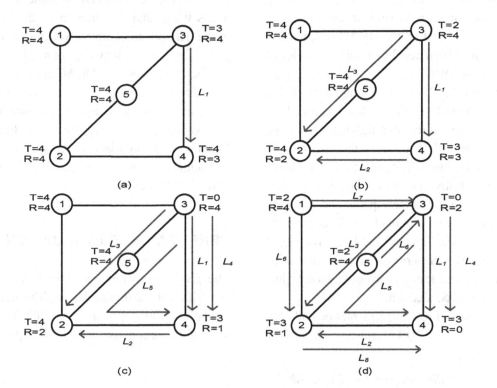

to lightpaths under the DRAC scheme. Therefore under the DRAC scheme Class 1 connections are more likely to be groomed onto new lightpaths and more grooming ports are more likely to be consumed. Connections with shared protection (Class 2) under the DRAC scheme are provisioned backup paths which traverse a sequence of lightpaths. However, under the DRAL scheme wavelengths used by the backup path are only reserved but they are not set up, and therefore, no additional grooming ports are required. Therefore for shared protection the DRAL scheme trades the bandwidth efficiency in routing each connection request for the savings in grooming ports usage.

Backup sharing: The DRAC scheme allows working and backup paths (of different connections) to be groomed into the same lightpath. Also the difference between backup paths of Class 2 under the two schemes creates a difference in the way they share the backup capacity. As a lightpath

may traverse multiple links, the reserved backup capacity on a lightpath, in case of the DRAC scheme, is less likely to be shared among multiple connections compared to the backup capacity reserved on a link in case of the DRAL scheme. Under the DRAL scheme the reserved wavelengths on a link act like a "pool" for all the failure scenarios and wavelength converters facilitate backup capacity sharing among different wavelengths on a link. However, under the DRAC scheme, sharing the backup capacity among different wavelengths on a link is not possible as the backup capacity is in the form of lightpaths and multiple lightpaths cannot share their reserved backup capacity. From above it is clear that for Class 2 connections the DRAC scheme provides flexibility in terms of grooming as it allows grooming working paths and backup paths (of different connections) onto the same lightpath, however the DRAL scheme is more flexible in terms of sharing backup capacity.

Rerouting: As discussed before, the DRAL scheme performs rerouting at an aggregate (lightpath) level, while the DRAC scheme performs rerouting at a per-flow (connection) level. Therefore the DRAL scheme affects more traffic than the DRAC scheme during the rerouting process as the DRAL scheme disconnects all the connections on the rerouted lightpath, while the DRAC scheme only disconnects the rerouted connection. Rerouting at the connection level results in more flexibility in terms of selecting rerouted connections and their new paths and allowing connections from different classes to be rerouted through the same lightpaths. Also, rerouting at connection level preserves the quality of service (QoS) and the traffic engineering (TE) constraints of the rerouted connection.

Complexity: From implementation and control point of view, the DRAL scheme is relatively simple compared to the DRAC scheme as it only needs the global information of the lightpaths in the network to compute working and backup paths for a new arriving connection request. On the other hand, the DRAC scheme requires, in addition to the global information of all the lightpaths, the detailed routing information of all the existing connections. Therefore, the DRAC scheme results in a larger signaling complexity and demands higher control bandwidth compared to the DRAC scheme.

PERFORMANCE EVALUATION

The proposed differentiated resilience scheme is verified through simulation. The high-speed Italian network topology (Ali, 2001), depicted in Figure 4, is considered as an example of a real

Figure 4. The Italian mesh network (Ali, 2001)

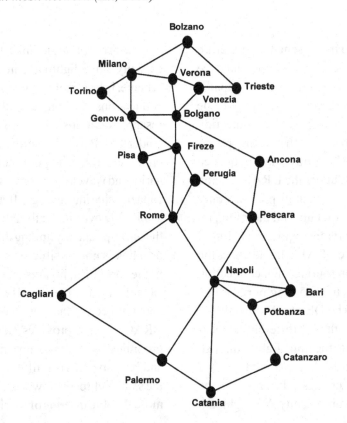

world network. The Italian network consists of 21 nodes and 36 bidirectional links. For the simulation scenario a ratio of 10% Class 1, 30% Class 2, and 60% Class 3 is assumed. Connections are assumed to be uniformly distributed among all the node pairs. The connection-arrival process is assumed to follow a Poisson distribution and the connection-holding time is assumed to be exponential distributed. Connections bandwidth requirements are assumed to be as follows: Class 1 and Class 2 requirements are assumed to be uniformly distributed in the range of (4-10) units. Requirements of Class 3 are assumed to be uniformly distributed in the range of (0-4) units. 100,000 connections are simulated in each simulation scenario. Each link supports 16 wavelengths and full wavelength conversion capability is assumed. The number of grooming ports μ_i at node i is set as $\mu_i = Wx\eta_i x\sigma$ where W is the number of wavelengths, n_i its nodal degree of node i and σ is a variable ($0 \leq \sigma \leq 1$) where $\sigma = 1$ implies that any incoming wavelength can be groomed (Ou, 2003). Failures are generated randomly. The inter-arrival time and holding time of failures are assumed to be exponentially distribution. Links are affected by failures according to a uniform distribution. We assume that there is an equal number of grooming-add and grooming-drop ports. Simulation results compare the performance of each traffic class under the DRAL scheme and the DRAC scheme with a large number of grooming ports ($\sigma = 1$) and a smaller number of grooming ports ($\sigma = 0.5$).

Figure 5 illustrates the connection request blocking probability of Class 1 (dedicated protection). With a large number of grooming ports ($\sigma = 1$), the DRAC scheme has lower blocking probability compared to the DRAL scheme. This is a result of the bandwidth efficiency in grooming at the connection level and the fact that the DRAC scheme allows working paths of different traffic classes to be groomed onto the same lightpath. However, when the number of grooming ports is small ($\sigma = 0.5$), the DRAL scheme has lower blocking probability. The DRAL scheme is not very sensitive to the changes in the number of grooming ports as it utilizes wavelengths more quickly than grooming ports. However, the DRAC scheme utilizes grooming ports more aggressively as it trades grooming ports for bandwidth efficiency in routing and grooming.

Results of Class 2 (shared protection) traffic are shown in Figure 6. The approach of provision-

Figure 5. Blocking probability of Class 1 traffic

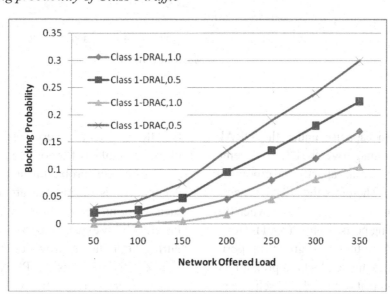

Figure 6. Blocking probability of Class 2 traffic

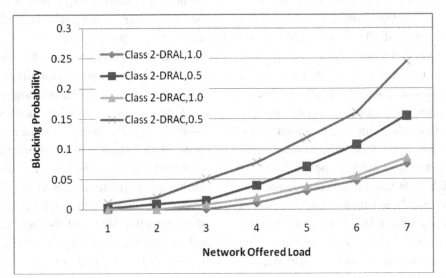

Figure 7. Blocking probability of working paths of Class 3 traffic

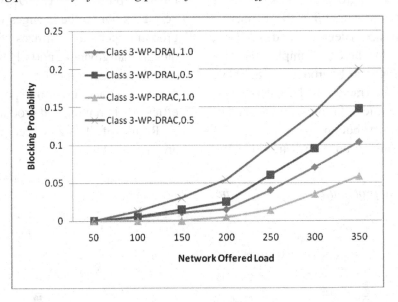

ing shared backup capacity under the DRAL scheme gives it advantage over the DRAC scheme in terms of sharing backup capacity and grooming ports consumption. Therefore the DRAL scheme slightly outperforms the DRAC scheme when the number of grooming ports is large (σ =1). However under a smaller number of grooming ports (σ =0.5) the DRAL scheme blocking probability is much lower than that of the DRAC scheme as

the DRAC scheme is very sensitive to the changes in the number of grooming ports.

The blocking probability experienced by Class 3 connections during the attempt to provision them working paths (Class3-WP) and the blocking probability experienced by Class 3 connections during the attempt to provision them backup paths in case of failure (Class3-BP) are shown in Figure 7 and Figure 8, respectively. When the number

Figure 8. Blocking probability of backup paths of Class 3 traffic

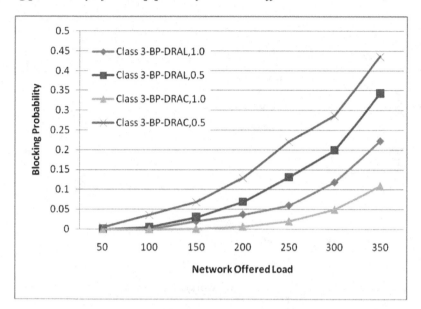

of grooming ports is large ($\sigma = 1$), the DRAC scheme outperforms the DRAL scheme as the DRAL scheme allows Class 3 connections to be groomed onto the same lightpaths with other Classes. Due to the DRAC scheme high sensitivity to change in the number of the grooming ports, the DRAL scheme outperforms the DRAC scheme when the number of grooming ports is smaller ($\sigma = 0.5$).

Figure 9 shows the rerouting probability experienced by Class 3 to allow Class 1 and Class 2 to establish their connections. As discussed before, the DRAL scheme affects more traffic than the DRAC scheme during the rerouting process as the DRAL scheme disconnects all the connections on the rerouted lightpath, while the DRAC scheme only disconnects the rerouted connection. Therefore, the rerouting probability of Class 3 is higher under the DRAL scheme. The same trend is noticed when the number of grooming ports is smaller ($\sigma = 0.5$).

To further reflect the difference between the two schemes, we define two parameters: wavelength efficiency (η_W) and grooming port efficiency (η_G). The two parameters measures how

efficiently the network utilizes wavelengths and grooming ports. They are defined as the load carried in the network carried load divided by the amount of allocated resources (wavelength or grooming ports). They are calculated as follows:

$$\eta_W = \frac{\sum_i l_i \times t_i}{\sum_i W_i \times t_i}$$

$$\eta_G = \frac{\sum_i l_i \times t_i}{\sum_i G_i \times t_i}$$

where t_i is the time period between two successive events, l_i is the network load during t_i, W_i is the number of wavelength links used during t_i, and G_i is the number of grooming ports used during t_i.

Figure 10 compares the wavelength efficiency of the DRAL scheme and the DRAC scheme for $\sigma = 0.5$ and $\sigma = 1$. We can see that the DRAC scheme has higher wavelength efficiency under different number of grooming ports as the DRAL scheme grooms traffic at the lightpath level and

Figure 9. Rerouting probability of working paths of Class 3 traffic

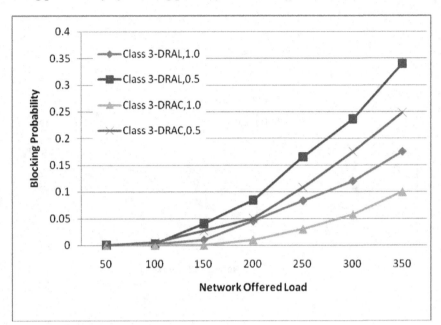

Figure 10. Wavelength efficiency

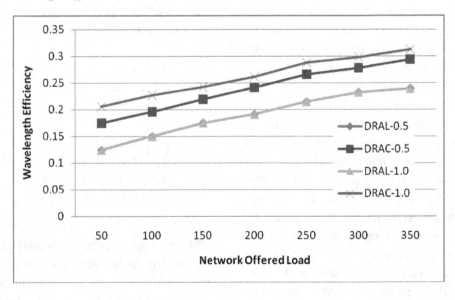

lightpaths may not be fully used. We can also see that the DRAL scheme wavelength utilization is not affected by changing the number of available grooming ports as the DRAL scheme utilizes wavelengths more quickly than grooming ports. However, the wavelength utilization of the DRAC scheme with $\sigma =1$ will be reduced by an average of 2% compared to $\sigma =0.5$ which support our previous observation that the DRAC scheme is sensitive to changes in the number of grooming ports.

Figure 11. Grooming-port efficiency

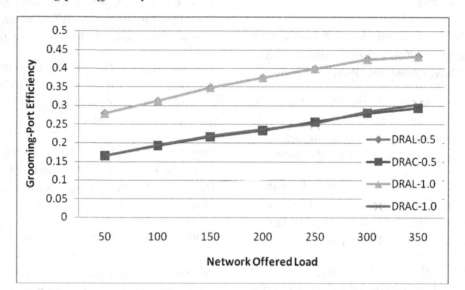

Figure 11 shows the grooming port efficiency of the DRAL scheme and the DRAC scheme for $\sigma =0.5$ and $\sigma =1$. As mentioned before the DRAC scheme utilizes grooming ports more aggressively therefore the DRAL scheme grooming-port efficiency is on average 14% higher than the DRAC scheme.

CONCLUSION

In this chapter, we have investigated dynamic traffic grooming with differentiated resilience for WDM mesh networks. We have studied two different schemes of provisioning different levels of resilience to connections with sub-wavelength requirements: DRAL and DRAC. The performance of different resilience classes (dedicated protection, shared protection and restoration) is evaluated through simulation. Simulation results show that traffic with dedicated protection requirements obtains a better blocking probability under the DRAC scheme when the number of grooming ports is large. However, when the number of grooming ports is small, the DRAL scheme results

in a lower blocking probability due to the high sensitivity of the DRAC scheme to the change in the number of grooming ports. The blocking probability experienced by traffic with shared protection requirements under the DRAL and the DRAC schemes are very close under a larger number of grooming ports. With a small number of grooming ports the DRAL scheme outperforms the DRAC scheme. The blocking probability experienced by connections with restoration requirements during the attempt to provision them working paths and during the attempt to provision them backup paths in case of failure is lower under the DRAC scheme when the number of grooming ports is large. With a small number of grooming ports the opposite trend is observed. The rerouting probability experienced by connections with restoration requirements to allow higher resilience classes to establish their connections are found to be higher under the DRAL scheme. Results also show that while the DRAC scheme utilizes wavelengths more efficiently, the DRAL scheme has better grooming-port utilization.

REFERENCES

Ali, M. (2001). *Transmission efficient design and management of wavelength-routed optical networks.* Kluwer Academic Publisher 2001.

Aneja, Y., Jaekel, A., & Bandyopadhyay, S. (2007). Some studies on path protection in WDM networks. *Photonic Network Communications, 14*, 165–176. doi:10.1007/s11107-007-0075-0

Autenrieth, A., & Kirstädter, A. (2002). Engineering end-to-end IP resilience using resilience- differentiated QoS. *IEEE Communications Magazine, 40*(1). doi:10.1109/35.978049

Bouillet, E. (2002). Stochastic approaches to compute shared mesh restored lightpaths in optical network architectures. *Proceedings - IEEE INFOCOM*, (June): 801–807.

Chlamtac, I., Faragó, A., & Zhang, T. (1996). Lightpath (wavelength) routing in large WDM networks. In *IEEE J. Select. Areas Commun., 14*, 909–913.

Cholda, P., et al. (2007). A survey of resilience differentiation frameworks in communication networks. *IEEE Communications Surveys and Tutorials, 9*(4).

Ou, C. K. (2003). Traffic grooming for survivable WDM networks--Shared protection. *IEEE Journal on Selected Areas in Communications, 21*(9), 1367–1383. doi:10.1109/JSAC.2003.818233

Ou, C. K. (2004). Traffic grooming for survivable WDM networks—-Dedicated protection. *OSA Journal of Optical Networking, 3*(1), 50–74. doi:10.1364/JON.3.000050

Shepherd, F. B. (2004). Lighting fibres in a dark network. *IEEE Journal on Selected Areas in Communications, 22*(9), 1583–1588. doi:10.1109/JSAC.2004.833850

Thiagarajan, S., & Somani, A. K. (2001). Traffic grooming for survivable WDM mesh networks. In *Proc. OptiCom'01, 2001,* (pp. 54-65).

Van Caenegem, B. (1998). Dimensioning of survivable WDM networks. *IEEE Journal on Selected Areas in Communications, 16*(7). doi:10.1109/49.725185

Vasseur, J. P. (2004). *Network recovery: Protection and restoration of optical, SONET-SDH, IP, and MPLS.* The Morgan Kaufmann Series in Networking, 2004.

Xiang, B., Wang, S., & Li, L. M. (2003). A traffic grooming based on shared protection in WDM mesh networks. In *Proc. IEEE PDCAT'03*, Cheng Du, China, (pp. 254-258), August 2003.

Yao, W., & Ramamurthy, B. (2004). Rerouting schemes for dynamic traffic grooming in optical WDM mesh networks. In *IEEE GLOBECOM'04*, vol. 3, (pp. 1793–1797).

Yao, W., & Ramamurthy, B. (2004). Survivable traffic grooming with differentiated end-to-end availability guarantees in WDM mesh networks. In *The 13th IEEE Workshop on Local and Metropolitan Area Networks, LANMAN 2004* (pp. 87–90).

Zhu, H., et al. (2002). Dynamic traffic grooming in WDM mesh networks using a novel graph model. In *Proc. IEEE GLOBECOM*, (pp. 2681–2685).

Zhu, H. (2003). A novel, generic graph model for traffic grooming in heterogeneous WDM mesh networks. *IEEE/ACM Transactions on Networking, 11*, 285–299. doi:10.1109/TNET.2003.810310

Zhu, K., & Mukherjee, B. (2002). Traffic grooming in an optical WDM mesh network. *IEEE Journal on Selected Areas in Communications, 20*, 122–133. doi:10.1109/49.974667

Zhu, K., & Mukherjee, B. (2002b). On-line approaches for provisioning connections of different bandwidth granularities in WDM mesh networks. In *Proc. OFC*, Mar. 2002.

Chapter 9
Robust Design and Management of Optical Networks:
Incorporating Availability–Awareness

Hussein T. Mouftah
University of Ottawa, Canada

Burak Kantarci
University of Ottawa, Canada

ABSTRACT

High capacity advantage of optical networks also introduces the risk of huge data loss in case of a service interruption, even if the outage lasts a short time. Therefore, survivable and reliable design and management of optical networks is urgent. However, deployment of efficient survivability policies does not always guarantee the continuity of the service. Long failure restoration delays, multiple failures, and lack of protection resources may lead to service unavailability. Hence, connection availability arises as a design constraint, and it is defined as the probability of a connection being in the operating state at any time. Availability-constrained optical network design and availability-constrained connection provisioning are two important problems to guarantee robustness of connections in a survivable network.

INTRODUCTION

As a result of the increasing bandwidth demand of the Internet applications, optical networking seems to be a strong candidate for the long haul communication. This great advantage of optical

networks also introduces the risk of huge data loss in case of a service interruption even though the outage lasts short. Therefore survivable and reliable design and management of optical networks is urgent (Mouftah and Ho, 2003). Various survivability schemes have been proposed to avoid data loss in case of a failure.

DOI: 10.4018/978-1-61350-426-0.ch009

Survivability schemes can basically be grouped in three main categories, i.e., ring protection, span protection and path protection. These survivability policies introduce the trade-off between restoration delay and resource redundancy which have also been widely studied in the literature. Existing survivability policies are derived from these categories, and these policies usually attempt to find a compromise for this trade-off. For instance, segment protection is a hybrid of span protection and path protection where failure detection delay is close to that of link protection, and backup redundancy is close to path protection (Krishna et al., 2000). Similarly, as another hybrid survivability scheme, pre-configured cycle (p-cycle) protection provides path/link protection-like backup resource consumption and ring protection-like restoration efficiency (Stamatelakis and Grover, 2000).

Figure 1 illustrates the survivability schemes that fall in the scope of this chapter. Figure 1.a illustrates the *Dedicated Path Protection (DPP)* scheme for two connections simply assuming that node failures are negligible. The connection from N1 to N7 uses the working path passing through the nodes, *N1-N3-N5-N7* on the wavelength channel λ_1 while the connection from N2 to N7 has the working path passing through the nodes, *N2-N4-N6-N7* on the wavelength channel λ_1. The former connection uses the backup path *N1-N4-N6-N7* on the wavelength channel λ_2 while the latter uses the backup path *N2-N1-N4-N5*-N7 on the wavelength channel λ_3. Thus, on the common link *N1-N4*, backup channels of the connection from N1 to N7 are not shared with the connection from N2 to N7.

Figure 1.b illustrates a *Shared Backup Path Protection (SBPP)* scenario for two connections. The connection from N1 to N7 switches to its backup path following the route *N1-N4-N6-N7* on wavelength λ_2 in case of a failure on its working path. The connection from N2 to N7 switches to its backup path following the route *N2-N1-N4-N5-N7* on wavelength channel λ_2. Thus, the two connections share the backup channel λ_2 on

the link *N1-N4* which will lead to unavailability of either of the connections in case of such a dual failure. Ramamurthy et al. presented a comprehensive discussion on the advantages and disadvantages of DPP and SBPP (Ramamurthy et al., 2003).

Figure 1.c illustrates a segment protection scenario which was presented by Krishna et al. (Krishna et al., 2000). Primary path of the connection follows the path formed by the links between the following nodes: *Source*-N1-N2-N3-N4-N5-N6-N7-N8-*Destination*, and it is protected by three overlapping backup segments. In the illustrated scenario, the link between the nodes N3 and N4 fails. Following the failure, the connection does not immediately switch to the backup resources but it forwards the traffic on a short segment of the working path from *Source* to *N2*. At node *N2*, the connection switches to the backup segment starting at *N2* and ending at *N6*. At node *N6*, the connection switches to the working path again and routes the traffic to the *Destination* node. In order to keep it simple, shared segment protection is not illustrated in the figure however, shared segment protection stands for the overlapping segment protection type where the backup resources on the backup segments are being shared by the connections.

Figure 1.d illustrates a *Pre-configured Protection Cycle (P-Cycle)*. In the figure, the bold lines denote the on-cycle links which form the main protection ring while the dashed lines represent the straddling links which can be protected by either half of the protection cycle in case of a failure. Thus, a p-cycle protects the links on the protection ring and further protects the off-cycle links whose both ends lie on the cycle. P-cycles offer mesh-like spare capacity and ring-like restoration time efficiency as stated by Stamatelakis and Grover (Stamatelakis and Grover, 2000).

Deployment of efficient survivability policies does not guarantee the continuity of the service. Therefore connection availability arises as a design constraint, and it is defined as the probability of

Figure 1. An illustration of the survivability schemes studied here (a) DPP (b) SBPP (c) Shared segment protection (d) P-Cycle protection

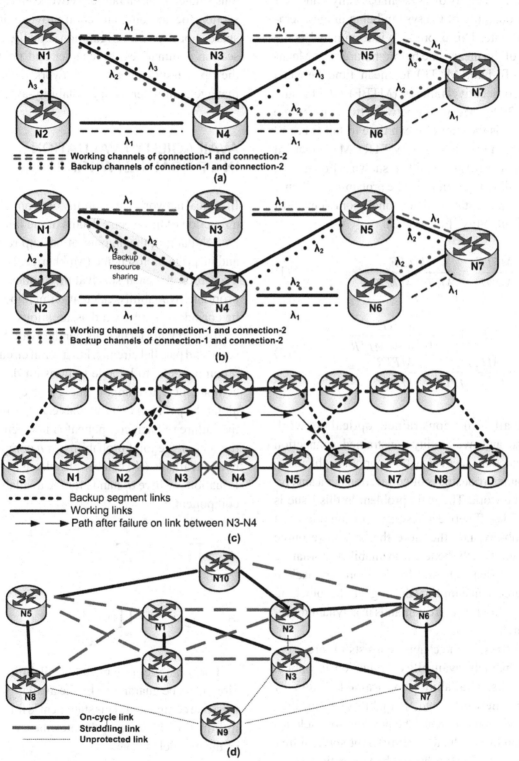

a connection being in the operating state at any time. Availability of a system basically stands for the probability of the system being in the operational state. This expression is formulated as the ratio of the mean time before failure (i.e., Mean-Time-To-Fail, MTTF) to mean time between two consecutive failures (MTBF) (Babbit and Best, 2006). Mean time between two consecutive failures is the sum of mean time to fail (MTTF) and mean time to repair (MTTR). Mathematical formulation of availability is shown in Equation 1. Unavailability stands for the probability of being in the failed state which is one's complement of the availability (Equation 2).

$$A = \frac{MTTF}{MTBF} = \frac{MTTF}{MTTF + MTTR} \tag{1}$$

$$U = 1 - A = 1 - \frac{MTTF}{MTTF + MTTR} = \frac{MTTR}{MTTF + MTTR} = \frac{MTTR}{MTBF} \tag{2}$$

Availability-constrained optical network design and availability-constrained connection provisioning seem as two important problems to guarantee robustness of connections in a survivable network. The main problem in this issue is the trade-off between resource consumption and availability, i.e., the more the backup resource consumption the better the availability (Tornatore et al., 2006a). Obviously, the more the backup resource consumption the higher the blocking probability for future demands in a dynamic traffic environment.

The rest of the chapter consists of three sections, namely availability evaluation in optical networks, availability-constrained design of optical networks and availability-constrained network management. The next section includes availability evaluation of different survivability schemes, and it is followed by the section which presents availability-constrained network planning studies in the literature. Network management section focuses on connection provisioning and spare capacity re-configuration. Finally, the last section summarizes the chapter and points out the open issues on robust network design and management in terms of availability-awareness.

AVAILABILITY EVALUATION IN OPTICAL NETWORKS

In optical networks, availability of a connection is a function of the precise details of the failures on its lightpath(s), the amount of backup resources and its protection scheme (Mukherjee, B. 2006). Therefore, for each survivability policy, an appropriate availability calculation method has to be considered under valid assumptions.

To start with, let us consider the availability of serial and parallel circuits. For a serial circuit with n components, to be able to transmit data from component C_1 to component C_n, each component along the path has to be available. Assuming that the failure of each component is independent of the others, the joint probability of n components being available (A_s) can be formulated as seen in Equation 3 where a_i stands for the availability of component, C_i.

$$A_s = \prod_{i=1}^{n} a_i \cong 1 - \sum_{i=1}^{n} (1 - a_i) \tag{3}$$

$$A_p = \sum_{i=1}^{n} a_i \cong 1 - \prod_{i=1}^{n} (1 - a_i) \tag{4}$$

For a parallel circuit, the availability of the circuit can be guaranteed by the availability of at least one component. This statement is formulated in Equation 4 where A_p denotes the availability of the parallel system.

In this section, availability evaluations of dedicated/shared path protection, p-cycles and shared segment protection are presented. For those schemes that have more than one availability analysis method, the unavailability formulations are distinguished by having the initials of the proposing authors as the subscripts of the unavailability function.

Dedicated Path Protection (DPP)

Arci et al. formulated the availability of a dedicated path protected connection as in Equation 5 (Arci et al., 2003). In Equation 5, availability of a 1:1 connection ($A^{(1:1)}$) is shown where A_w and A_b represent the availability of the working and the backup paths, respectively. As seen in the equation, a 1:1 connection is available if its working path is available or its backup path is available when its working path is unavailable.

$$A^{(1:1)} = A_w + (1 - A_w).A_b = A_w + A_b - A_w A_b$$
(5)

Shared Backup Path Protection (SBPP)

Arci et al. formulated the availability of a shared backup path protected connection ($A_{DA}^{(c)(SBPP)}$) as seen in Equation 6 where c denotes the index of the connection (Arci et al., 2003). In the formulation, S_c represents the *sharing* group of c, which is the set of connections sharing at least one backup resource with connection c while A_{w_m} denotes the working path of connection m which is in S_c. Thus, for a connection to be available, one of the following two conditions has to hold: 1) The working path of the connection is available, 2) If the working path of the connection is unavailable, its backup path has to be available AND the working paths of all connections in its sharing group are available.

$$A_{DA}^{(c)(SBPP)} = A_w + (1 - A_w).A_b.\prod_{m \in S_c} A_{w_m}$$
(6)

As seen above, Equation 6 provides a conservative bound for the connection availability due to the assumption that the connection is unavailable when concurrent failures affecting its working path and the working path(s) of any connection(s) in its sharing group. Zhang et al. proposed a more relaxed formulation assuming that the connection can still survive with a probability of δ^k in case of k concurrent failures on the working paths (Zhang et al., 2007). Equation 7 shows the availability calculation of a shared backup protected connection ($A_{JZ}^{(c)(SBPP)}$) based on the related reference. In the equation, ρ_k represents the probability of k concurrent failures on the working paths while δ_c^k stands for the probability of connection c to survive in case of k concurrent failures on the working paths.

$$A_{JZ}^{(c)(SBPP)} = A_w + (1 - A_w).A_b.\sum_{k=0}^{K} (\rho_k.\delta_c^k)$$
(7)

Mello et al. proposed an analytical approach under dual failure assumption where the failure state of the network is represented by a Markov chain (Mello et al., 2005). Each node (state) of the Markov chain has either one index or two indices such that node-i represents the state where link-i is the only failed link in the network while node-ij denotes the failure of link-j following the failure of link-i. Node-0 stands for the state where all links are available, i.e., zero-failure state. Steady state probabilities (π_i, π_{ij}) can be obtained by the solution of the balance equations by using the means of linear algebra as explained by the authors. These steady state probabilities are further used to calculate the availability of a connection based on its survivability policy

This approach is based on a realistic assumption of dual failures as the dominant factors on connection unavailability. A connection which is provisioned with respect to the SBPP is unavailable if one of the following two conditions holds: 1) Failure of a link on the working path followed by the failure of a link on the backup path, or 2) Failure of a link on the backup path or a link on the working path of a connection in the sharing group (SW_c) followed by the failure of a link on the working path. Equation 8 formulates the availability of a shared backup path protected connection with respect to the matrix-based unavailability calculation approach ($A^{(c)(SBPP)}_{DAM}$). The total of the two terms inside the parenthesis denotes the unavailability of connection c. The first term in the parenthesis represents the first condition while the second term stands for the second unavailability condition stated above.

$$A^{(c)(SBPP)}_{DAM} = 1 - \left[\sum_{\substack{i \in WP_c \\ j \in B\hat{P}_c}} \pi_{ij} + \sum_{i \in WP_c, \ j \in BP_c \cup SW_c} \pi_{ji} \right] \tag{8}$$

Although this approach was proposed for SBPP, Mello et al. have also formulated the availability calculation of an unprotected connection ($A^{(c)(Unp)}_{DAM}$) by using the steady state probabilities of the link failure states. As seen in Equation 9, an unprotected connection is unavailable if one of the following three conditions holds: 1) Failure of a link on the working path is followed by the failure of a link which is not on the working path, or 2) Failure of a link which is not on the working path is followed by the failure of a link on the working path, or 3) Dual failure on the working path.

$$A^{(c)(Unp)}_{DAM} = 1 - \left[\sum_{\substack{i \in WP_c \\ j \notin W\hat{P}_c}} \pi_{ij} + \sum_{\substack{i \in WP_c \\ j \notin W\hat{P}_c}} \pi_{ji} + \sum_{\substack{i \in WP_c \\ j \in W\hat{P}_c}} \pi_{ij} \right] \tag{9}$$

Shared Segment Protection

Availability calculation of shared segment protection is more complicated than dedicated/shared path protection. Here, in order to compute the connection availability, the term, *protection domain* has to be presented. Protection domain stands for a working path segment together with its protection segment. Thus, in shared segment protection, a connection can be considered as a series of protection domains. Therefore for a connection to be in the operating state, each protection domain has to be available as shown in Equation 10 where A_i denotes the availability of a working path link, and Λ_{Si} stands for the segment protection for link i (Kantarci et al., 2008c).

$$A^{(c)(Seg)} = \prod_{i \in W_c} (A_i + A_{S_i} - A_i . A_{S_i}) \tag{10}$$

Since a link can be protected by more than one segment, A^c_{Si}, availability of the segment protection of a working link (say, link i) of connection c has to be considered as shown in Equation 11 where A^c_S stands for the availability of segment s in the protecting segment list of the working link i of connection c (Seg_i). According to the equation, segment protection of link i is available if at least one backup segment in its segment protection list is available.

$$A^c_{S_i} = \sum_{s \in Seg_i} A^c_S - \sum_{p \in Seg_j} \sum_{\substack{s \in Seg_i \\ i \neq j}} (A^c_S . A^c_p)$$

$$+ \sum_{\substack{r \in Seg_k \\ j \neq r}} \sum_{p \in Seg_j} \sum_{\substack{s \in Seg_i \\ i \neq j}} (A^c_S . A^c_p . A^c_r) - \dots \tag{11}$$

$$\cong 1 - \prod_{s \in Seg_i} (1 - A^c_S)$$

Since availability of the segment protection of a link is a function of the availability of the backup segments, shared backup segment availability computation has to be formulated as well. Equation 12 formulates the availability of the backup segment *s*. According to the formulation, a backup segment of connection *c* is available if each link along the segment is available and one of the following two conditions holds: 1) Working paths of all connections sharing at least one backup resource with connection *c* are available, or 2) If condition-1 fails, and connection *c* is still able to switch its traffic to the backup path with a probability of δ ($\delta<1$) where SG_z stands for the set of the connections sharing backup resources with connection *c*. These two conditions provide a conservative bound for the availability calculation due to the assumption of unavailability of connection *c* in case of any failure on the working paths of the connections in its sharing group. Since a connection can be protected by more than one backup segment, a failure on the working path of connection *z* can be recovered by switching the traffic to another backup segment which is not utilized by connection *c*. Therefore, unavailability of each connection in the sharing group is normalized by the number of the backup segments protecting that connection ($|Seg_z|$).

$$A_S = (\prod_{k \in S} A_k).$$
$$\left[\prod_{z \in SG_s} \left[1 - \frac{(1 - \prod_{j \in W_z} A_j)}{|Seg_z|} \right] + \left[1 - \prod_{z \in SG_s} \left[1 - \frac{(1 - \prod_{j \in W_z} A_j)}{|Seg_z|} \right] \right] . \delta \right]$$

(12)

Substitution of Equation 12 into Equation 11, and then into Equation 10 gives the open form of the connection availability in shared segment protection.

Pre-Configured Protection Cycles (P-Cycles)

According to the availability evaluation proposed by Cloqueur and Grover, a connection is protected by a series of protection domains where each protection domain consists of a p-cycle (Cloqueur and Grover, 2005). At each protection domain,

a connection is assumed to pass through either the straddling spans or the on-cycle spans of the protection domain. Four span sets are defined for cycle *x* and a working path *p* of connection *c* as follows: S_x^c represents the set of the spans on cycle *x* while S_x^s stands for the set of straddling spans of cycle *x*. $S_{p,x}^c$ and $S_{p,x}^s$ stand for the spans of the path *p* on cycle *x* and the spans of path *p* crossing the straddling spans of cycle *x*, respectively. Six dual failure scenarios leading to connection unavailability are considered for availability calculation. Unavailability of each scenario is summed up to obtain end-to-end unavailability.

Mukherjee, D. S. et al. proposed an enhanced availability analysis approach for the connections protected by p-cycles (Mukherjee, D. S. et al., 2006). Here, the working path of a connection can have both an on-cycle span and a straddling span on a protection domain. Like the availability analysis defined above, a survivable connection is assumed to consist of a series of protection domains each of which stands for a pre-computed p-cycle. The four span sets defined above are used to define the six failure categories (scenarios) as summarized below:

- **Category-1:** Dual failure affecting S_x^c and S_{px}^c.
- **Category-2:** Dual failure affecting S_{px}^c and S_x^s.
- **Category-3:** Dual failure affecting S_{px}^c and S_{px}^s.
- **Category-4:** Dual failure affecting S_x^c and S_{px}^s.
- **Category-5:** Dual failure affecting two spans in S_{px}^s.
- **Category-6:** Dual failure affecting S_x^s and S_{px}^s.

For each category, an outage probability is calculated as shown by Equation 13-Equation 18. For Category-1, in case of a dual failure affecting a straddling span and an on-cycle span on the working path, the connection is unavailable regardless of the order of the errors. For Category-2, if a straddling span on the working path of the connection fails before a straddling span out of the working path, the connection can switch its traffic onto the p-cycle-*x*. If the failures occur in reverse order, the connection becomes unavailable. Therefore, the factor ½ is included in the formulation. Category-3 represents a situation where both of the failures are on the working path leading to an absolute outage. Category-4 considers two possible failure patterns where the first failure pattern consists of the failure of an on-cycle span which is not on the working path followed by the failure of a straddling span on the working path, and the second failure pattern consists of the same failures in reverse order. The first pattern can occur with a probability of 50%, and it leads to unavailability of the connection. In the second pattern, the straddling span can switch its traffic onto one of the two halves of the p-cycle however the second failure of this failure pattern which is supposed to be on an on-cycle span can occur on the corresponding backup path with a probability of 50%, and it may lead to unavailability of the protection of the straddling span. Therefore the joint probability of the connection to survive is 25% which contributes the unavailability formulation with the factor ¾ as seen in Equation 16. Category-5 represents the situation where two straddling spans on the working path fail. In this case the connection can still survive if the failed straddling spans are not in the *crossing situation*. Therefore, the connection is unavailable with a probability of 50%. Conversely, in Category-6, the connection can still survive unless the two straddling spans are not in the *crossing situation*.

Therefore, its formulation has a factor of ½ to represent this possibility.

$$U_{DSM}^{k,Category-1} = |S_x^c| \cdot |S_{px}^c| U^2 \tag{13}$$

$$U_{DSM}^{k,Category-2} = \frac{1}{2}|S_{px}^c| \cdot |S_x^s| U^2 \tag{14}$$

$$U_{DSM}^{k,Category-2} = \frac{1}{2}|S_{px}^c| \cdot |S_x^s| U^2 \tag{15}$$

$$U_{DSM}^{k,Category-4} = \frac{3}{4}|S_{px}^p| \cdot |S_x^c| U^2 \tag{16}$$

$$U_{DSM}^{k,Category-5} = \frac{1}{2}|S_{px}^s| \cdot |S_{px}^s - 1| U^2 \tag{17}$$

$$U_{DSM}^{k,Category-6} = \frac{1}{2}|S_{px}^s| \cdot |S_x^s| U^2 \tag{18}$$

Unavailability of a domain of a connection can be calculated as the sum of the outage probabilities (unavailability) caused by each failure category as seen in Equation 19. Finally, connection unavailability (U_{DSM}^k) can be expressed as shown in Equation 20 (see Table 1) where X is the set of the protection domains of connection k.

$$U_{DSM}^{k,Domain-x} = \sum_{i=1}^{6} U_{DSM}^{k,Category-i} \tag{19}$$

Table 1. Equation 20

$$U_{DSM}^k = \sum_{x \in X} U_{DSM}^{k,Domain-x} = \sum_{x \in X} \begin{bmatrix} |S_x^c| \cdot |S_{px}^c| U^2 + \frac{1}{2}|S_{px}^c| \cdot |S_x^s| U^2 + |S_{px}^c| \cdot |S_{px}^s| U^2 \\ +\frac{3}{4}|S_{px}^p| \cdot |S_x^c| U^2 + \frac{1}{2}|S_{px}^s| \cdot |S_{px}^s - 1| U^2 + \frac{1}{2}|S_{px}^s| \cdot |S_x^s| U^2 \end{bmatrix} \tag{20}$$

AVAILABILITY-CONSTRAINED PLANNING OF OPTICAL NETWORKS

Availability-constrained planning stands for determining the location and the amount of physical resources of the network in order to maximize the availability introduced per routing and fiber and wavelength assignment (RFWA) configuration. A robust availability-constrained design has to be validated by an accurate availability estimation and realistic assumption of optical component reliability characteristics. This section introduces various approaches for the availability-constrained design and planning of optical WDM networks.

Dedicated/Shared Backup Path Protection

Tornatore et al. proposed a two-step approach to provision the static connection demands (Tornatore et al., 2006b). The first step is called Maximum Connection Availability Design (MCAD) while the second step is called Availability-Constrained Physical Resource Optimization (ACPRO). MCAD starts with an over-provisioned network, and attempts to provision the connection demands with maximum availability where the network topology is represented by a multi-layered graph with directed arcs assigned the WDM channel unavailability as the cost metrics for path selection algorithm.

Maximum Connection Availability Design with Dedicated Path Protection (MCAD-DPP)

For a connection demand, MCAD-DPP aims to find a circle with the minimum cost where each arc cost is assigned the WDM channel unavailability on its corresponding optical link. It runs two-step Dijkstra's shortest path algorithm and Bhandari's one-step cycle search algorithm (Bhandari, 1998), then, selects the result leading to the minimum unavailability for the connection demand; i.e.,

$U_c = U_w U_b$ where U_w and U_b denote the unavailability of the working and the backup path of the connection, respectively.

Maximum Connection Availability Design with Shared Backup Path Protection (MCAD-SBPP)

Circle selection algorithm cannot be applied to SBPP since backup resources have to be shared, rather than being dedicated to a single demand. Therefore, MCAD-SBPP computes the working and backup lightpaths separately. It sorts the connections that are not provisioned yet in a list, C, and for each connection k in the list, it maintains a vector where the connections sharing at least one working link with connection k are stored. In the algorithm, the costs of the links assigned to the protection path are set to zero in order to force the links to be preferred as a backup resource by the connections that will be provisioned after connection k, i.e., forcing those links to be shared.

Availability-Constrained Physical Resource Optimization (ACPRO)

The second step of the availability design scheme defined by Tornatore et al. aims to minimize the fiber deployment by keeping the connection availability in a pre-defined margin, and it is called Availability-Constrained Physical Resource Optimization (ACPRO) (Tornatore et al, 2006b). ACPRO is transparent to the protection scheme, i.e., it can run by using either of the results of MCAD-DPP or MCAD-SBPP. Starting from the least utilized fiber, ACPRO goes through each K-fiber where K-fiber stands for a fiber with K wavelengths utilized as backup or primary resources. At each step, a K-fiber is disabled temporarily upon releasing the connections passing through that fiber. If an alternate RFWA configuration can be found (based on MCAD-DPP/SBPP) for each released connection within the unavailability tolerance margin, new configurations are stored while the disabled fiber is removed.

Dynamic Adjustment of the Sharing Degree with MCAD/ACPRO

Further enhancement possibility on connection availability is considered, and a dynamic adjustment scheme is presented on top of MCAD and ACPRO defining a trade-off function based on the collected data from the network (Kantarci et al., 2008b). The trade-off function takes the resource-overbuild and average connection unavailability as the input parameters, and it runs upon provisioning every N connections. Equation 21 defines the trade-off metric for the $(n+1)^{th}$ period of run where U_n stands for the average connection unavailability at the end of the nth period of run while RO_n denotes the average resource overbuild at the end of the nth period. Resource overbuild stands for the ratio of total backup channels (Λ^b_n) to total working channels (Λ^w_n) in the network.

$$T(n+1) = RO_n, U_n = \frac{\Lambda^b_n}{\Lambda^w_n} . U_n \qquad (21)$$

According to the sharing degree update procedure (Kantarci et al., 2008b), the trade-off function uses currently calculated and previously stored values of the trade-off metric, and based on the change in the value of the metric, it determines to either increment or decrement the sharing degree. If current value is less than the previously calculated value, previous action on the sharing degree is repeated. If the previous action was incrementing the sharing degree, sharing degree is incremented, and vice versa. If current value of the trade-off metric is greater than the previously calculated value, previous action on the sharing degree is reversed, i.e., if previous action was to increment, sharing degree is decremented, and vice versa.

A similar adaptive modification approach on adjusting the availability offered by the network was also introduced in order to improve the network performance which was defined as a function of availability and connection acceptance rate (Lin et al., 2005). Here, the research presented by Kantarci et al. adopts the adaptive adjustment approach in order to find a compromise between availability and resource overbuild by determining the feasible sharing degree on the wavelength channels (Kantarci et al., 2008a, Kantarci et al., 2008b). In the first step of this scheme, working path selection is the same as the working path selection in MCAD. Backup path selection uses a different cost metric for the arcs as shown in Equation 22. The cost of an arc on the working path of the connection is assigned to infinity. A channel which is reserved as a backup path resource with a number of sharing connections less than currently calculated sharing degree is assigned a small cost value which is obtained by multiplying the channel unavailability with the number of sharing connections and dividing by a large number, K. If a channel does not have these two conditions, its unavailability is assigned as the cost.

$$C^{backup}_i = \begin{cases} \infty & c(i) = 1 \vee \exists k, i \in w_k \\ U_i.c(i) / K & c(i) > 0 \wedge c(i) < SH \wedge \exists k, i \in b_k \\ U_i & else \end{cases} \qquad (22)$$

At the end of the second step, all empty fibers are removed, and the scheme proceeds with the second step. The second step is also based on ACPRO. Thus, starting from the least utilized fiber, all K-fibers are searched and disabled temporarily one-by-one. Working path search is done by finding the maximum reliable path. Backup path search is done on a single-layered graph, and each link cost is set to the unavailability of the WDM channels on it. Upon finding a backup path, the first free channel is attempted to be assigned to the connection. If there is not a free channel, a backup channel which is shared by a number of connections less than the final feasible sharing degree calculated in the first step is searched. The search operation on each fiber can be completed in $O(W)$ time. For each probed fiber, if the reconfigured connection can have a better availability compared to its previous configuration, its new RFWA configuration is accepted.

According to the referred research, increasing the number of WDM channels per fiber (W) beyond 16 does not have a significant effect on average connection availability (Tornatore et al., 2006b, Kantarci et al., 2008b). On the other hand, dynamic sharing can lead to an enhancement in terms of availability with less channel utilization than that of the dedicated path protection (Kantarci et al., 2008b).

Shared Segment Protection

Kantarci and Mouftah evaluated performance of availability-constrained provisioning schemes under different wavelength/fiber scenarios and illustrated the results in terms of resource consumption and connection availability (Kantarci and Mouftah, 2010). Although this work is done under dynamic environment, it addresses the impact of wavelength/fiber ratio on connection availability under shared segment protection. According to the results, the number of wavelengths per fiber does not have a significant impact on average connection availability until moderate load levels which are defined to be beyond 100 Erlang under US Nationwide Backbone Topology. The details of the related work and the results are presented in the next section under the subsection entitled shared segment protection.

P-Cycles

Majority of the related work in availability design of optical networks with p-cycles are based on lin-

ear models optimizing the capacity placement and routing in p-cycles. The first two linear models are *Selectively Enhanced Availability Capacity Placement* (*SEACP*) and *p-cycle Multi-Restorability Capacity Placement (PC-MRCP)* (Cloqueur and Grover, 2005).

Selectively Enhanced Availability Capacity Placement (SEACP)

Based on the availability analysis in the previous subsection, both schemes rely on the fact that availability of a connection with on-cycle working spans is less than the availability of a connection with straddling working spans. SEACP aims to optimize the routes and the spare capacity on the protection domains so that each working path is protected against single span failure. SEACP computes n shortest paths between the source and the destination. Table 2 presents the summary of the constants, inputs and the variables used in the formulations (Cloqueur and Grover, 2005).

Integer Linear Programming (ILP) formulation of *SEACP* is shown in Table 3 by Equations 23 through Equation 30. Two service classes are defined here with respect to their protection requirements as follows: 1) *Gold* service class, and 2) *Gold-plus* service class. Since *Gold-plus* service class requires higher availability, the connections of this class are forced to be routed over the straddling spans while those of the *Gold* service class category can be routed over either the on-cycle or the straddling spans.

The objective function in Equation 23 stands for minimizing the total allocated capacity on the spans. The first constraint shown in Equation 24 guarantees that for all connections, regardless of their service class, sufficient capacity exists on the working routes in order to supply their capacity requirements. Equation 25 forces each working span to allocate appropriate working capacity for the *Gold* service class. Equation 26 is the same as Equation 25 except it stands for the allocated working capacity for the *Gold-plus* service

Table 2. Summary of constants, inputs and variables used in formulations

S:	Set of all spans in the network. In the formulation, i and k denote the indices of a failed span and a regular span, respectively.
M:	Set of capacity module types in the network
Z^m:	Capacity of the module-m
D^g:	Set of the demands of the *Gold* service class
D^{g+}:	Set of the demands of the *Gold-plus* service class
D^p:	Set of the demands in the *Platinum* class
d^r:	Demand units for the demand r
Q^r:	Set of the candidate working paths for demand r
$\zeta_k^{r,q}$:	A binary variable such that; 1 if route q for the demand r crosses the span k, and 0, otherwise.
X:	Set of the cycles where the p-cycles can be placed
π_k^x:	A binary variable such that; 1 if the cycle x crosses the span k, and 0, otherwise
ρ_k^x:	2 if k is a straddling span, 1 if k is an on-cycle span, and 0, otherwise.
$g^{r,q}$:	Working flow on route q by which demand r is served.
w_k^a:	Number of working capacity units allocated on span k for the service class a.
n^x:	Number of p-cycles on cycle x
s_k:	Number of spare capacity units allocated on span k
n_k^m:	Number of *m-type* capacity modules required on span k

class. Equation 27 provides a lower bound for the number of p-cycles to be able to protect all working traffic from both service classes. Since service class *Gold-plus* have to be routed over the straddling spans, Equation 28 provides a lower bound for the number of p-cycles where span i is a straddling span while serving as a working span for the *Gold-plus* service class. Next constraint in Equation 29 forces each span to be allocated enough spare capacity to cover all p-cycles that it crosses. The last constraint in Equation 30 guarantees that on each span, there exist enough capacity modules to supply the working flows of the two service classes as well as the allocated spare capacity.

P-Cycle Multi-Restorability Capacity Placement (PC-MRCP)

Similar to SEACP, PC-MRCP computes n shortest paths between source and destination. PC-MRCP offers two options for protection as follows: 1) Two working channels on a straddling span are protected by two on-cycle backup paths, or 2) A single working channel on a straddling span is protected by two on-cycle backup paths. PC-MRCP considers three service classes with respect to their protection requirements. *Gold* class service also exists here by having the same requirements as defined in SEACP. *Platinum* service is defined additionally which represents the connections requiring two protection paths for a single working channel. Thus, *Platinum* service class is expected

Table 3. Equations 23-30

Minimize $\displaystyle\sum_{k\in S}\sum_{m\in M} C_k . n_k^m$		(23)
Subject To		
$\displaystyle\sum_{q\in Q^r} g^{r,q} = d^r$	$\forall r \in D^g \cup D^{g+}$	(24)
$w_k^g = \displaystyle\sum_{r\in D^g}\sum_{q\in Q^r} \zeta_k^{r,q} . g^{r,q}$	$\forall\, k \in S$	(25)
$w_k^{g+} = \displaystyle\sum_{r\in D^{g+}}\sum_{q\in Q^r} \zeta_k^{r,q} . g^{r,q}$	$\forall\, k \in S$	(26)
$\displaystyle\sum_{x\in X} \rho_i^x . n^x \geq w_i^g + w_i^{g+}$	$\forall\, i \in S$	(27)
$\displaystyle\sum_{x\in X} \rho_i^x . (1-\pi_i^x) . n^x \geq w_i^{g+}$	$\forall\, i \in S$	(28)
$s_k \geq \displaystyle\sum_{x\in X} \pi_k^x . n^x$	$\forall\, k \in S$	(29)
$\displaystyle\sum_{m\in M} n_k^m . Z^m \geq w_k^g + w_k^{g+} + s_k$	$\forall\, k \in S$	(30)

to have full restoration and survivability against dual failure.

The proposed design models for SEACP and PC-MRCP show two important results regarding to p-cycle protection and availability: 1) Deployment of shorter p-cycles introduces better availability, 2) Routing on straddling links is more advantageous when compared to routing on on-cycle links (Cloqueur and Grover, 2005).

AVAILABILITY-CONSTRAINED NETWORK MANAGEMENT

This section focuses on availability-guaranteed connection provisioning and spare capacity reconfiguration. At the end of the section, presented schemes are briefly summarized and compared in terms of protection policy, failure type, granularity, optimization and restorability.

Connection Provisioning

In the following two subsections, availability-constrained connection provisioning schemes are investigated for shared backup path protection and shared segment protection. Although dedicated path protection seems to be excluded from this section, it is implicitly included in the connection provisioning schemes that are proposed for shared path protection as follows. Since connection requests arrive with pre-specified availability requirements, a demand with high availability requirement is provisioned with respect to DPP whenever SBPP is detected to violate the required availability level.

Shared Backup Path Protection

There are several availability-constrained connection provisioning schemes that are proposed for shared backup path protection. In this section, we study five of them starting with *Compute A Feasible Solution (CAFES)* which forms a basis for reliable connection provisioning (Ou et al., 2004). Then we move to the enhancements to CAFES, namely *Availability Guaranteed Differentiated Service Provisioning (AGDSP)* (Song et al., 2007), *Global Differentiated Availability-aware Connection Provisioning (G-DAP)* (Kantarci et al., 2009) and *Link-by-Link Availability-Aware Connection Provisioning (LBL-DAP)* (Kantarci et al., 2009). Finally we present *Shared Backup Path Protection with Differentiated Reliability with Node Failures (SBPP-DiR-NF)* (Pándi et al., 2006) which differs from the previous work in terms of its assumption on the failure characteristics of the network components.

Compute-A-Feasible-Solution (CAFES): A Reliable Provisioning Approach

Ou et al. proposed the first reliable connection provisioning approach where K candidate working paths are selected. Then their corresponding backup paths are selected so that the sharing of backup resources without violating the risk group constraints is maximized (Ou et al., 2004). In CAFES, while searching a backup path for a connection demand, a link can be in one of the following three states: shareable, not shareable but allowed to be utilized, neither shareable nor

allowed to be utilized. The parameters and variables used in the algorithm and the formulations are as follows.

- $C_w^c(e)$: Cost assigned to link e while searching a working path for connection c
- $C_b^c(e)$: Cost assigned to link e while searching a backup path for connection c
- λ_f^e: Number of free wavelength channels on link e
- $v_e^{e'}$: Number of working paths passing through link e' and protected by link e
- v_e^*: Number of wavelength channels to be reserved as spare resources on link e
- w_c': Set of links on the candidate working path (k^{th} minimal path) of connection c
- w_c: Set of links on the working path of connection c

Basically, link cost assignment for the backup search is done with respect to the current state of the link so that a shareable link is assigned a negligible cost, a non-shareable but utilizable link is assigned a full cost, and a link that is neither shareable nor utilizable is assigned an infinite cost (Tornatore et al., 2005). This is formulated in Equation 31 (see Table 4) where ε is a negligible value close to zero.

The algorithm starts with selecting $|K|$ minimal working path candidates with respect to various cost metrics. For each working path in the set K, a corresponding backup path is searched by using the cost assignment function in Equation 2.1. The path pair leading to the minimum cost is selected

Table 4. Equation 31

$$C_b^c(e) = \begin{cases} \infty & e \in w_c' \vee \left[\lambda_f^e = 0 \wedge (\exists e' \in w_c', v_e^{e'} = v_e^*) \right] \\ \in . C_w^c(e) & \forall e' \in w_c', v_e^{e'} < v_e^* \\ C_w^c(e) & else \end{cases} \qquad (31)$$

to provision the connection. Here, CAFES can run as an availability guaranteeing connection provisioning algorithm by setting $C_w^c(e)$ to either u_e or to $-log(a_e)$ while u_e and a_e denote the unavailability and the availability of link e, respectively.

Availability Guaranteed Differentiated Service Provisioning (AGDSP)

AGDSP is proposed to run in an environment where connections arrive with differentiated availability requirements (Song et al., 2007). The parameters used in the formulations and the algorithm are given below while the parameters presented in the definition of CAFES are also used with the same symbols here.

- B(e): Number of backup (spare) wavelengths reserved on link e.
- $\alpha(e)$: $(N(e)+1)/B(e)$ where $N(e)$ denotes the number of connections whose backup paths pass through link e.

AGDSP starts assigning a working path (w_c) to the arriving connection c with respect to the *Most Reliable Path (MRP)* fashion. MRP stands for the path with the maximum availability from source to destination. In order to compute the MRP, each link e, is assigned the cost of $-log\ (a_e)$ where a_e denotes the availability of the link. Then, the path leading to the minimum cost is selected as the MRP. Equation 32 (see Table 5) is used as the cost assignment function when searching for a backup path. If link-e is traversed by the working path of connection c or all channels in link e

are neither shareable nor utilizable to connection c, then the link is avoided to be selected by the backup path. If at least one channel on link e is shareable to connection c, then, link cost is reduced by a negligible value proportional to the average sharing index ($\alpha(e)$) and the unavailability of link e. Otherwise, if there is at least one free channel on link e, then it is assigned to a greater cost value as seen in the last line of the formulation. Thus, the links with shareable channels are preferred to the links with free channels by AGDSP.

Global Differentiated Availability Aware Provisioning (G-DAP)

Relying on the trade-off between resource-overbuild and availability, G-DAP attempts to minimize the value of the "unavailability vs. resource overbuild" trade-off by running a heuristic which aims to control the sharing degrees on the link for each availability class (Kantarci et al, 2009).

Here, the trade-off function in Equation 21 is modified to work for each service class and used as shown in Equation 33 where $T^k(n+1)$ is the trade-off value for the service class k to be used to determine its feasible sharing degree (SH_k) in the $(n+1)^{th}$ period. U_n^k denotes the average unavailability of the connections of the service class k at the end of the nth period, and RO_n stands for the average resource overbuild at the end of the nth period.

$$T^k(n+1) = RO_n, U_n^k = \frac{\Lambda_n^b}{\Lambda_n^w} \cdot U_n^k \qquad (33)$$

Table 5. Equation 32

$$C_b^c(e) = \begin{cases} \infty & e \in w_c \vee \lambda_f^e = 0 \wedge \exists e' \in w_c, v_e^{e'} = B(e) \\ \varepsilon.\alpha(e).(-\log(a_e)) & \forall e' \in w_c, v_e^{e'} < B(e) \\ 1 + \varepsilon.\alpha(e).(-\log(a_e)) & \lambda_f^e > 0 \end{cases} \qquad (32)$$

G-DAP runs a trade-off update function to modify the sharing degree for each availability class. The procedure defined in the dynamic sharing algorithm (Kantarci et al., 2008b) is modified to work for each service class separately, and it runs periodically. Here, a period is defined as the arrival of N connection requests. G-DAP is mainly based on CAFES. It selects three working path candidates with respect to the following criteria: Path-1: Path with the least cost, Path-2: Path with the least cost after removing the link with the highest availability on Path-1, Path-3: Path leading to the shortest hop count. For Path-1 and Path-2, G-DAP assigns a trade-off cost to each link as shown in Equation 34 where λ_b^e and λ_w^e denote the number of backup and the number of working channels on link e, respectively, and a_e stands for the availability of link e.

$$C_w(e) = \frac{\lambda_b^e}{\lambda_w^e} \cdot (1 - a_e) \qquad (34)$$

If one of these paths satisfies the availability requirement of the connection, G-DAP quits. Otherwise, for each working path, it searches a backup path by assigning the link cost with respect to the cost assignment function shown in Equation 35. According to the function, a link on the working path or having neither free nor spare capacity is assigned infinite cost. If a link has at least one spare channel that is being shared by a number of connections less than the feasible sharing degree for the service class (SH_k) of the corresponding connection request c, it is assigned $(1/SH_k)^{th}$ of the cost assigned while searching the working path for the request. Otherwise, backup search procedure uses the same cost metric as the working path search procedure. In the equation, $\lambda^e(\omega)$ denotes the number of connections sharing the channel ω on link e.

$$C_b^{k,c}(e) =$$
$$\begin{bmatrix} \infty & e \in w_c \vee \lambda_s^e + \lambda_f^e = 0 \\ \dfrac{1}{SH_k} \cdot C_w(e) & \lambda_s^e > 0 \wedge \exists \omega, \lambda^e(\omega) < SH_k \\ C_w(e) & else \end{bmatrix}$$
$$(35)$$

Once G-DAP finds the three working/backup path pairs for the connection request, it computes a trade-off index (T_p) for each candidate pair. As seen in Equation 36, the trade-off index leads to the product of the connection unavailability $(1 - A_c^P)$ under the path pair P and a resource consumption factor. The resource consumption factor is the sum of the number of working channels ($|w_p|$), and the backup channels ($|b_p|$) divided by the actual average sharing degree for the corresponding service class along the path. In the resource consumption factor, actual average sharing degree is summed and divided by the number of backup lightpaths along the path, the numerator of the resource consumption factor gains another factor leading to the backup path hop count which contributes as a square operation.

$$T_P = (1 - A_c^P) \cdot \left| |w_P| + \frac{|b_P|^2}{\displaystyle\sum_{e \in b_P} (\frac{1}{\lambda_s^e} \cdot \sum_{\omega \in e} \lambda^e(\omega))} \right|$$
$$(36)$$

Among the three candidate path pairs, the one leading to the lowest trade-off index is selected for the connection request c. In order to assign a wavelength to the connection, G-DAP first groups the wavelength channels on each backup link in three categories as working, spare and free channels. Then it sorts the spare channels with respect to their sharing degrees. On each link, it attempts to assign a spare wavelength starting from the one with the lowest sharing degree without violating the availability requirements of the existing connections. If the connection cannot be assigned a spare channel, a free channel on the corresponding link is assigned to the connection. However, if a free channel cannot be found, the connection request is blocked. At the end of wavelength assignment phase, if the availability of the connection is below the availability requirement of its service class, the connection request is blocked as well.

G-DAP uses the availability calculation in proposed by Mello et al. (Mello et al., 2005) however, it allows the connections that have common links in their working paths to share backup links unless this situation does not violate their availability requirements. Therefore, as seen in Equation 37, one more term is added to the availability calculation where W_{Sc} is the set of the working links of the connections sharing at least one backup resource with connection c. Thus, a connection is unavailable if a backup link fails following the failure of a working link (first term), or a working link fails following the failure of a backup or a working link of the sharing group of connection c (second term), or a working link of a sharing group connection which has a common link on its working path with connection c fails (third term).

$$A_c = 1 - \left[\sum_{\substack{i \in WP_c \\ j \in BP_c}} \pi_{ij} + \sum_{i \in WP_c, \ j \in BP_c \cup SW_c} \pi_{ji} + \sum_{j \in (WP_c \cap W_{Sc})} \pi_j \right]$$
(37)

In the link cost assignment phase, G-DAP has a complexity of $O(|E|.|W|)$ where E and W stand for the set of links and the wavelengths per link, respectively. In the wavelength assignment phase, it first groups the wavelengths, then, checks the wavelengths one-by-one, therefore, for each link it takes $O(W)$ in the worst case. Performance of G-DAP under the NSFNET topology can be seen in the next subsection with comparisons to availability aware implementation of CAFES and the scheme presented in the next subsection (also in Kantarci et al., 2009).

Link-By-Link Differentiated Availability Aware Provisioning (LBL-DAP)

LBL-DAP builds a linear trade-off function with the inputs of availability and resource consumption. An ILP formulation runs periodically, and attempts to minimize the value of this trade-off function. The outputs of the formulation are used to control the sharing degrees on the wavelength channels in each link for each availability class (SH_k^e) (Kantarci et al., 2009).

LBL-DAP re-defines the trade-off function in terms of resource gain and average availability of the service class k as seen in Equation 38. Here, resource gain for the service class (RG_k) stands for the possible idle capacity after forming the lightpaths with respect to the obtained feasible sharing degrees on the links for the corresponding class. As seen in the equation, the trade-off value to be used to determine the feasible sharing degrees is calculated as the product of the resource gain and average connection availability of the service class k. To introduce high availability to the connections, more resources have to be reserved on the backup paths, leading to lower sharing degrees and less idle resources. Thus, the increase in A_k leads to the decrease in RG_k, and vice versa.

$$T_{LBL-DAP}^k = RG_k . A_k$$
(38)

Since RG_k and A_k are both functions of the feasible sharing degrees in the links, the trade-off function cannot be used in an Integer Linear Program (ILP) formulation as is. Therefore, an approximation is taken to get the trade-off function linear as seen in Equation 39 (see Table 6) which is the objective function of the ILP model formulated to be run by LBL-DAP periodically. In the objective function, A_k and RG_k denote the average connection availability and resource gain for service class k, respectively while $\hat{A}_k$ and $\hat{RG}_k$ stand for the values of these parameters that were stored at the end of the last period.

Table 6. Equations 39-47

Objective	
$$\min \sum_k A_k \hat{R}G_k + \hat{A}_k . RG_k, \quad k = 1, 2, 3$$	(39)
Subject to	
$$A_c + \sum_{i \in w_c, j \in b_c} (\pi_{ij} + \pi_{ji}).SH_k^j = 1, \quad \forall c, k \mid c \in C^{(k)}$$	(40)
$$A_k = \frac{1}{\mid C^{(k)} \mid} \cdot \sum_{c \in C^{(k)}} A_c, \qquad \forall k$$	(41)
$$SH_k^{AVG} = \frac{1}{\mid E_b \mid} \cdot \sum_{e \in E_b} SH_k^e, \qquad \forall k$$	(42)
$$\lambda_s^e.SH_k^e \geq e_s^{(k)}, \qquad \forall e \in E, k$$	(43)
$$\mid E_b \mid .SH_k^{AVG} - H_k . \mid C^{(k)} \mid = RG_k, \qquad \forall k$$	(44)
$$SH_k^i = SH_k^j, \quad \forall i, j \in E_b \mid \frac{\rho_i}{\lambda_s^i} = \frac{\rho_j}{\lambda_s^j}$$	(45)
$$SH_{\min} \leq SH_k^e \leq SH_{\max}$$	(46)
$$RG_k \geq 0, \qquad \forall k$$	(47)

The constraints of the ILP model are formulated from Equation 40 to Equation 47. Equation 40 defines an approximation for connection availability such that, the connection is assumed to be unavailable in case of a dual failure affecting a working link and a backup link. It is further assumed that unavailability of a connection of the service class k increases with the sharing degrees on the links for its service class, therefore dual failure probability by a working and a backup link is multiplied by the sharing degree for service class k on the corresponding backup link (SH_k^j). In the equation, $C^{(k)}$ represents the set of the connections from service class k. Equation 41 stands for the average connection availability for the service class k which is calculated straightforward by dividing the total connection availability by the number of connections in the set $C^{(k)}$. Similarly, Equation 42 represents average feasible sharing degree for each service class which is denoted by SH_k^{AVG}. Equation 43 is a capacity constraint to guarantee that a backup link, e has enough spare capacity so that it can still accommodate all connections of a service class utilizing link e as a backup link although each spare channel of the link is shared with a number of connections as much as the feasible sharing degree for that class on link e. Here, $e_s^{(k)}$ represents the set of the connections of service class k utilizing a channel on link e as a backup resource. Equation 44 is the definition of resource gain for service class k where E_b and H_k stand for the set of backup links and the average hop count of the backup paths for the service class k, respectively. Equation 45 forces the feasible sharing degree on two links with the same backup utilization to be equal. Backup utilization is formulated as the ratio of the number of connections utilizing a channel on link i as a backup resource (ρ_i) to the number of total backup channels on link i (λ_s^i). Feasible sharing degree for a service class on a link is bounded below and above by a minimum and a maximum value, respectively as shown in Equation 46. Finally Equation 47 formulates the positivity constraint for RG_k.

Once the ILP formulation is built and solved at the end of a period, feasible sharing degrees in each link for each service class (SH_k^j) are obtained. As in G-DAP, LBL-DAP uses these values when searching for a backup path for a candidate working path out of the three. Three working path candidates are computed the same as in G-DAP and their corresponding backup paths are searched by assigning the link costs with respect to the function shown in Equation 48. Based on the function, for a connection of the service class k, the links on which there exists at least one spare wavelength channel that has a sharing degree which is less than the feasible sharing degree in it for class k is prioritized in backup path search process. Wavelength assignment process in LBL-DAP is the same as in G-DAP.

$$C_b^{k,c}(e) = \begin{Bmatrix} \infty & e \in w_c \vee \lambda_s^e + \lambda_f^e = 0 \\ \dfrac{1}{SH_k^e} \cdot \lambda_s^e & \lambda_s^e > 0 \wedge \exists \omega, \lambda^e(\omega) < SH_k \\ SH_{min} & else \end{Bmatrix} \quad (48)$$

In the referred research (Kantarci et al., 2009), performance evaluation of LBL-DAP, G-DAP and CAFES is presented in terms of blocking probability, resource overbuild and average connection unavailability in an heterogeneous environment where link availabilities are evenly distributed in the set {99.9%, 99.99%, 99.999%} and five service classes are assumed with the availability requirements of 98%, 99%, 99.9%, 99.99% and 99.999%. An incoming connection request can belong to one of these service classes (class-1, class-2, class-3, class-4, class-5) with the probabilities of 12%, 25%, 32%, 23% and 8% such as the most expensive class contracts less amount of subscribers. Figure 2-4 illustrate the performance evaluation under the NSFNET topology where 16 wavelengths per fiber exist and connections arrive with respect to the Poisson distribution.

Figure 2 illustrates the blocking probability under different load levels for the three schemes. According to the results, G-DAP can decrease blocking probability until the heaviest loads of the simulations due to prioritizing the links with feasible sharing degree in the backup path selection. LBL-DAP takes the advantage of determining the link-by-link feasible sharing degrees for each service class through periodically run ILP formulation, and it achieves to decrease the blocking probability at each load level.

In Figure 3, resource overbuild of LBL-DAP, G-DAP and CAFES is shown. As seen in the figure, G-DAP can achieve the same resource overbuild as CAFES as the load gets heavier. Furthermore, LBL-DAP can achieve a compromise between availability and resource consumption by the objective function in its ILP formulation therefore it can decrease resource overbuild as well as the blocking probability.

Figure 4 illustrates a further test on the performance of these schemes in terms of average connection unavailability. According to the figure, G-DAP and LBL-DAP do not lead to an increase in the average connection unavailability of CAFES.

Figure 2. Blocking probability of LBL-DAP, G-DAP and CAFES under the NSFNET topology (Kantarci et al., 2009)

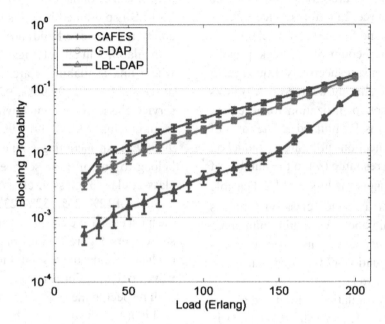

Figure 3. Resource overbuild of LBL-DAP, G-DAP and CAFES under the NSFNET topology (Kantarci et al., 2009)

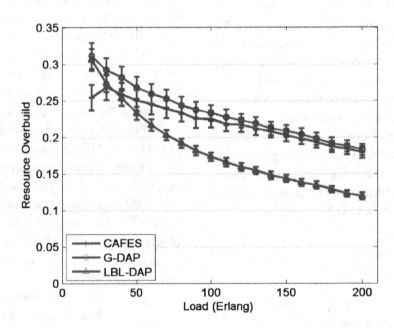

Figure 4. Average connection unavailability introduced by LBL-DAP, G-DAP and CAFES under the NSFNET topology (Kantarci et al., 2009)

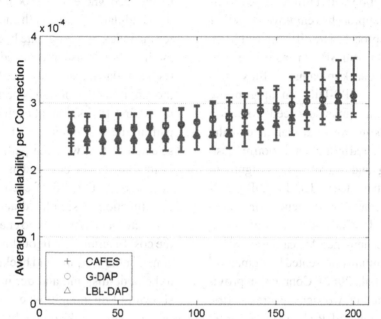

Availability-Constrained Provisioning under Consideration of Node Failures

In this chapter, before this subsection, a network failure is used to refer a link failure. Pándi et al. proposed an availability model by considering link and node failures together (Pándi et al., 2006) Here, we refer this scheme as *SBPP-DiR-NF*. A lower bound for the end-to-end availability is proposed, and it is used together with *Disjoint Path Pair Matrix (DPPM)* (Tacca et al., 2004) scheme which provides a sub-optimal solution for the Shared-Path Protection with Differentiated Reliability problem. In this work, a connection demand is defined by the following components $c = \{s, d, t_a, t_h, A_c^{req}\}$ where s and d denote source and destination, and t_a and t_h stand for the arrival and the holding times of the connection, respectively. A_c^{req} represents the availability requirement of connection c. A conservative lower bound is computed for connection availability as

seen in Equation 49 where $\kappa^{(c)}$ stands for the set of unprotected components along the working path of connection c ($W^{(c)}$) consisting of the set of nodes, n_c^w and the links, e_c^w from source to destination, and $\rho^{(c)}$ denotes the set of the protected components along the working path. In the equation, E and V represent the set of links and the set of nodes, respectively. According to the equation, a connection experiences an outage in case of unavailability of any unprotected component along the lightpath (first term in Equation 49) or in case of dual failure where a protected working path link fails concurrently with any link out of the working path (second term in Equation 49). For the sake of fast end-to-end computation, the effect of *sharing group* concept is neglected here.

$$A_c = (1 - U(\kappa^{(c)})) \cdot (1 - U(\rho^{(c)}) \cdot U((E \cup V) \setminus W^{(c)})$$

$$(49)$$

In order to guarantee connection availability, Shared Path Protection with Differentiated Reliability (SPP-DiR) approach is employed together with the DPPM approach. The difference of SPP-DiR from the traditional SBPP is as follows: In SBPP working and backup paths are fully link-disjoint however in SPP-DiR, an appropriate portion of the working links may need protection while the rest can be unprotected guaranteeing the availability level desired by the connection request. However, routing and wavelength assignment with availability guarantee under SPP-DiR is NP-Complete (Pándi et al., 2006). Hence, simulated annealing is applied to find a feasible solution for an arriving request. Simulated Annealing approach for SPP-DiR was initially presented by Fumagalli et al. (Fumagalli et al., 2002). Connection provisioning is performed in two steps where the first step is the conventional SBPP by using *Disjoint Path Pair Matrix (DPM)*, and the second step is the modification of the SBPP by setting some of the working links unprotected. Second step consists of iterative steps run by simulated annealing. At iteration step of simulated annealing, the algorithm gets the topology (G), the set of the list of unprotected links of each demand (F) and the current demand set (C). At each iteration step, the algorithm selects a random demand. If all working links of the demand are protected, then a working link is selected on random basis, temporarily set unprotected, and if the connection availability can still meet the desired level, the link is set unprotected permanently and added to the list $F^{(i)}$. If a connection has both protected and unprotected links, then either of the following two steps is taken with equal probability: 1) A random link is selected and removed from the list $F^{(i)}$, 2) A protected link is chosen randomly, and it is added to $F^{(i)}$ if setting the link unprotected does not violate the availability requirement.

Tacca et al. compute working and backup lightpaths of the connections by using the DPPM prior to the simulated annealing approach (Tacca et al., 2004). At the beginning, for each source-destination pair, a DPPM matrix of size $k_1 x k_2$ is maintained where i ($0 \leq i < k_1$) stands for the index of a candidate working path and j ($0 \leq j < k_2$) stands for the index of a candidate backup path. A slight modification is also introduced to the second step (i.e., simulated annealing step) as follows: Once the initial SBPP solution is obtained with respect to the first-fit fashion, at each iteration step, one path in the first column of DPMM is randomly selected so that working path of the connection is attempted to be changed as well as the protection path. At the end of each iteration step, the cost function of simulated annealing ($C_{SA}(c)$) is computed, and if there is no change in the value of the cost function, the algorithm quits. Pándi et al. consider the working and backup candidate paths to be node-disjoint, and define the cost function shown in Equation 50 where $S^{(c)}$ denotes the set of wavelength links on the backup lightpath that are being shared by c and other connections for protection (Pándi et al., 2006). In the related research, the two-step provisioning approach runs for each demand separately under a dynamic environment where connection requests arrive and release following Poisson and exponential distributions, respectively.

$$C_{SA}(c) = \frac{|W^{(C)}| + |\rho^{(C)}| + |S^{(C)}|}{2} + A_c - A_c^{(req)}$$

(50)

Pándi et al. tested proposed connection provisioning scheme under metropolitan, national and continental area networks by the deployment of two different OXC technologies, namely, planar tilt mirror-type Micro-Opto-Electro-Mechanical Systems (MOEMS) and the indium phosphide-based (InP) waveguide switches. It is shown that, when node failures cannot be neglected, optical node technology selection has a significant impact on the availability performance, and MOEMS-based switching technology leads to better availability guarantee when compared to Inp-based switching technology (Pándi et al., 2006).

Shared Segment Protection

This section presents an adaptive scheme called *SLA-Aware Protection Switching (SAPS)* which switches between sub-path protection and shared segment protection (Kantarci and Mouftah, 2010). Numerical results will also provide some idea on the effect of the physical infrastructure of the network (such as number of wavelengths per fiber) on the performance of availability-guaranteeing shared segment protection.

SLA-Aware Protection Switching (SAPS)

SAPS categorizes the connections with respect to availability requirements as priority-class (Category-P) and economy-class (Category-E) connections (Kantarci and Mouftah, 2010). Priority-class connections are protected by shared segment protection while economy-class connections are protected by shared sub-path or shared backup path protection. For those require shared segment protection, the provisioning scheme applies a conventional segment selection algorithm, namely Generalized Segment Protection (GSP) (Ou et al., 2005), and link costs are assigned considering the availability requirements and feasible sharing degrees for the corresponding classes in order to minimize the availability vs. resource-overbuild trade-off defined earlier.

According to the segment protection, in a directed graph, a segment intersects the working path of a connection at two nodes, i.e., start and end nodes of the segment. Hence, an appropriate arrangement of costs of the links intersecting the working path can enhance the availability offered by the segment protection. In a previous work, it was shown that the more the number of segments the better the availability (Kantarci et al., 2008c). Moreover, since the baseline algorithm GSP finds

overlapping segments, a working link where two protection segments overlap has two alternate protection routes increasing the restorability of the corresponding portion of the working path.

SAPS is based on the following assumption: Category-E connections do not need high availability as Category-P connections do therefore those connections can be protected by sub-path protection while segment protection can be used for the connections requiring high availability, i.e., Category-P. For the service classes forming Category-P and Category E, feasible sharing degrees are determined by running the trade-off function in Equation 33 and modifying the sharing degrees by the procedure defined in the previous work (Kantarci et al., 2008a, Kantarci et al., 2008b) for each service class separately.

Upon the arrival of a Category-E connection, SAPS switches to the sub-path protection mode, searches three candidate working paths and their corresponding sub-path protection segments. While searching the backup protection segments, feasible sharing degree calculated for the service class of the incoming request is used as a prioritization mechanism between the candidate backup links. Thus, Equation 35 is used to assign link costs on the possible backup links.

Provisioning a Category-P connection with shared segment protection is as follows: Three working paths are selected based on the criteria presented in the previous section. Here, least cost path leads to the MRP. If MRP is able to meet the availability requirement of the connection request, SAPS quits and provisions the connection as unprotected. Otherwise, for each working path, a corresponding segment protection is searched. Prior to link cost assignment, GSP modifies the links intersecting the working path as follows: If the working path traverses the links between the following nodes $s, n_1, n_2, n_3 \ldots n_i, n_{i+1}, \ldots d$, a link between n_j and n_{i+1} is moved temporarily to traverse the nodes n_j and n_i where the working path does not traverse the node n_j. Another modification is reversing the direction of the links on the work-

Table 7. Equation 51

$$C_b^{k,c}(e) = \begin{cases} \infty & e \in \underline{w_c} \vee \lambda_w^e = W \\ -C_w(e) & e \in w_c \\ \varepsilon.(-SH_k).C_w(e) & \lambda_s^e > 0 \wedge e \cap w_c \neq \varphi \wedge \exists \omega \mid \lambda^e(\omega) < SH_k \\ \varepsilon.C_w(e) & \lambda_s^e > 0 \\ C_w(e) & else \end{cases}$$
(51)

ing path as seen in the second line of Equation 51 (see Table 7). For those specifying the following three criteria, link cost is assigned to a low and a negative value to force the link to be selected: 1) There is at least one spare channel in the link, 2) the corresponding link intersects the working path at a node, 3) there exists at least one spare channel whose sharing degree is less than the

calculated feasible sharing degree for the service class of the connection c.

Once SAPS finds the segment protections for the corresponding candidate working paths, it attempts to assign wavelengths starting from the spare wavelength channel with the least sharing degree. Availability calculation is done based on

Figure 5. The mesh topology used to evaluate shared segment protection-based schemes

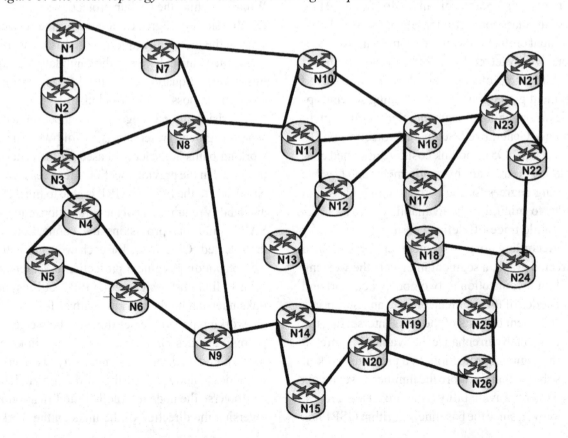

Equation 10. Among the three solutions, the one leading to the highest availability is selected.

Performance of SAPS is evaluated in terms of blocking probability, resource overbuild and average connection availability, and it is compared to *Availability-Constrained Generalized Segment Protection (AC-GSP)* which is availability-guaranteed adaptation of the GSP algorithm (Kantarci and Mouftah, 2010). The mesh topology in Figure 5 is used to evaluate the performance of SAPS. Link availability values are distributed in {99.9%, 99.99%} with respect to the fiber lengths. Fiber lengths are taken from the US nationwide topology used in a related work (Song et al., 2007). In the performance evaluation, three service classes are assumed where the connections with availability requirements of 99.99% and 99.999% form Category-P and those with the availability requirement of 99.9% form Category-E.

As seen in Figure 6, under resource rich environment where 32 wavelengths exist on each fiber, SAPS can decrease blocking probability significantly. On the other hand, under resource-scarce environment where there are 16 wavelengths per fiber, decrease in blocking probability with SAPS can be experienced until heavy loads since under heavy loads SAPS starts to suffer from resource limitation.

In Figure 7, resource overbuild of SAPS and AC-GSP is illustrated. Here, it is seen that SAPS leads to a slight increase in resource overbuild due to the attempt of protecting Category-P connections with more number of segments.

Spare Capacity Re-Configuration

Partial restorability of the failure-impacted connections in Generalized Multi-Protocol Label Switching (GMPLS) networks under dual failure assumption is a different approach for availability-constrained connection provisioning. Guo et al. presented the partial restorability concept as well as *Linear Programming (LP)* models to re-provision the spare capacity in the links under different scenarios (Guo et al., 2007, Guo et al., 2010).

Figure 6. Blocking probability of SAPS and Availability Guaranteed GSP (Kantarci and Mouftah, 2010)

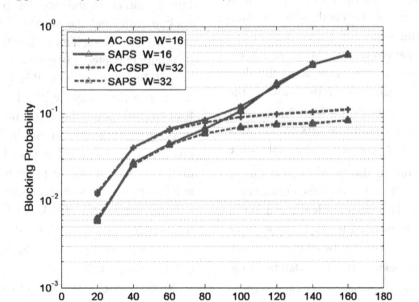

Figure 7. Resource Overbuild of SAPS and Availability Guaranteed GSP (Kantarci and Mouftah, 2010)

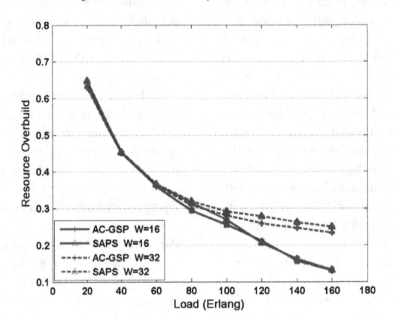

Let us assume R to be a set of failure patterns in the network. Partial restoration can be explained by two scenarios: 1) Working path of a connection is protected by at least one backup path where the connection is restorable under a certain set of failure patterns, and 2) The spare capacity allocated along the backup lightpath is some fraction (θ_c) of the bandwidth allocated along the working path.

Four failure pattern sets are considered as follows. A failure pattern r_{mn} represents the dual failure situation where link n fails following the failure of link m. Similarly, a failure pattern r_m stands for the single failure of link m. The first failure pattern set R^c_{wp} consists of the dual failure patterns where one of the failed links is on the working path and the other is on the backup path of connection c. This situation leads to a 100% availability impairment of the connection since the connection does not have the chance to switch its traffic on the backup path. $R^c_{w\backslash p}$ stands for the set of failure patterns consisting of a single failure on the working path or a dual failure where the

first failed link is on the working path and the other one is not on the backup path of connection c. A failure pattern in this set leads to an availability impairment of $(1-\theta_c)$ due to restoring θ_c of the working traffic of the connection. R^c_{w2p} denotes the set of dual failure patterns where the first failed link is neither on the working path nor on the backup path, and the second failure is on the working path. Here, if the connection c can restore its failure impacted traffic on the backup path, its availability impairment will be $(1-\theta_c)$. However, the connection may switch its traffic on the backup path unless the first failed link is on the working path of a connection which shares at least one backup channel with connection c. The probability of such a failure is neglected here, and availability impairment of connection c is assumed to be approximately $(1-\theta_c)$. The final failure pattern set, R^c_w denotes the list of failure patterns which do not affect the availability of connection c (Guo et al., 2007).

Spare Capacity Reconfiguration runs as follows: Upon the arrival of each connection request, the request is routed with respect to the SBPP by using the Successive Survivable Routing algorithm the details of which are presented by Liu et al. (Liu et al., 2005). If the calculated availability is less than the desired level, the connection is dropped. If at some point during the system run, the network traffic distribution demonstrates a significant difference, this is named as a network event, and a network event calls a new *Spare Capacity Allocation (SCA)* module if the last SCA process is not running. Upon solving the LP formulation for SCA, spare capacities on the links are updated, and the system keeps waiting for the new connection requests. In this section, two SCA models are presented in the following two subsections.

Failure-Independent Spare Capacity Allocation (FID-SCA)

The objective of failure independent spare capacity allocation is minimizing the total spare capacity to be allocated to satisfy the end-to-end avail-

Table 8. Equations 52-57

B_c:	Bandwidth allocated for working traffic of connection c	
π_r:	Probability of the failure pattern r to occur	
$A_c^{(req)}$:	Availability requirement of connection c	
V_j:	Spare capacity allocated on link j.	
S_{j,r_m}:	Spare capacity allocated on link j when the single error failure occurs on link m	
$S_{j,r}$:	Spare capacity allocated on link j in case of the failure pattern r	
θ_c:	Protection level of connection c	
A_c:	Availability of connection c	
$S_{j,r}$:	Spare capacity allocated on link j under the failure pattern r	
$\theta_{c,r}$:	Protection level of connection c under the failure pattern r	
Objective $\min \sum\limits_{j \in E} V_j$		(52)
Subject to		
$V_j = \max S_{j,r_m} \quad \forall j \in E$		(53)
$S_{j,r_m} = \sum\limits_{\forall c\,s.t.\,m \in w_c, j \in b_c} B_c \cdot \theta_c \quad \forall j \in E, r_m \in R, m \neq j$		(54)
$A_c = 1 - \sum\limits_{r \in R^c_{wp}} \pi_r - \sum\limits_{r \in (R^c_{w2p} \cup R^c_{w \backslash p})} (1 - \theta_c) \cdot \pi_r \quad \forall c \in C$		(55)
$A_c \geq A_c^{(req)} \quad \forall c \in C$		(56)
$0 \leq \theta_c \leq 1 \quad \forall c \in C$		(57)

ability of the connections as shown in Equation 52 where V_j is the spare capacity allocated on link j on a topology with $|E|$ edges (see Table 8). Spare capacity allocation on a link is the maximum of the spare capacity allocation values on the link under the failure pattern r_m leading to an availability impairment for the active connections as seen in Equation 53. Spare capacity allocated on link j in case of a single failure scenario is shown in Equation 54. According to the constraint, spare capacity allocated on a link j has to be the total of the restored capacities of all connections that have link j as a backup resource. Connection availability is formulated in Equation 55 where availability impairment due to the failure pattern set R^c_{w2p} is simply assumed to be $(1-\theta_c)$. The constraint in Equation 56 guarantees that availability requirement of each connection is not violated by spare capacity reconfiguration on the links. Finally, the last constraint in Equation 57 assures that the protection level, θ_c has a rational value in order to represent a fraction.

Failure-Dependent Spare Capacity Allocation (FD-SCA)

Apart from the previous approach, Failure-Dependent Spare Capacity Allocation aims to obtain protection level per failure pattern r for each connection ($\theta_{c,r}$). Furthermore, spare capacity allocation values on the links are obtained for each failure pattern. The LP formulation of FD-SCA is similar to that of FID-SCA as it can be seen in Equation 58-Equation 63 (see Table 9). As seen in the formulation, spare capacity allocation on each link is formulated for each restorable pattern. On the other hand, protection level is determined for each restorable pattern as seen in Equation 60 while connection availability is calculated considering each failure pattern as shown in Equation 61.

As stated by Guo et al., using the protection level as a key parameter enhances the performance of the GMPLS network in terms of capacity utilization when compared to conventional SBPP with 100% restorability (Guo et al., 2010). Another advantage of employing the space capacity

Table 9. Equations 58-63

Objective $\min \sum_{j\in E} V_j$	(58)
Subject to	
$V_j = \max S_{j,r} \quad \forall j \in E$	(59)
$S_{j,r} = \sum_{\forall c s.t. m\in w_c, j\in b_c, r\in(R^c_{w2p}\cup R^c_{w\backslash p})} B_c \cdot \theta_{c,r} \quad \forall j \in E, r \in R, m \neq j$	(60)
$A_c = 1 - \sum_{r\in R^c_{wp}} \pi_r - \sum_{r\in(R^c_{w2p}\cup R^c_{w\backslash p})} (1-\theta_{c,r}) \cdot \pi_r \quad \forall c \in C$	(61)
$A_c \geq A^{(req)}_c \quad \forall c \in C$	(62)
$0 \leq \theta_{c,r} \leq 1 \quad \forall c \in C$	(63)

re-configuration is shown to be the satisfaction of the availability requirements by assuring a protection level less than 100% while selecting appropriate spare capacities on the links.

Table 10 gives a brief and a comparative summary of the availability-constrained connection provisioning schemes presented in this chapter. As seen in the table, the provisioning schemes are compared with respect to their protection policy, failure assumption, granularity, optimization requirement, and restorability guaranteed by each scheme.

SUMMARY AND DISCUSSION

As seen in the previous sections, robust network design and management appears as a key issue in optical networking. This chapter has presented three main sections after a detailed Introduction providing a general view on survivability policies

and the availability concept. In the first section, we have presented various availability analysis approaches for the optical network survivability schemes, namely dedicated path protection (DPP), shared backup path protection (SBPP), shared segment protection and p-cycles. In the second section, based on each availability analysis scheme for each survivability policy, we have presented planning models for optical networks, and discussed their advantages as well as the performances. In the third section, we have focused on availability-constrained connection provisioning in optical networks under the survivability schemes whose end-to-end connection availability calculations were presented in the chapter. In the last subsection of the third section, we have presented and discussed connection provisioning with the spare capacity re-configuration in GMPLS networks, and explained the schemes existing in the literature. In most of the surveyed schemes, we have pointed the trade-off between resource

Table 10. Summary of the availability-constrained provisioning schemes studied in the chapter

Provisioning Scheme	Protection Policy	Failure type	Granularity	Optimization	Restorability
CAFES (Ou et al., 2004)	SBPP, DPP	Link failures	Wavelength	NO	100%
AGDSP (Song et al., 2007)	SBPP, DPP	Link failures	Wavelength	NO	100%
G-DAP (Kantarci et al., 2009)	SBPP, DPP	Link failures	Wavelength	NO	100%
LBL-DAP (Kantarci et al., 2009)	SBPP, DPP	Link failures	Wavelength	ILP-based	100%
SBPP-DiR-NF (Pándi et al., 2006)	SBPP-DiR	Link + Node Failures	Wavelength	NO	100%
SAPS (Kantarci and Mouftah, 2010)	Shared Segment + Sub-path Protection	Link failures	Wavelength	NO	100%
FD-SCA (Guo et al., 2007)	SBPP, DPP	Link failures	Multi-granular	LP-based	Partial
FID-SCA (Guo et al., 2007)	SBPP, DPP	Link failures	Multi-granular	LP-based	Partial
SEACP (Cloqueur and Grover, 2005)	P-Cycles	Span failures	Wavelength	ILP-based	100%
PC-MRCP (Cloqueur and Grover, 2005)	P-Cycles	Span failures	Wavelength	ILP-based	100%

overbuild and availability, blocking probability and availability, and network capacity and availability. Therefore, availability-constrained planning and provisioning solutions have to consider these aspects in order to minimize the trade-off.

Since scalability is one of the important design parameters, incorporating availability design and management in multi-granular optical networking and the employment of appropriate architectures supporting availability-constrained design and provisioning are still the open issues for the researchers in this field. Furthermore, consideration of Quality-of-Service and availability-guarantee together in connection provisioning stands as a possible future extension. Availability-constrained designs of storage area networks and optical grids seem as other open issues to complement the studies in this area.

REFERENCES

Arci, D., Pattavina, A., Petecchi, D., & Tornatore, M. (2003). Availability models for protection techniques in WDM networks. *IEEE Workshop on Design of Reliable Communication Networks (DRCN)*, (pp. 159-166).

Babbit, J., & Best, R. (2006). *Maintaining availability in an optical backbone network*. Paper presented at the National Fiber Optic Engineers Conference / Optical Fiber Communication Conference (NOEC/OFC), Anaheim, CA, USA.

Bhandari, B. (1998). *Survivable networks: Algorithms for diverse routing*. Kluwer Academic Publishers.

Cloqueur, M., & Grover, W. (2005). Availability analysis and enhanced availability design in p-cycle-based networks. *Springer Science. Photonic Network Communications, 10*(1), 55–71. doi:10.1007/s11107-005-1695-x

Fumagalli, A., Tacca, M., Unghvary, F., & Farago, A. (2002). Shared path protection with differentiated reliability. *IEEE International Conference on Communications (ICC)*, (pp. 2157-2161).

Guo, Q., Ho, P.-H., Yu, H., & Mouftah, H. T. (2007). Availability-constrained shared backup path protection (SBPP) for GMPLSBased Spare Capacity Reprovisioning. *IEEE International Conference on Communications (ICC)*, (pp. 2186-2191).

Guo, Q., Ho, P.-H., Yu, H., Tapolcai, J., & Mouftah, H. T. (2010). Spare capacity re-provisioning for high availability shared backup path protection connections. *Elsevier Computer Communications, 33*(5), 603–611.

Kantarci, B., & Mouftah, H. T. (2010). SLA-aware protection switching in optical WDM networks. *25th Queen's Biennial Symposium on Communications,* (pp. 230-233).

Kantarci, B., Mouftah, H. T., & Oktug, S. (2008a). Connection provisioning with feasible shareability determination for availability-aware design of optical networks. *International Conference on Transparent Optical Networks (ICTON), vol. 3* (pp. 19-22).

Kantarci, B., Mouftah, H. T., & Oktug, S. (2008b). Arranging shareability dynamically for the availability-constrained design of optical transport networks. *IEEE Symposium on Computers and Communications (ISCC),* (pp. 68-73).

Kantarci, B., Mouftah, H. T., & Oktug, S. (2008c). *Availability analysis and connection provisioning in overlapping shared segment protection for optical networks.* Paper presented at International Symposium on Computer and Information Sciences (ISCIS), Istanbul, Turkey.

Kantarci, B., Mouftah, H. T., & Oktug, S. (2009). Adaptive schemes for differentiated availability-aware connection provisioning in optical transport networks. *IEEE / OSA. Journal of Lightwave Technology, 27*(20), 4595–4602. doi:10.1109/JLT.2009.2025246

Krishna, G. P., Pradeep, M. J., & Murthy, C. S. R. (2000). A segmented backup scheme for dependable real-time communication in multi-hop networks. *IEEE International Workshop on Parallel and Distributed Real-Time Systems,* (pp. 678-684).

Lin, R., Wang, L., Li, L., & Guo, L. (2005). A new network availability algorithm for WDM optical networks. *International Conference on Computer and Information Technology,* vol. 1 (pp. 480-484).

Liu, Y., Tipper, D., & Siripongwutikorn, P. (2005). Approximating optimal spare capacity allocation by successive survivable routing. *IEEE/ACM Transactions on Networking, 13*(11), 198–211. doi:10.1109/TNET.2004.842220

Mello, D. A., Schupke, D., & Waldman, H. (2005). A matrix-based analytical approach to connection unavailability estimation in shared backup path protection. *IEEE Communications Letters, 9*(9), 844–846. doi:10.1109/LCOMM.2005.1506722

Mouftah, H. T., & Ho, P.-H. (2003). *Optical networks: Architecture and survivability.* Kluwer Academic Publishers.

Mukherjee, B. (2006). *Optical WDM networks.* Springer.

Mukherjee, D. S., Assi, C., & Agarwal, A. (2005). An alternative approach for enhanced availability analysis and design methods in p-cycle-based networks. *IEEE Journal on Selected Areas in Communications, 24*(12), 23–34. doi:10.1109/JSAC.2006.258220

Ou, C., Rai, S., & Mukherjee, B. (2005). Extension of segment protection for bandwidth efficiency and differentiated quality of protection in optical/MPLS networks. *Elsevier Optical Switching and Networking, 1*(1), 19–33. doi:10.1016/j.osn.2004.10.002

Ou, C., Zhang, J., Zhang, H., Sahasrabuddhe, L. H., & Mukherjee, B. (2004). New and improved approaches for shared-path protection in WDM mesh networks. *IEEE/OSA. Journal of Lightwave Technology, 22*(5), 1223–1232. doi:10.1109/JLT.2004.825346

Pándi, Z., Tacca, M., Fumagalli, A., & Wosinska, L. (2006). Dynamic provisioning of availability-constrained optical circuits in the presence of optical node failures. *IEEE/OSA. Journal of Lightwave Technology, 24*(9), 3268–3279. doi:10.1109/JLT.2006.879505

Ramamurthy, S., Sahasrabuddhe, L., & Mukherjee, B. (2003). Survivable WDM mesh networks. *OSA Journal of Lightwave Technology, 24*(4), 870–883. doi:10.1109/JLT.2002.806338

Song, L., Zhang, J., & Mukherjee, B. (2007). Dynamic provisioning with availability guarantee for differentiated services in survivable mesh networks. *IEEE Journal on Selected Areas in Communications, 25*(3), 35–43. doi:10.1109/TWC.2007.024505

Stamatelakis, D., & Grover, W. D. (2000). Theoretical underpinnings for the efficiency of restorable networks using preconfigured cycles (p-cycles). *IEEE Transactions on Communications, 48*(8), 1262–1265. doi:10.1109/26.864163

Tacca, M., Monti, P., & Fumagalli, A. (2004). The disjoint path-pair matrix approach for online routing in reliable WDM networks. *IEEE International Conference on Communications (ICC),* (pp. 1187-1191).

Tornatore, M., Maier, G., & Pattavina, A. (2006a). Capacity versus availability trade-offs for availability-based routing. *Journal of Optical Networking, 5*, 858–869. doi:10.1364/JON.5.000858

Tornatore, M., Maier, G., & Pattavina, A. (2006b). Availability design of optical transport networks. *IEEE Journal on Selected Areas in Communications, 24*(8), 1520–1532.

Tornatore, M., Ou, C. S., Zhang, J., Pattavina, A., & Mukherjee, B. (2005). An efficient shared-path-protection strategy based on connection-holding-time awareness. *IEEE/OSA. Journal of Lightwave Technology, 23*(10), 3138–3146. doi:10.1109/JLT.2005.856174

Zhang, J., Zhu, K., Zang, H., Matloff, N. S., & Mukherjee, B. (2007). Availability- aware provisioning strategies for differentiated protection services in wavelength-convertible WDM mesh networks. *IEEE/ACM Transactions on Networking, 15*(5), 1177–1190. doi:10.1109/TNET.2007.896232

ADDITIONAL READING

AlSukhni, E. M., & Mouftah, H. T. (2010). A Framework for Distributed Provisioning Availability-Guaranteed Least-Cost Lightpaths in WDM Mesh Networks, *IEEE Symposium on Computers and Communications,* (pp. 180-183).

Cholda, P., Tapolcai, J., Cinkler, T., Wajda, K., & Jajszczyk, A. (2009). Quality of resilience as a network reliability characterization tool. *IEEE Network, 23*(2), 11–19. doi:10.1109/MNET.2009.4804331

Ho, P.-H., Tapolcai, J., & Haque, A. (2008). Spare Capacity Reprovisioning for Shared Backup Path Protection in Dynamic Generalized Multi-Protocol Label Switched Networks. *IEEE Transactions on Reliability, 57*(4), 551–563. doi:10.1109/TR.2008.2006037

Huang, S., Martel, C. & Mukherjee, B. (2010). Adaptive Reliable Multipath Provisioning in Survivable WDM Mesh Networks. *IEEE/OSA Journal of Optical Communications and Networking, 2(6),* 368-380.

Kiaei, M. S., Ranjbar, A., Jaumard, B., & Assi, C. (2009). An Improved Analysis for Availability-Aware Service Provisioning in p-Cycle-Based Mesh Networks. *IEEE/OSA. Journal of Lightwave Technology, 27*(20), 4424–4434. doi:10.1109/JLT.2009.2024167

Naser, H., & Mouftah, H. T. (2004a). Enhanced pool sharing: a constraint-based routing algorithm for shared mesh restoration networks. *OSA Journal of Optical Networking, 3*(5), 303–323. doi:10.1364/JON.3.000303

Naser, H., & Mouftah, H. T. (2004b). A Multilayer Differentiated Protection Services Architecture. *IEEE Journal on Selected Areas in Communications, 22*(8), 1539–1547. doi:10.1109/JSAC.2004.830465

Staessens, D., Colle, D., Lievens, U., Pickavet, M., Demeester, P., & Colitti, W. (2008). Enabling high availability over multiple optical networks. *IEEE Communications Magazine, 46*(6), 120–126. doi:10.1109/MCOM.2008.4539475

Tapolcai, J., Ho, P.-H., Verchere, D., Cinkler, T., & Haque, A. (2008). A New Implementation of Shared Segment Protection Method for Guaranteed Recovery Time. *IEEE Transactions on Reliability, 57*(2), 272–282. doi:10.1109/TR.2008.923480

Chapter 10

New Dimensions for Survivable Service Provisioning in Optical Backbone and Access Networks

Paolo Monti
Royal Institute of Technology, Sweden

Lena Wosinska
Royal Institute of Technology, Sweden

Cicek Cavdar
Royal Institute of Technology, Sweden

Andrea Fumagalli
The University of Texas at Dallas, USA

Jiajia Chen
Royal Institute of Technology, Sweden

ABSTRACT

Originally, networks were engineered to provide only one type of service, i.e. either voice or data, so only one level of resiliency was requested. This trend has changed, and today's approach in service provisioning is quite different. A Service Level Agreement (SLA) stipulated between users and service providers (or network operators) regulates a series of specific requirements, e.g., connection set-up times and connection availability that has to be met in order to avoid monetary fines. In recent years this has caused a paradigm shift on how to provision these services. From a "one-solution-fits-all" scenario, we witness now a more diversified set of approaches where trade-offs among different network parameters (e.g., level of protection vs. cost and/or level of protection vs. blocking probability) play an important role.

This chapter aims at presenting a series of network resilient methods that are specifically tailored for a dynamic provisioning with such differentiated requirements. Both optical backbone and access networks are considered. In the chapter a number of provisioning scenarios - each one focusing on a specific Quality of Service (QoS) parameter - are considered. First the effect of delay tolerance, defined as the amount of time a connection request can wait before being set up, on blocking probability is investigated when Shared Path Protection is required. Then the problem of how to assign "just-enough" resources to meet each connection availability requirement is described, and a possible solution via a Shared Path Protection Scheme with Differentiated Reliability is presented. Finally a possible trade off between deployment cost and level of reliability performance in Passive Optical Networks (PONs) is investigated.

DOI: 10.4018/978-1-61350-426-0.ch010

The presented results highlight the importance of carefully considering each connection's QoS parameters while devising a resilient provisioning strategy. By doing so the benefits in terms of cost saving and blocking probability improvement becomes relevant, allowing network operators and service providers to maintain satisfied customers at reasonable capital and operational expenditure levels.

INTRODUCTION

Wavelength Division Multiplexing (WDM) enables optical networks to transport hundreds of wavelength channels through a single optical fiber, with a capacity that currently varies from 10 Gbit/s to 40 Gbit/s for each channel, and that is expected to reach 100 Gbit/s in the near future (Ray, 2010). Morcover, one single fiber cable consists of a large number of optical fibers, and an accidental single cable cut may lead to the interruption of a very large number of optical connections with the likely interruption of an enormous amount of services. For this reason it is extremely important to provide efficient survivability mechanisms in optical networks. With this regard a lot of work can be found in the literature that addresses the resiliency problem in both optical core (Mukherjee, 2006) and access networks (Chen, Mas Machuca, Wosinska & Jaeger, 2010; Yeh & Chi, 2007; Chan, Chan, Chen & Tong, 2003).

The term *core* refers to the backbone infrastructure of a network that usually interconnects large metropolitan areas, and may span across nations and/or continents (Figure 1). Usually interconnected in a mesh pattern the backbone nodes aggregate and transmit traffic from and to the peripheral areas of the network (i.e., the metro/access segment). The term *access* refers to the so called *last mile* or segment of a network where central offices (COs) and remote nodes (RNs) provide connectivity, using tree topologies, between the end users and the rest of the network infrastructure. Depending on the reach of the access segment core and access may or may not be interconnected via a metro infrastructure. With short reach access solutions (i.e., the CO is placed a few tens of kilometers from the end

users) the traffic from the end users is aggregate at the metro level before being sent to the core. With long reach access solutions (i.e., the CO is more than one hundred kilometers from the end user) the traffic goes directly from the access into the core segment.

Most of the attention was earlier devoted to reliability methods that were able to provide resiliency to all optical channels, or *lightpaths*, indistinctly. This was motivated essentially by the fact that in the absence of survivability mechanisms the first priority was to develop solutions that provide uninterrupted services in the case of network link or node failures. Another reason for this flat architecture was the nature of the services carried over the lightpaths. Historically, networks were engineered to provide only one type of service, i.e. either voice or data, so only one level of resiliency was needed.

This trend has changed now and today's approach in providing network connections is quite different. Network operators and service providers integrate an increasing number of services with different resilience requirements in the same network. These services are different in nature, e.g., real time versus background data transfer, as well as in their scope, e.g., critical financial transactions versus recreational activities (Cholda, Mykkeltveit, Helvik, Wittner & Jajszczyk, 2007). Examples of this differentiated scenario are optical networks with dynamic connection provisioning where specifics services, e.g., Video-on-Demand (VoD) requests to corporation and backup virtual private networks (VPNs), may require bandwidth capacity during specific time intervals with flexible or strict connection set-up times and differentiated reliability requirements. Another example is bandwidth on demand (BoD) services

Figure 1. Telecom network hierarchy example: core, metro and access segment

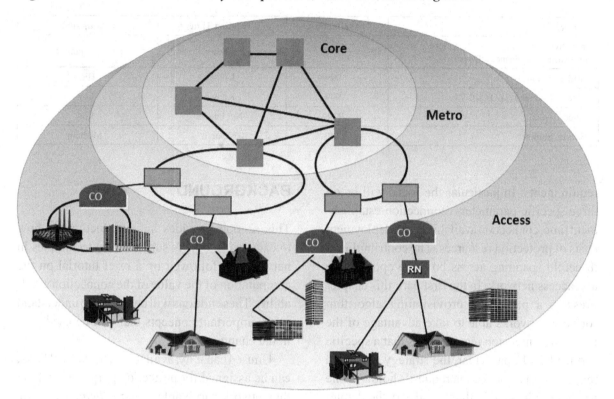

with differentiated reliability requirements that enables the customer to order and receive the desired connectivity within hours or minutes of the request. Such services are already provisioned by a large US carrier in the form of Real-Time BoD and Scheduled BoD (Liu & Chen, 2007). Along the same concept, another large US carrier's On-time Provisioning service guideline specifies a deadline for each service order and gives the customer the right to withdraw the request if the carrier cannot set-up the required service within the specified amount of time (ATT, 2009). Table 1 presents a few examples of mapping of specific services with their respective requirements. These requirements are often specified as part of the Service Level Agreement (SLA) between the client and the operator and it is clear that given such a plethora of requirements a "one-solution-fits-all" approach for network resilience is not efficient.

This strong focus on strict end-to-end requirements for the provisioned services has also triggered a growing interest on how and up to which level resiliency is provided in optical access networks. It is known that fiber access networks without any protection are characterized by poor reliability (Tran, Chae & Tucker, 2005; Wosinska & Chen, 2008). Therefore, some type of protection should be provided to satisfy the resiliency requirements of the network services. Obviously, adding redundant components and systems will improve network reliability. However, in the access the network costs are shared by a limited number of users. Therefore, both system deployment cost and network management cost should be minimized.

The objective of this chapter is to present a series of network resilience methods that are specifically tailored to offering a dynamic provisioning scenario with such differentiated

Table 1. Service differentiation for sample services

Services	Availability	Holding Time	Set-up time
Sensitive Services (Telemedicine, Financial Trans.)	.99999	Known	Medium
Grid computing	.999	Known	High
Video on demand (VoD), IPTV	.999	Known	Low
Voice Trunks	.9999	Not known	Low
Backup Storage	.99	Flexible	Medium

requirements. In particular the focus will be on three specific parameters: connection establishment time, connection availability and deployment costs of protection resources. The contribution is threefold spanning across both the optical core and access network. In the first part, this chapter presents a protection provisioning algorithm for core networks able to take advantage of the temporal dimension requirements that a specific service has. In particular, the strategy presented makes use of a connection request's holding time and delay tolerance. In the second part, the chapter still focuses on core networks and will investigate a protection strategy that assigns spare resources to connections based on their specific survivability level requirements. More specifically the presented protection algorithm explores the tradeoff between the service availability level and how efficiently network resources are used, measured in terms of connection blocking probability. Finally, the chapter addresses the problem of resiliency in optical access networks with particular focus on deployment cost. A series of protection schemes are presented and compared in terms of level of protection provided versus cost per user.

The chapter is organized as follows. First some background information about survivability techniques and connection availability computation is provided. Then each contribution is presented in separate subchapters. Finally some concluding remarks are provided.

BACKGROUND

This section provides an introduction on how to categorize today's survivability techniques in networking, followed by a brief tutorial on the computation of the value of the connection availability. These notions will be helpful to understand a few important concepts that will be used later in the chapter.

Unused capacity, available in the optical links, can be assigned for protection purposes, making the network survivable, i.e., *resilient*. There are two ways of protecting traffic: *path protection* and *link (or segment) protection*. In path protection schemes the traffic disrupted by a fault is rerouted along a different path between the source and destination nodes. Therefore, each node pair requires an additional link or node disjoint path depending on the type of failure the connection needs to be protected from. In link protection schemes the traffic is rerouted around the failed link only.

Network survivability schemes can be classified in two groups, i.e., *protection* and *restoration* (Mukherjee, 2006). Protection refers to pre-provisioned backup resources allocated for failure recovery. Protection schemes are typically fast and they can offer recovery time below 50 ms. They can offer various protection levels ranging from 1+1 and 1:1 (i.e., *dedicated* protection), to M:N (i.e., *shared* protection). In 1+1 protection, the traffic is transmitted simultaneously on two distinct paths from the source to the destination.

The destination node selects from which path it receives the incoming traffic. In case of a fiber or node failure, the destination node has to switch over to the other path to avoid interruption in data reception. In 1:1 protection, there are also two separate paths between the end nodes. In this case, the transmission takes place only on one path, the working path. In case of a fiber cut, both nodes have to switch to the other path, the protection path. With the M:N protection N working paths share M protection paths. Only single fiber or node failures can be protected while, in the event of multiple failures, survivability is not guaranteed. Protection techniques, however, can be quite expensive due to the need for extra network equipment. Restoration on the other hand refers to the rerouting of traffic around the point of failure if there are resources available. The alternative route is discovered or reserved on the fly. For this reason restoration usually takes longer time than protection. If sufficient network resources are not available upon failure, restoration is not possible.

A parameter often used to define and differentiate the level of protection is the connection asymptotic *availability*, which is referred to as the probability that a connection is up at an arbitrary point in time. The computation of this parameter is not always simple, i.e., as the complexity of a network increases analytical availability calculation becomes more and more time consuming. It is often very hard or even impossible to include all parameters from a real network in the analytical availability calculation. There are two methods that can be used to compute availability: Markovian models and Monte Carlo simulations. They are briefly described next.

The basic assumption for Markovian models is the exponential distribution of time between failures and reparation time. This approximation reflects the real behavior of electronic and photonic component failures during their operational time. The availability for a structure is derived using state transition diagram devised for a certain network.

A working state of a component is changed to a non-working state by the occurrence of a failure and the opposite transition occurs as a consequence of a repair action. The state of a connection in the network is evaluated from the component states according to logical expressions that describe the relationship between component events (failure/repair) and the state of a connection (working or non-working state). Basic parameters for each Markov availability model are the component failure rate and the reparation rate.

Monte Carlo simulation can be used to generate the times to failure (*TTF*) and the time to repair (*TTR*) of components in the network. Each *TTF* and *TTR* is derived from a random number generator with a defined probability density function (PDF) that is component related. Statistical data related to the occurrence of a specific component failure is collected during the component life-test or by measuring *TTF*s for already deployed systems. By monitoring real optical links one can distinguish between failures of cables and failures of optical/electronic devices. By monitoring the maintenance data from the field, the PDF for *TTR* can be estimated. The mean time to failure and the mean time to repair can then be calculated as the mean value of the corresponding PDFs. Each component changes randomly from a working to a non working state. The impact of each component state change is analyzed and a decision is then made whether the connection state is affected by the component state change or not. The connection mean uptime T_{up} and mean downtime T_{down} are then cumulatively calculated. When the simulation is completed the asymptotic connection availability A is computed as:

$$A = \frac{T_{up}}{T_{up} + T_{down}}.$$

(1)

The availability calculation based on Monte Carlo simulation also introduces a *simulation error* but the number of simulation iterations can bound this error. Unfortunately, desirable accuracy may require long simulation runs. In addition, the time complexity of the simulation is a function of the number of network elements and the level of network redundancy. In a highly redundant network some network events are very rare and require many single or multiple element failures, including dependent failures, to be simulated before being able to measure the desired outcome.

DYNAMIC SCHEDULING OF SURVIVABLE CONNECTIONS IN OPTICAL WDM NETWORKS

User-controlled, large-bandwidth, on-demand services with differentiated timing requirements will play an important role in the future Internet. Connections are set up and released for specific time durations, with sliding or fixed set-up times, for applications such as video-on-demand, IPTV, backup storage, grid computing, and collaborative solutions in finance and R&D. With the development of (*i*) new and agile switching devices, and (*ii*) control and management plane architectures such as Automatically Switched Optical Networks (ASON) and Generalized Multiprotocol Label Switching (GMPLS), optical WDM networks are now able to provide dynamic circuits to meet the high bandwidth requirements of these dynamic services.

To characterize resource requirements for such applications, scheduled traffic models have been proposed by Cavdar *et al.* (2010). In this subchapter, we focus on dynamic scheduling of survivable connections with flexible set-up times. After a customer issues a connection request, the customer waits for a response. The request is accepted or rejected according to the network operator's ability to provide the required level of service quality. If the connection request cannot be satisfied and set up within a certain amount of time, say t_d, the customer withdraws the request. We call t_d the *delay tolerance* of the customer, which describes a customer's patience, i.e., the maximum duration a customer is willing to wait until the connection is set up (Figure 2). Delay tolerance of a connection request can be defined as a service-level specification (SLS) stated in a contract known as the service-level agreement (SLA), which is explained in detail by Clemente *et al.* (2005).

The network performance can be improved by exploiting the various SLA terms in temporal dimension. In particular, this subchapter discusses the use of a connection request's holding time and delay tolerance, where holding time defines the time duration of the service. More specifically, we consider the dynamic scheduling of survivable connections with delay tolerance. We study the performance of a dynamic scheduling approach on shared-path protection (SPP) for

Figure 2. Connection set-up time can slide until the end of the delay tolerance

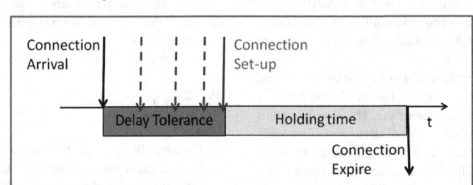

efficient capacity usage. The performance of different scheduling algorithms is compared and discussed, giving priority to requests according to their (*i*) arrival rates, (*ii*) delay tolerances, or (*iii*) holding times.

There has been a substantial amount of research on survivable connection provisioning in optical networks. In what follows we will categorize the existing work according to traffic models focusing mainly on scheduled traffic models. In general, traffic models can be classified into two groups: *unscheduled* and *scheduled.* In unscheduled models, time-domain specifications, such as holding time of a connection, are ignored. Connections are provisioned at the time they arrive according to the current network state, without considering the connection duration. But scheduled models consider the holding time of connections so that provisioning algorithms can optimize resources in both space and time. Both unscheduled and scheduled traffic models can be either static or dynamic. In a static traffic model, the set of traffic demands (unscheduled or scheduled) is known in advance. In contrast, for a dynamic traffic model, the arrival time and holding time of requests are generated randomly, based on certain distributions.

Set-up and tear-down times for scheduled traffic demands can be fixed, e.g., Li & Wang (2006), or they can be allowed to slide within a larger time window, e.g., Jaekel & Chen (2007), in which case they are called, respectively, fixed scheduled and sliding scheduled traffic demands. In a sliding scheduled traffic model, setup time slides within a time window, where the arrival time, holding time, and maximum end-time of the window are given. Tanwir *et al.* (2008) consider survivable routing and wavelength assignment for a sliding scheduled traffic model and use restoration to provide survivability.

Delay tolerance gives us the time difference between the window size and the holding time and can be used as a measure of flexibility of the time window. As the performance of provisioning algorithms with sliding scheduled demands

is dependent on this flexibility, a larger ratio of delay tolerance to holding time can allow more effective temporal sliding and may lead to more efficient resource utilization. Delay tolerance, proposed first by Cavdar, Tornatore & Buzluca (2009) as a connection oriented metric, is defined by each connection request which allows sliding scheduling of the demands.

Significant work has been done for dynamic unscheduled traffic with shared-path protection (SPP), e.g., by Ou, *et al.* (2004) and SPP with differentiated reliability, e.g., Fumagalli, Tacca, Unghvary & Farago (2002). Moreover, with fixed set-up times, Tornatore *et al.* (2005) considers the a-priori knowledge of holding time for SPP and Cavdar *et al.* (2007) study holding time aware availability-guaranteed connection provisioning with SPP under dynamic traffic demands. SPP is also studied for static scheduled demand models with fixed window by Li & Wang (2006) and sliding window by Jaekel & Chen (2007). Dynamic provisioning of SPP has been studied with sliding scheduled connection requests by Cavdar, Tornatore & Buzluca (2009) with availability guarantee and by Cavdar *et al.* (2010) with the comparison of different scheduling policies. In this subchapter we will explain the problem of dynamic scheduling of shared-path protected connections with delay tolerance (SDT) and discuss the performance of different dynamic scheduling policies on SPP.

Shared Path Protection with Delay Tolerance (SDT)

This section first provides a formal definition of the shared path protection problem with delay tolerance (SDT), then three different algorithmic solutions are presented as a solution of the problem.

Problem Statement

Given: a) Physical topology of a network represented by a graph G with a set of links and nodes; W specifies the number of wavelengths on each link; b) a connection request $R=\{s, d, t_a, t_h, t_d, n\}$, between source-destination pair (s,d) with arrival time (t_a), holding time (t_h), delay tolerance (t_d), and counter for retrials (n) to count each attempt to set up the connection request; c) a threshold (T) to restrict the number of retrials.

Output: A shared-path-protected connection comprehensive of a working path (l_w), a backup path (l_b), and setup time (t_s).

Objective: Minimize backup resource consumption and overall network blocking probability.

In SDT, if the network cannot provide a path pair for a specific request, the request is either delayed by sliding the set-up time for the delay tolerance duration, or it is rejected when the delay tolerance expires.

Backup Routing

Our reference algorithm finds primary and shared backup paths using a version of CAFES, which was proposed by Ou, *et al.* (2004). CAFES is a two-step, edge-disjoint path-pair algorithm. In the first step, a minimal cost working path (l_w) is computed, and then the link costs are updated to find a link-disjoint backup path (l_b) with minimal costs.

To keep track of backup resource utilization, we associate a conflict set v_e with a link e. To identify the sharing potential between backup paths, $v_e^{e'}$ denotes the number of backup wavelengths reserved on link e to protect primary paths passing through link e'. B(e) = number of wavelengths in the backup pool where shared wavelengths are reserved on link e, N(e) = number of connections which share wavelengths in the backup pool on link e, f(e) = number of free wavelengths on link e, and d(e) = distance of link e. In this study it is assumed that d(e)=1.

To calculate minimal-cost routes for backup paths, the link-cost calculation method proposed by Ou, *et al.* (2004) has been used for shared-path protection (SPP). As a primary objective, SPP encourages shareability, and minimizes hop distance. Therefore, cost C(e) for a candidate backup link e is calculated as follows:

$$C(e) =$$
$$\begin{cases} \infty, & \text{if } e \in p \text{ or if } f(e) = 0 \text{ and } \exists e' \in p, v_e^{e'} = B(e); \quad (2) \\ \varepsilon \times d(e), & \text{if } \exists e' \in p, v_e^{e'} < B(e); \quad (3) \\ d(e), & \text{otherwise, if } f(e) > 0. \quad (4) \end{cases}$$

Case (2) (infinite cost) corresponds to insufficient resources on a link to set up the backup path. Case (3) (negligible cost) corresponds to the case of a shareable backup pool where there is no need to allocate extra spare capacity for the incoming connection. Case (4) (full cost) gives the cost of the link where a new wavelength needs to be added in the backup pool.

Different SDT Algorithms

To solve the SDT problem, we introduce three different algorithms, which give priority based on arrival rate, delay tolerance, or holding time of a connection request. Here, the requests that normally would be blocked by a conventional SPP approach are rescheduled by putting the request back into the queue for another set up attempt. For the details, the interested reader is referred to Cavdar, *et al.* (2010).

A request is rescheduled only if an existing connection in the network departs within the current request's delay tolerance (t_d). The connection request is then rescheduled immediately after the departure, and t_d is updated. The resources released with the departure of a connection request change the network state and provide an opportunity to find available resources. A rescheduling algorithm is then needed to assign priorities when more than one connection request are to be delayed after a departure. Three different scheduling strategies are considered:

Figure 3. Network topologies used during the performance evaluation phase

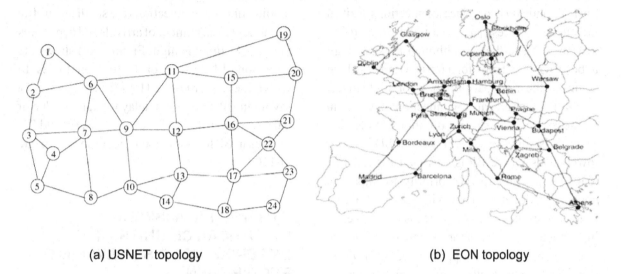

(a) USNET topology   (b) EON topology

- **Algorithm SDT_ar:** the main strategy in SDT_ar is prioritizing connections according to their arrival time, which is the traditional first-come-first-served (FCFS) queuing policy.
- **Algorithm SDT_dt:** gives priority to the impatient connection requests with smaller t_d in the queue.
- **Algorithm SDT_ht:** gives priority to requests with smaller holding time in the queue.

Illustrative Numerical Examples

For performance evaluation of three scheduling policies, a dynamic network environment is simulated. Connection arrivals follow a Poisson process with exponentially distributed holding time and delay tolerance, with each connection requiring one wavelength unit of bandwidth. Average delay tolerance (D) is normalized to the holding time, while the average holding time (H) has average equal to one. Therefore, offered network traffic load in Erlangs equals the arrival rate. In this study, we used 2 different network topologies: USNET with 24 nodes representing a backbone

topology in US (Figure 3(a)); and EON with 28 nodes, representing a pan-European backbone network (Figure 3(b)). In both cases, each link has 16 bidirectional wavelength channels. In each experiment, 100000 unicast connection requests, symmetric and uniformly distributed among all node pairs, are considered. Each plotted value has a 95% confidence level, with confidence interval not larger than 0.05 of the plotted value except in case of very small value of blocking probability (BP).

Figures 4(a) and 4(b) compare, in terms of BP versus arrival rate, the three different scheduling algorithms with CAFES, applied to USNET and to EON network topologies for D = 0.5. In order to have a fair comparison with CAFES, all three scheduling algorithms are based on the same routing strategy used in CAFES: two-step primary-backup routing without wavelength continuity constraint. Significant savings in BP are achieved at all load levels by applying SDT_ar, SDT_dt and SDT_ht in both topologies. The savings are larger for lower load values (e.g., 50% at arrival rate of 150 compared to 70% at 100 connections per time unit in USNET for SDT_ht). An asymptotic decrease of the gain is reached when the network load increases over a specific value in SDT_ar and SDT_dt, since connections wait in the queue for a restricted amount of time, delay tolerance. Nevertheless, although requests wait to be provisioned only for a limited time, SDT_ht achieves the best performance for high loads, because it gives priority to connections that remain in the system for shorter durations. Note that SDT_dt is superior to SDT_ar and SDT_ht for lower loads where BP is less than 8%. At low loads, it is better to give priority to impatient connections over requests with smaller holding times versus the priority that needs to be given at higher loads. For the more detailed results the reader is referred to the paper by Cavdar, *et al.* (2010) where it is also shown that there is no significant change in resource overbuild (which measures the usage of backup resources over primary resources) by applying SDT, except a slight decrease which occurs due to the increase in provisioning success rate. As a result, all three SDT algorithms bring significant gain in BP, without sacrificing resources for spare capacity usage.

Another important aspect is the duration of the delay tolerance (in our examples so far, D is normalized to the holding time of the connection). Longer delay tolerance allows more opportunities

for re-scheduling and re-routing, especially if the holding times of connections are small and traffic dynamicity, i.e., number of arrivals and departures in a unit of time, is high. Figure 4 (c) shows the reduction of BP vs. D for the two topologies. In USNET, even for D=0.2·H, SDT achieves a 50% saving in BP. Even for a delay tolerance value of 0.05, the algorithm SDT_dt achieves around 7% savings in BP for both topologies at an arrival rate of 150.

DYNAMIC PROVISIONING OF OPTICAL CIRCUITS WITH DIFFERENTIATED SURVIVABILITY REQUIREMENTS

WDM networks can be made survivable by means of path protection schemes implemented at the WDM layer (Mukherjee, 2006). A path protection scheme requires allocation of spare (or standby) resources that can be used in the event of a fault. For a lightpath, a path protection scheme consists of assigning a working and a protection path between the source and the destination. The working path carries the offered traffic during normal network operation. When the working

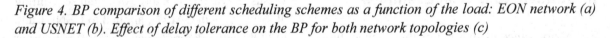

Figure 4. BP comparison of different scheduling schemes as a function of the load: EON network (a) and USNET (b). Effect of delay tolerance on the BP for both network topologies (c)

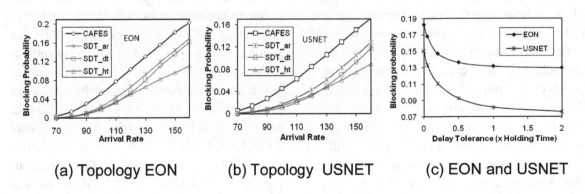

(a) Topology EON   (b) Topology USNET   (c) EON and USNET

path is disrupted by a fault, the affected traffic is rerouted over the protection resources.

Even though the aim of a protection scheme is straightforward, the amount of spare capacity a lightpath should be allocated to guarantee the required level of resiliency is a question without a univocal answer. For example, should a lightpath always be protected against any single fault regardless of the reliability requirement for the specific service, or it can tolerate some downtime if possible? Conventional protection schemes are not able to answer this type of questions because they are meant to guarantee either a full protection in the presence of a network fault or no protection at all. These approaches are very simple and have proven to be a valid solution in many network scenarios. However, their simplicity comes with a cost in terms inefficiency in using the network resources. For example, with the Dedicated Path Protection (DPP) scheme (Mukherjee, 2006) the resources reserved for the protection are dedicated to a specific connection. In order to have a better resource efficiency multiple working paths may be allowed to share resources reserved for protection, i.e., the Shared Path Protection (SPP) scheme (Mukherjee, 2006). Nonetheless, both DPP and SSP lack the ability to adapt to the different protection requirements, and may not be adequate in those scenarios where over-reservation of redundant network resources is not acceptable.

This problem can be addressed by applying a concept called Differentiated Reliability (DiR). The DiR approach leverages on the intuition that different connections may require different protection levels, e.g., backup storage may sustain some brief interruptions while, for example, bank transactions cannot tolerate any disruption at all. The validity of such intuition is supported by the strong role concepts such as Quality of Service (QoS) and Differentiated Services have in today's communication networks. According to the DiR paradigm, each arriving connection request comes with a specific reliability requirement that must be

met by the protection scheme and accordingly is assigned a certain protection level. This assumption makes it possible to reserve the minimum amount of network resources that are necessary to meet the level of protection required by a connection. In fact, the DiR approach focuses only on the protection level offered to each individual connection. There are several ways to express the level of protection. One option is to have the service protection level defined in terms of conditional failure probability, referred to as the probability that, once established, the connection survives a single fault in the network. Another option is to define an asymptotic connection availability, as specified for example by the Service Level Agreement (SLA).

The DiR concept was first introduced in the work by Fumagalli & Tacca (2006) where it was applied to provide different degrees of protection level in networks with a ring topology. The same concept was then extended to be used in more general mesh topologies under single (Fumagalli, *et al.*, 2002) and dual (Tacca, Fumagalli & Unghvary, 2003) link failure scenario. In all these studies the DiR approach was able to yield a significant reduction of the total number of network resources that are needed to accommodate a given set of lightpaths, i.e., a static provisioning scenario. In this subchapter the DiR problem is studied while considering a WDM network with dynamic traffic provisioning. The contribution is twofold. The first part of this subchapter focuses on describing how the SPP scheme can be combined with the DiR concept in a dynamic provisioning environment (resulting in the so called SPP-DiR scheme) when each connection protection level is described in terms of maximum conditional failure probability, assuming a single link failure scenario. The second part of the subchapter illustrates how to apply the SPP-DiR concept to account also for the impact of node failure on the connection survivability. In this part of the study, the level of protection of each lightpath is specified in terms of asymptotic connection availability. The failure scenario considered is also more general where multiple link/node failures are assumed.

The SPP-DiR Problem with Dynamic Provisioning

This section presents the SPP-DiR problem applied in dynamic provisioning scenario under the assumption of a single link failure. Consider a WDM network with an arbitrary mesh topology, where wavelength conversion is not available. It is assumed that only link failures are possible, and the probability that two or more links are down at the same time is considered to be negligible (Mukherjee, 2006). The WDM mesh network is modeled as a graph $G(N,L)$, where N represents the set of network nodes and L the set of network links. Each link $(m,n) \in L$ is characterized by the value of its conditional failure probability, $P_f(m,n)$. Based on the single failure assumption, the *conditional link failure probability* is the probability that a link is failed, given that a single link failure has occurred in the network. By assuming a single link failure scenario, the link failure probability is given by the product between the conditional link failure probability and the probability of having a single failure. For example, assuming a uniform distribution of faults among all the links, the *conditional link failure probability* can be expressed as:

$$P_f(m, n) = \frac{1}{|L|} \ \forall (m, n) \in L. \tag{5}$$

In such a scenario, it is assumed that each arriving connection request is characterized by a *Maximum Conditional Failure Probability (MCFP)* value. MCFP represents the maximum acceptable probability that, given the occurrence of a network link failure, the service data flow will be permanently disrupted.

With this rationale in mind, it is possible to select a set of links of the working path for which an arriving connection request d will not need to resort to the protection path. This set must be selected to satisfy the required protection level, formally expressed by the connection's MCFP. Notice that with SPP-DiR two (or more) connections whose working paths have a common link may also share a link and a wavelength for their respective protection paths. This option is available when at least one of the two connections can afford to be permanently disrupted upon the failure of the link that is shared by the working paths. By the same reasoning, it is also possible to have a working path completely unprotected if the working path failure probability still satisfies the reliability requirement indicated by the connection's MCFP. These last options are not supported by the conventional SPP scheme where 100% protection against any single failure is offered, i.e., SPP supports MCFP = 0 only. The following example illustrates how the SPP-DiR scheme works in a dynamic provisioning scenario.

Assuming a uniform link failure distribution, the link conditional failure probability for the network in Figure 5 is $P_f(m,n)=1/7$, $\forall (m,n) \in L$. Three connection requests are shown. d_1 arrives first and requires $MCFP_{d1} = 0$. The chosen working path is C-B. The protection path is C-E-B. d_2 arrives next and requires $MCFP_{d2} = 0$. The chosen working path is D-E-A. The protection path is therefore D-C-B-A. Finally, d_3 arrives and requires $MCFP_{d3} =1/7$. According to its reliability requirement of d_3 can sustain the failure of one link along its path. Taking advantage of this possibility, the working path is routed along D-E-B. The protection path for d_3 is D-C-B and is used only in the case of a failure on link (E,B), leaving link (D,E) unprotected, i.e., should link (D,E) fail d_2 will revert to its protection path since it cannot sustain any link failure. As shown in the example, protection resources along link (C,B) can be shared between connections d_2 and d_3 even though their working paths are not route

Figure 5. SPP-DiR problem, an example

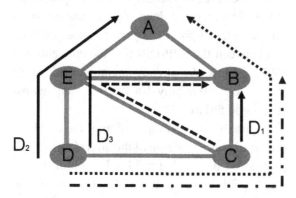

disjoint. Notice that by requiring a higher protection level, i.e., $MCFP_{d3} < 1/7$, connection d_3 is then blocked due to the lack of available wavelengths in the network. Although manually constructed, this example serves the purpose of showing that the SPP-DiR scheme has the potential to yield better resource utilization when compared to the conventional SPP scheme, while still guaranteeing each connection request sufficient resources to satisfy its protection requirement.

SPP-DiR Problem Definition and Solution

Let $H_{w,d}$ be the set of wavelength links used by the working lightpath to accommodate connection request d and $H_{p,d}$ be the set of wavelength links used by the protection lightpath assigned to connection request d. To guarantee the availability of a protection lightpath when the working lightpath is affected by a failure, working and protection lightpaths must be route-disjoint:

$$H_{w,d} \cap H_{p,d} = \varnothing, \qquad (6)$$

Let U_d be the set of unprotected links along the working lightpath of d, the conditional failure probability of d can be calculated as:

$$P_{f,d} = \sum_{(i,j) \in U_d} P_f(i,j) \le MCFP_d, \qquad (7)$$

where $P_f(i,j)$ is the conditional link failure probability. Let $H_{s,d} \subseteq H_{p,d}$ be the set of links, used by the protection lightpath of d, that share resources with other protection lightpaths already routed in the network. Based on the routing for both working and protection lightpaths, a cost function measuring the goodness of the choice for the routing is defined as:

$$C_d = |H_{w,d}| + |H_{p,d}| + |H_{s,d}| + (MCFP_d - P_{f,d}). \qquad (8)$$

The cost function measures the amount of resources provisioned to a connection. In addition it measures the excess of reliability $(MCFP_d - P_{f,d})$ that d receives. Solving the SPP-DiR problem means to provision each connection request with enough resources to satisfy equation (7), while minimizing the cost function defined in equation (8), which in turn has an impact on the overall blocking probability.

In order to solve the SPP-DIR problem presented above a two-step algorithm is used. The approach works as follows. In the first step, called SPP-DiR-FF, the algorithm aims at solving the Routing and Wavelength Assignment (RWA) problem for each connection request d while guaranteeing that the MCFP requirement is met by the protection scheme. The protection strategy is based on a modified version of the conventional Shared Path Protection (SPP) scheme where the DiR concept is applied with a larger granularity, i.e., connections can be fully protected or fully unprotected only. In the second step, called SPP-Dir-SA, the algorithm aims at reducing the reliability degree of the connection requests provisioned in the first step, with the intent of reducing the output of the cost function defined in equation (8). This is accomplished by selecting a subset of links along the working lightpath for which protection is not required. These links are chosen using a meta-heuristic algorithm based on Simulated Annealing (SA). For more information about the presented two-step strategy the interested reader is referred to Monti, Tacca & Fumagalli (2004).

SPP-DiR Performance Study

This section presents a collection of results obtained while solving the SPP-DiR dynamic provisioning problem presented in the previous section. Since the number of candidate paths between a source and a destination grows exponentially with the network size, to reduce the search complexity the candidate paths for each connection request are generated using the disjoint path-pair matrix (DPM) approach (Monti, Tacca & Fumagalli, 2004). DPM uses the first k_1 shortest paths as candidates for the working path. For each working path candidate, the first k_2 shortest paths found when the links in the primary path are removed are used as candidates for backup paths. The European optical network (Batchelor *et al.*, 2000) with 19 nodes and 39 bidirectional links is used as reference. It is assumed to have one fiber for each direction of propagation, with 32 wavelengths per fiber. The conditional link failure probability is obtained assuming a uniform distribution of failures over all links i.e., $P_f(i,j)$ = 1/39 $\forall(i,j) \in L$. The connection requests arrive according to a Poisson process with arrival rate λ. Source and destination nodes of each connection are randomly chosen using a uniform distribution over all possible node pairs. Unless otherwise specified, each connection request is assigned a reliability degree requirement of MCFP = 0.03, i.e., in the network topology under consideration each connection may be able to have up to one working link that is unprotected. Once established, a connection remains in the system for a time that is exponentially distributed with parameter $1/\mu$ =

1, i.e., the value of the arrival rate is equal to the value of the network load. The results shown in Figures 6 and 7 provide a performance comparison between the SPP-DiR and the conventional SPP schemes. As already mentioned, the SPP scheme can offer only 100% protection against any single failure.

Figure 6 (a) shows the value of the blocking probability as a function of the arrival rate λ. The plot shows that with a mild reduction of the offered reliability degree, i.e., MCFP = 0.03, the SPP-DiR scheme is able to decrease the blocking probability when compared to the SPP scheme. The reduction is more significant in the presence of multiple candidate paths, i.e., k_1=20, k_2=10, since the algorithm is able to find path options that better match the reliability requirement of each service. Figure 6 (b) plots the value of the average number of shared protection links versus λ. Results obtained for both the SPP-DiR and SPP schemes are shown. In the case under study, it is found that by closely matching the service's reliability requirement, the SPP-DiR scheme improves the number of shared protection links by 49% when compared to SPP.

Figure 7 (a) shows the normalized average excess of reliability as a function of the arrival rate. The excess of reliability of a connection, defined in equation (8) is averaged over all the provisioned connection requests, and normalized to MCFP=0.03. The excess of reliability obtained is always below 20%, a considerable reduction when considering that the SPP scheme has a value for the excess of reliability always equal to 100%. Figure 7 (b) shows the value of the blocking probability versus MCFP. The plots indicate the existing trade-off between the reliability degree that is guaranteed and the blocking probability. The values shown at MCFP = 0 represent the blocking probability of the SPP scheme. These results confirm that by attempting to closely match the connection's reliability requirement, the SPP-DiR scheme is successful in reducing the average

Figure 6. Performance evaluation: blocking probability (a) and average number of shared links (b) as a function of the arrival rate

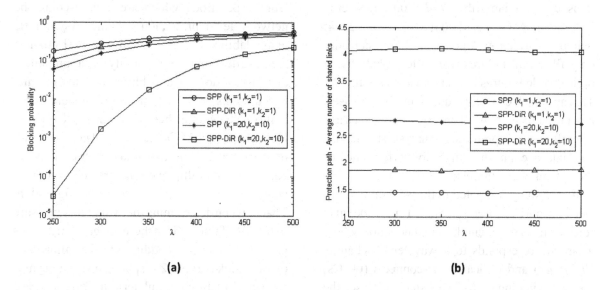

(a)   (b)

amount of network resources that must be reserved to establish a newly arrived connection. In turn, this fact may significantly reduce the value of the blocking probability.

Impact of Optical Node Failures on Network Reliability Performance

This section extends the previously presented study where only link failures were considered. Network survivability scenario with only link

Figure 7. Performance evaluation: excess of reliability as a function of the arrival rate (a) and blocking probability as a function of the maximum conditional failure probability (b)

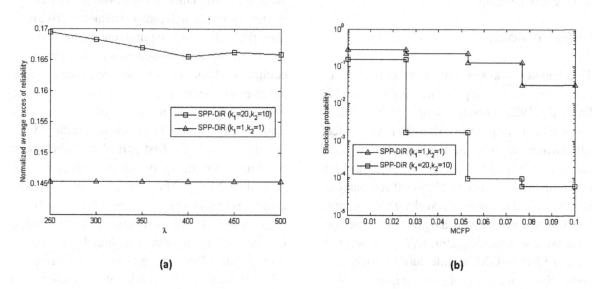

(a)   (b)

failure assumption is in line with what can be found in the literature where most of the contributions, e.g. (Ramamurthy, *et al.*, 2003; Doucette, Coloqueur & Grover, 2003; Ou, *et al.*, 2004; Schupke, Gruber & Autenrieth, 2002), consider only fiber link failures while the probability of optical node failures is assumed to be negligible. This approach can be valid in a number of cases, but probably not in all the possible scenarios. Therefore, a comprehensive analysis should also take into account the reliability performance of the optical nodes in the network.

In the study presented in this section we consider an optical circuit switched network (also referred to as wavelength routed network) where a circuit corresponds to a wavelength channel (*lightpath*) and optical cross-connects (OXCs) are the switching nodes. This study analyses the impact of an optical node failure on the end-to-end lightpath provisioning in survivable WDM networks by combining both the node level and the network wide reliability calculations in a single reliability provisioning framework. The node level reliability calculations based on the models in Wosinska (1993) are embedded in a network level protection scheme, i.e., the SPP-DiR approach, making it possible to study the effect of node reliability performance on end-to-end service provisioning.

Node Reliability

To calculate the availability of the optical nodes in the network we adopt the models presented in Wosinska (1993) where the number of wavelengths on each fiber is equal to 4, 32 and 64. The optical MEMS technology has been selected for this study due to the relatively low energy consumption, high reliability performance and low cost compared to an OXC based on tunable wavelength converters (TWC) and a passive wavelength selective device, i.e. arrayed waveguide grating (AWG). Moreover, we consider two OXC architectures, namely with and without inherent protection (Figure 8).

Table 2 shows the unavailability values for OXCs in Figure 8 with different number of wavelengths per fiber, which are later used in the network level protection scheme. Asymptotic unavailability (U) is a reliability performance measure denoting the probability that a component or system is down at an arbitrary instance of time. Calculations are based on the component availability values published in Wosinska, Thylen & Holmstrom (2001). Devices used at nodes for link termination, i.e., splitters, transmitters, receivers, couplers, and multiplexers, are treated as components connected to links in series configuration. That is, when determining the availability measure of links the failure rate of these optical components is also considered in addition to the failure rate of fiber links. Table 2 displays also the asymptotic unavailability of link terminations at nodes, derived using the same technique as in the case of the node level availability results. The difference of the unavailability figures of link terminations at nodes between protected and unprotected node architectures comes from the optical power splitter used at the input fibers in the protected architecture.

Network Level Reliability

We assume a dynamic network scenario where incoming connection (lightpath) requests arrive with specified reliability requirements. The network topology, the link capacities and the switching equipment deployed at nodes are given as input parameters. Incoming connection requests arrive with specified availability requirements and a centralized decision mechanism, similar to the one described in the first part of the subchapter, is utilized for lightpath setup. Lightpaths are provisioned only when there are enough free resources in the network to meet their reliability requirements. In order to obtain efficient resource utilization in the network the shared path protection scheme (SPP) is combined with differentiated reliability (DiR). This combination is referred to

Figure 8. Considered OXC architectures: with protection (a), and without protection (b)

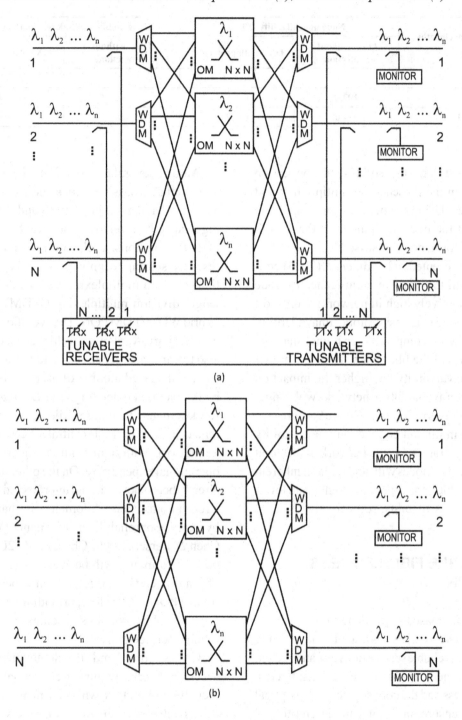

as SPP-DiR (Monti, Tacca & Fumagalli, 2004). The results of SPP-DiR are compared with the ones obtained with dedicated path protection scheme (DPP). In order to provide suboptimal solutions in polynomial time, a heuristic technique is utilized, which makes use of a time-efficient

Table 2. Node and link termination reliability performance

Number of wavelengths per fiber	Node unavailability (10^{-6})		Unavailability of link terminations (10^{-6})	
	OXC without protection	OXC with protection	OXC without protection	OXC with protection
4	24.0	0.00036	0.6	0.9
32	192.0	0.018	2.4	2.7
64	384.0	0.072	4.8	5.1

method to estimate the end-to-end connection availability in the presence of multiple link and node failures. Using the heuristic presented, the influence of the node equipment on the overall network performance is assessed.

Results presented in Pandi, *et al.* (2006) confirm the intuition that the significance of node failures is relatively high in networks with short fiber links since the asymptotic unavailability of nodes may be comparable or higher than the unavailability of the fiber links. Thus, the higher the node unavailability the higher the impact on the connection availability in networks with longer fiber links. This is reflected in Figure 9 where U_{min} denotes the minimum connection unavailability that can be guaranteed in the network of different size and based on nodes with and without inherent protection, which corresponds to higher and lower value of node availability, respectively.

SURVIVABLE FIBER ACCESS NETWORK

Due to the increased dependency on electronic services all over society and due to the growing importance of reliable service delivery, an efficient fault management strategy has to be considered in both the access and the core segment of an optical network to ensure an uninterrupted end-to-end service provisioning. However, in contrast to the core segment, access networks are very cost sensitive due to low sharing factor of the network infrastructure.

Among several existing fiber access network architectures, passive optical networks (PONs) are considered as an important candidate to offer high capacity at relatively low cost. Three types of PON solutions, each one utilizing different resource sharing technologies can be identified: time-division multiplexing (TDM) PON, wavelength-division multiplexing (WDM) PON and hybrid WDM/TDM PON. The evolution of PON is progressing towards not only higher bandwidth but also towards a larger coverage of the access areas and an increased number of users. This is driven by the fact that extending the PON reach from a few kilometers to hundred kilometers enables the replacement of multiple central offices (COs) with a single one, with significant saving in capital and operational expenditure. On the other hand, it has already been shown that an unprotected PON with a reach up to twenty kilometers is characterized by a very poor reliability performance (Wosinska, Chen & Larsen, 2009; Chen, *et al.*, 2010). Such poor performance will be even worse in the case of long-reach (i.e. up to 100 km expected in the future) PONs. Therefore, providing protection in future PON installations becomes essential for a reliable service delivery.

On the other hand, the deployment of fiber access networks require a considerable investment from operators, while, as mentioned before, in this network segment cost is a very important factor. Therefore, operators may choose to provide at first mostly unprotected services. Up to now, only business users are including in their Service Level Agreement (SLA) some penalties

Figure 9. Minimum connection unavailability as function of the network size and as a function of different switch architectures

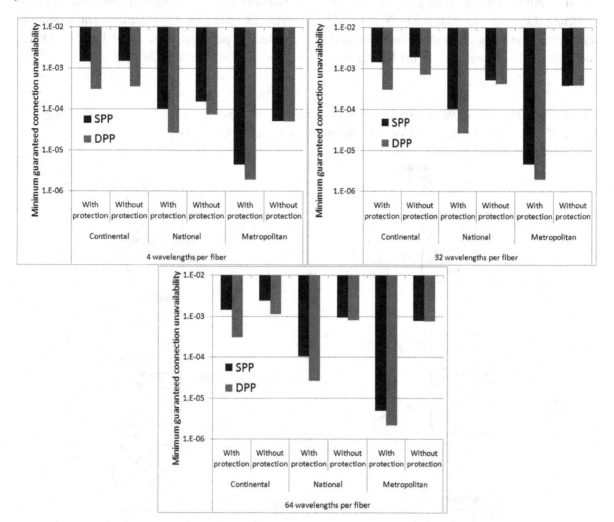

to be paid for those service interruptions that exceed an agreed time threshold. However, new services (e.g. telehealth) may in the future extend this requirement to the private users. In such a case, operators will be facing the need to upgrade their access networks with protection resources to improve reliability performance in order to avoid penalties for service interruptions.

This subchapter focuses on survivable PON architectures aiming at comparing the cost and reliability performance of some representative approaches. First, different PON protection schemes are reviewed, including both solutions proposed in the standards and in the literature. Then the cost and reliability performance are evaluated for all the presented reliable PON architectures.

Protection Schemes in PON

Tree is the most commonly used topology in fiber access networks. Among the various options, trees with a single splitting point (Figure 10 (a)) are the most commonly used configuration in PONs. In such configuration one single fiber, called feeder fiber (FF), connects an optical line terminal (OLT) at the CO to a remote node (RN), which is an intermediate node where the optical signal is split to reach each and every optical network unit (ONU) deployed in the PON. An ONU represents a termination point at the user side, and each ONU is connected to the RN through a separate fiber, called distribution fiber (DF). The drawback in having a tree structure is that it requires additional fiber deployment to provide protection paths between the OLT and the ONUs to be used in case of a fiber cut.

Figure 10. TDM PON architectures: (a) basic, (b) Type A, (c) Type B, (d) Type C, (e) neighboring protection (Chen, Chen & He, 2006), and (f) ring protection (Yeh & Chi, 2007)

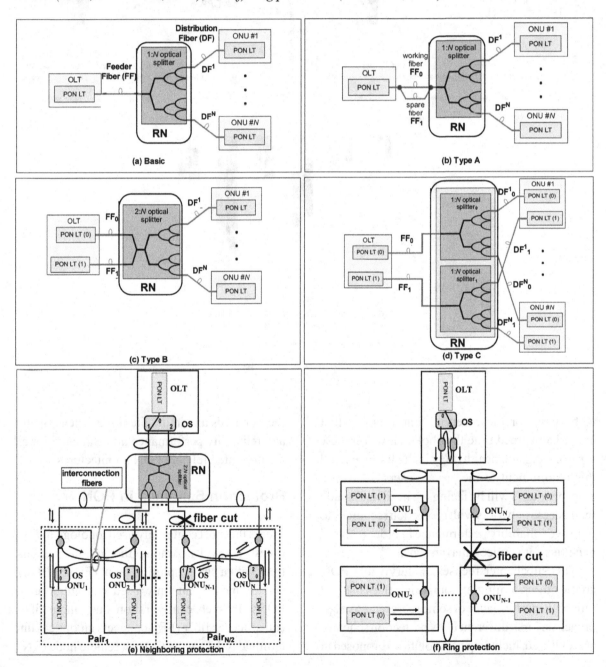

In order to provide a more reliable service delivery over the PON infrastructure several protection architectures have been proposed. In the late 90s, some standard protection architectures were defined by ITU-T (ITU-T, 1997). The ITU-T standard proposes a simple and straightforward concept, i.e., provision of duplicated components for the parts that need to be protected. Figures 10 (b), (c) and (d) show three standard protection architectures defined by ITU-T. They are based on the duplication of network resources and are referred to as type A, B, and C. In Type A (Figure 10 (b)) only the FF is duplicated. Type B protection (Figure 10 (c)) duplicates the shared part of the PON (i.e., both FF and the line terminals at the OLT). Type C protection (Figure 10 (d)) represents 1+1 dedicated path protection with full duplication of the PON resources. In addition to the protection schemes just explained, ITU-T also defined a scheme referred to as Type D protection where the FF and the DFs can be duplicated independently. This additional protection scheme enables network operators to offer different reliability levels to different users. Type D protection provides end users with either full or partial protection referred to as Type D1 or D2 respectively.

Type C and Type D1 protection schemes are able to offer a relatively high reliability performance but they require duplication of all network resources to realize their protection function. This may result in deployment costs that are too high. Therefore, a lot of research work has been done to develop cost-efficient and reliable access network architectures. The work presented in Chan, *et al.* (2003), Chen, Chen & He (2006), Chen & Wosinska (2007), and Chen, Wosinska & He (2008), proposed neighboring protection (NP) schemes where two adjacent ONUs protect each other using interconnection fibers (see Figure 10 (e)). In this way, the cost invested in burying redundant disjoint DFs to each ONU can be saved and, consequently, the deployment cost can be significantly reduced. Figure 10 (e) shows a neighboring protection architecture for TDM PON

proposed in Chen, Chen & He (2006) where two geographically disjoint fibers provide dedicated protection against a FF cut between OLT and RN, and adjacent ONUs are paired to realize dedicated protection for DFs. Figure 10 (e) provides also an example on what would happen with the presented protection scheme should a fiber failure occur between ONU_{N-1} and the RN, where N denotes the total number of ONUs. ONU_{N-1} detects the loss-of-light and a control signal is generated to trigger the optical switch (OS) from port 1 (the normal state) to port 2 (the protection state). The corresponding interconnection fiber, which connects port 2 of the OS in ONUN-1, works for both the upstream and downstream traffic flows associated with ONU_{N-1}. This NP scheme can also be used with WDM PON (Chan, *et al.*, 2003) and with hybrid WDM/TDM PON (Chen & Wosinska, 2007; Chen, Wosinska & He, 2008).

On the other hand, ring topologies are able to offer resiliency with a minimum number of links. Therefore a protection scheme based on rings can also offer a cost-effective solution for PON by reducing the fiber deployment cost. A ring protection for TDM PON proposed in Yeh & Chi (2007) is shown in Figure 10 (f). During normal operation, the downstream signal from the OLT is transmitted counterclockwise to the line terminal LT(0) at each ONU, while the upstream signal from line terminals LT(0) at each ONU is transmitted clockwise. Figure 10 (f) provides an example on what would happen when a fiber cut occurs between ONU_{N-1} and ONU_N, where N denotes the total number of ONUs in the TDM PON. In the presence of a fiber cut, the ONUs where LT (0) loses its connection to the OLT will start using LT (1) to reconnect the OLT. In the meantime, the OLT will also switch the direction of its optical switch to the port number 2. In this way, the downstream signal will be separated to pass counterclockwise and clockwise simultaneously. For the upstream signal, LT (0) at each ONU not affected by the fiber cut maintains its clockwise transmission while LT (1) at the ONUs in protection mode sends the signal counterclockwise.

Reliability Performance

This section studies the availability of a connection between the OLT and each ONU using the method presented in Wosinska, Chen & Larsen (2009). The analysis is done for the resilient PON architectures presented in the previous section where different values of the total length of the FF plus the DF (referred to as *reach*) are assumed. Two values of the reach are considered, namely 20 km and 100 km. The other input data (e.g., failure rate, mean time to repair) used for the presented availability calculations is obtained from Chen, *et al.* (2010). The results for connection unavailability and their relative deployment cost per user are shown in Table 3. The cost is computed as a function of the equipment cost value, the cost for the fiber infrastructure and installation cost (Chen, *et al.*, 2010) and it normalized to the cost of the non protected (i.e., Basic) TDM PON case.

In future, it is expected that network operators will need to offer 5 nines connection availability (i.e. connection availability greater than 99.999%),

which corresponds to a connection downtime of less than 6 minutes per year. From Table 3, it can be seen that basic TDM PON without any protections shows reliability performance lower than 99.999% for both 20 km and 100 km reach. Therefore, it is necessary to provide protection in PONs in order to improve their reliability performance. On the other hand, Type C and D1 schemes with neighboring protection and ring protection can offer very high connection availability (higher than 99.999%) for both 20 km and 100 km reach. However, the comparison of deployment cost per user shows that neighboring and ring protection schemes are much more cost-efficient than Type C and D1. In addition, it can be seen that ring protection has the lowest deployment cost per user while maintaining an acceptable reliability performance. On the other hand, ring protection has a problem with the power budget. When the optical signal passes through several ONUs, it becomes degraded and attenuated. It restricts the total number of ONUs that can be connected to the ring. Therefore, compared with cost efficient NP scheme, ring protection cannot be applied to PONs deployed in dense populated areas. Furthermore, it can be observed that the

Table 3. Reliability performance results

Network Architectures		Unavailability		Relative deployment cost per user (%)	
		Reach=20km	Reach=100km	Reach=20km	Reach=100km
Basic	TDM PON	2.76E-04	1.37E-03	100%	100%
Standard protection (TDM) (ITU-T, 1997)	Type A	7.20E-05	7.36E-05	101%	106%
	Type B	7.03E-05	7.19E-05	103%	108%
	Type C	7.64E-08	1.88E-06	200%	200%
	Type D1	4.72E-08	1.70E-06	200%	200%
	Type D2	6.92E-05	7.08E-05	104%	111%
Neighboring protection (NP)	TDM	5.22E-06	6.86E-06	121%	127%
	WDM	7.50E-06	9.14E-06	126%	132%
	Hybrid I	6.42E-06	8.06E-06	121%	123%
	Hybrid II	4.80E-06	6.44E-06	121%	123%
Ring protection	TDM	2.41E-06	4.305E-06	65%	87%

reliability performance for a PON with NP scheme does not suffer as much from reach extension, while connection unavailability for basic PON, Type C and D1 increases significantly (5 to 25 times more).

CONCLUSION

This chapter presented a series of resiliency strategies that can be used in network scenarios where operators and service providers have to accommodate services with different resilience requirements. The common denominator of these strategies is their ability to leverage the different requirement levels imposed by each service to make a more efficient use of the network resources and to reduce the network deployment costs.

The first part of the chapter demonstrated a new dimension for the shared-path protection (SPP) scheme called Shared Path Protection with Delay Tolerance (SDT). Exploiting the flexibility provided by this QoS specification, SDT is able to significantly decrease blocking probability without sacrificing spare capacity utilization. The focus of the presented study was on different scheduling strategies for SPP that can be used in a dynamic provisioning scenario. It was shown that significant reduction of blocking probability is achievable in typical backbone network topologies independently of the load condition. Delaying connection requests even for a short duration brings approximately 50% reduction of blocking probability.

In the second part of the chapter an on-line Shared Path Protection scheme with differentiated Reliability (SPP-DiR) was described. SPP-DiR dynamically reserves network resources to set up incoming connection requests, with the objective of guarantee the required availability with the minimum possible redundancy. Two cases were analyzed. In the first case only single link failures were considered. In the second case the impact of nodes faults was also taken into account in a

scenario where multiple simultaneous network failures are possible. It was shown that when compared to the conventional SPP scheme, the presented SPP-DiR algorithm reduces the overall blocking probability by making use of the spare protection wavelengths, while guaranteeing the required availability of each connection. The presented results also confirm that the widely used assumption of negligible node failures may not be acceptable in networks with relatively large number of nodes and short fiber links where the asymptotic unavailability of nodes may be comparable or higher than the unavailability of the fiber links.

In the last part, the chapter provided an overview of recent advances in protection schemes for different types of PONs along with an assessment of some representative approaches in terms of reliability and deployment cost.

FUTURE TRENDS

This section briefly explores how the protection concepts presented in this chapter can be extended to account for other critical aspects of optical networks.

One interesting issue that can be investigated is the possibility of exploiting sub wavelength granularities. All the results presented in this chapter for the backbone segment assume that each lightpath uses the bandwidth of a full wavelength. It would be interesting to explore how the proposed mechanisms (i.e., both SPP-SDT and SPP-DiR) perform when they have to protect sub wavelength channels. Another interesting aspect to consider is how the presence of optical physical impairments will influence the differentiated QoS protection mechanisms presented in the chapter. Degradation of optical signal due to the physical layer phenomena limits the maximum span of a lightpath and influence the choices made during the routing phase. Their effect, in terms of reduced lightpath reach, is expected to be even more critical in the provisioning phase of protection lightpaths that are, on average, longer then their respective primary lightpaths.

In one of the sections this chapter highlighted a trade-off between the deployment cost (CAPEX) and the level of reliability performance in fiber access networks. Since economical aspects are most critical in the access part of the networks, the future trend will migrate towards minimizing the operational expenditures (OPEX) during the access network operation time in order to minimize the total cost of ownership (TCO) for the operator. In this respect, the work presented in this chapter can be extended to include consideration of failure related OPEX, such as service interruption penalty and reparation cost.

REFERENCES

ATT. (2009). *AT&T high speed internet business edition service level agreements*. Retrieved May 2009, from http://www.att.com/gen/general?pid=6622

Batchelor, P., Daino, B., Heinzmann, P., Hjelme, D. R., Inkret, R., & Jäger, H. A. (2000). Study on the implementation of optical transparent transport networks in the European environment-Results of the research project COST 239. *Photonic Network Communications*, 2(1), 15–32. doi:10.1023/A:1010050906938

Cavdar, C., Song, L., Tornatore, M., & Mukherjee, B. (2007). *Holding-time-aware and availability-guaranteed connection provisioning in optical WDM mesh networks. High-Capacity Optical Networks and Enabling Technologies (HONET), Dubai*. UAE.

Cavdar, C., Tornatore, M., & Buzluca, F. (2009). *Availability-guaranteed connection provisioning with delay tolerance in optical WDM mesh networks*. Optical Fiber Communication Conference and Exposition, San Diego, CA, USA.

Cavdar, C., Tornatore, M., Buzluca, F., & Mukherjee, B. (2010). Shared-path protection with delay tolerance (SDT) in optical WDM mesh networks. *IEEE/OSA. Journal of Lightwave Technology*, 28(14), 2068–2076. doi:10.1109/JLT.2010.2051414

Chan, T., Chan, C., Chen, L., & Tong, F. (2003). A Self-protected architecture for wavelength division multiplexed passive optical networks. *IEEE Photonics Technology Letters*, 15(11), 1660–1662. doi:10.1109/LPT.2003.818657

Chen, J., Chen, B., & He, S. (2006). Self-protection scheme against failures of distributed fiber links in an ethernet passive optical network. *OSA Journal of Optical Networks*, 5(9), 662–666. doi:10.1364/JON.5.000662

Chen, J., Mas Machuca, C., Wosinska, L., & Jaeger, M. (2010). Cost vs. reliability performance study of fiber access network architectures. *IEEE Communications Magazine*, 48(2), 56–65. doi:10.1109/MCOM.2010.5402664

Chen, J., & Wosinska, L. (2007). Analysis of protection schemes in PON compatible with smooth migration from TDM-PON to hybrid WDM/TDM PON. *OSA Journal of Optical Networks*, 6(5), 514–526. doi:10.1364/JON.6.000514

Chen, J., Wosinska, L., & He, S. (2008). High utilization of wavelengths and simple interconnection between users in a protection scheme for passive optical networks. *IEEE Photonics Technology Letters*, 20(6), 389–391. doi:10.1109/LPT.2007.915655

Cholda, P., Mykkeltveit, A., Helvik, B. E., Wittner, O., & Jajszczyk, A. (2007). A survey of resilience differentiation frameworks in communication networks. *IEEE Communications Surveys and Tutorials*, 9(1-4), 32–55. doi:10.1109/COMST.2007.4444749

Clemente, R., Bartoli, M., Bossi, M. C., D'Orazio, G., & Cosmo, G. (2005). *Risk management in availability SLA*. Italy: Design of Reliable Communication Networks, Island of Ischia.

Doucette, J., Coloqueur, M., & Grover, W. D. (2003). On the availability and capacity requirements of shared backup path-protected mesh networks. *SPIE Optical Networking Magazine, 4*(6), 29–44.

Fumagalli, A., & Tacca, M. (2006). Differentiated reliability (DiR) in wavelength division multiplexing rings. *IEEE/ACM Transactions on Networking, 14*(1), 159–168. doi:10.1109/TNET.2005.863708

Fumagalli, A., Tacca, M., Unghvary, F., & Farago, A. (2002). *Shared path protection with differentiated reliability*. International Conference on Communications, New York, NY, USA.

ITU-T. (1998). *Recommendation G983.1*.

Jaekel, A., & Chen, Y. (2007). Demand allocation without wavelength conversion under a sliding scheduled traffic model. *International Conference on Broadband Communications, Networks, and Systems (Broadnets)*, Raleigh, NC, USA.

Li, T., & Wang, B. (2006). *Approximating optimal survivable scheduled service provisioning in WDM optical networks with iterative survivable routing*. International Conference on Broadband Communications, Networks, and Systems (Broadnets), San Jose, CA, USA.

Liu, S., & Chen, L. (2007). *Deployment of carrier-grade bandwidth-on-demand services over optical transport networks: A Verizon experience*. Optical Fiber Communication Conference and Exposition, San Diego, CA, USA.

Monti, P., Tacca, M., & Fumagalli, A. (2004). Resource-efficient path-protection schemes and online selection of routes in reliable WDM Networks. *OSA Journal of Optical Networking, special issue on Next-Generation WDM Network Design and Routing, 3*(4), 188-203.

Mukherjee, B. (2006). *Optical WDM networks*. New York, NY: Springer.

Ou, C., Zhang, J., Sahasrabuddhe, L. H., & Mukherjee, B. (2004). New and improved approaches for shared-path protection in WDM mesh networks. *IEEE/OSA. Journal of Lightwave Technology, 22*(5), 1223–1232. doi:10.1109/JLT.2004.825346

Pandi, Z., Tacca, M., Fumagalli, A., & Wosinska, L. (2006). Dynamic provisioning of availability-constrained optical circuits in the presence of optical node failures. *IEEE/OSA. Journal of Lightwave Technology, 24*(9), 3268–3279. doi:10.1109/JLT.2006.879505

Ramamurthy, S., Sahasrabuddhe, L., & Mukherjee, B. (2003). Survivable WDM mesh networks. *IEEE/OSA. Journal of Lightwave Technology, 21*(4), 870–883. doi:10.1109/JLT.2002.806338

Ray, M. (2010). *100G DWDM optical networking transport: The telecom industry prepares*. Retrieved December 2010, from http://searchtelecom.techtarget.com/feature/100G-DWDM-optical-networking-transport-The-telecom-industry-prepares

Schupke, D. A., Gruber, C. G., & Autenrieth, A. (2002). *Optimal configuration of p-cycles in WDM networks*. International Conference on Communications, New York, NY, USA.

Tacca, M., Fumagalli, A., & Unghvary, F. (2003). *Double-fault shared path protection scheme with constrained connection downtime*. Banff, Alberta, Canada: Design of Reliable Communication Networks.

Tanwir, S., Battestilli, L., Perros, H., & Karmous-Edwards, G. (2007). Dynamic scheduling of network resources with advance reservations in optical grids. *International Journal of Network Management, 18*(2), 79–105. doi:10.1002/nem.680

Tornatore, M., Ou, C. S., Zhang, J., Pattavina, A., & Mukherjee, B. (2005). PHOTO: An efficient shared-path protection strategy based on connection-holding-time awareness. *IEEE/OSA. Journal of Lightwave Technology, 23*(10), 3138–3146. doi:10.1109/JLT.2005.856174

Tran, A. V., Chae, C., & Tucker, R. S. (2005). Ethernet PON or WDM PON: A comparison of cost and reliability. *TENCON*, IEEE Region 10.

Wosinska, L. (1993). Reliability study of fault-tolerant multiwavelength nonblocking optical cross connect based on InGaAsP/InP laser-amplifier gateswitch arrays. *IEEE Photonics Technology Letters, 5*(10), 1206–1209. doi:10.1109/68.248429

Wosinska, L., & Chen, J. (2008). Reliability performance analysis vs. deployment cost of fiber access networks. 7th International Conference on Optical Internet, Tokyo, Japan.

Wosinska, L., Chen, J., & Larsen, P. C. (2009). Fiber access networks: Reliability analysis and Swedish broadband market. *IEICE Transactions on Communications. E (Norwalk, Conn.), 92-B*(10), 3006–3014.

Wosinska, L., Thylen, L., & Holmstrom, R. P. (2001). Large-capacity strictly nonblocking optical cross-connects based on microelectrooptomechanical systems (MEOMS) switch matrices: Reliability performance analysis. *IEEE/OSA. Journal of Lightwave Technology, 19*(8), 1065–1075. doi:10.1109/50.939785

Yeh, C., & Chi, S. (2007). Self-healing ring-based time-sharing passive optical networks. *IEEE Photonics Technology Letters, 19*(15), 1139–1141. doi:10.1109/LPT.2007.900155

ADDITIONAL READING

Chen, J., & Wosinska, L. (2010). *Efficient next-generation optical networks - design and analysis of fiber access and core networks*. VDM Verlag, Dr Muller Aktiengesellschaft & Co. KG.

Tacca, M., Fumagalli, A., Paradisi, A., Unghvary, F., Gadhiraju, K., & Lakshmanan, S. (2003). Differentiated reliability in optical networks: theoretical and practical results. *IEEE/OSA. Journal of Lightwave Technology, 21*(11), 2576–2586. doi:10.1109/JLT.2003.819554

Tacca, M., Monti, P., & Fumagalli, A. (2004). The disjoint path-pair matrix approach for online routing in reliable WDM networks. *International Conference on Communications*, Paris, France, 2004.

Wosinska, L. (1999). *A study of the reliability of optical switching nodes for high capacity telecommunications networks* (TRITA-MVT Report 1999:4). Ph.D. dissertation, Royal Institute of Technology, Stockholm, Sweden, 1999.

KEY TERMS AND DEFINITIONS

Availability: The probability of a network resource to be in an operating state at a random time t in the future.

Delay Tolerance: The maximum time that a customer can wait after issuing the connection request to have the connection set-up.

Holding Time: The time duration between the set-up time and the teardown time of a connection.

Service Level Agreement: A contract signed between bandwidth provider and customer, which defines different level of service specifications to be met during the connection's life-time or set-up.

Set-up Time: The time it takes to establish a connection since the connection request is issued.

Chapter 11
Distributed Quality of Service Based Provisioning Framework for Survivable Optical Networks

Emad M. Al Sukhni
University of Ottawa, Canada

Hussein T. Mouftah
University of Ottawa, Canada

ABSTRACT

This chapter provides new distributed frameworks to support Quality of Service (QoS) differentiation. These frameworks provide differentiated protection services to meet customers' availability requirements effectively. We describe the availability-analysis for connections with different protection schemes. Through this analysis, we show how connection availability is affected by resource sharing. Based on the availability analysis, the proposed framework provisions each connection in which an appropriate level of protection is provided according to its predefined availability requirement. We consider the networks without wavelength conversion capability as well as dynamic traffic environment. In these distributed frameworks we propose several distributed schemes to provision and manage connections cost-effectively while satisfying the existing and new connections availability requirements.

INTRODUCTION

All-optical networks are potential candidates for future wide-area backbone networks. Such networks provide high throughput of the order of terabits per second. They display low error rates,

and are characterized by minimum delay. Due to those features, they can satisfy the emerging applications such as supercomputer visualization, medical imaging, and distributed CPU interconnect. Optical network provides a large number of wavelengths per fiber; and present technology allows transmission rates of up to 10 Gbps per channel. The optical network consists

DOI: 10.4018/978-1-61350-426-0.ch011

of wavelength cross-connects (OXCs) interconnected by point-to-point fiber links in an arbitrary mesh topology. In these networks, a connection is referred to as a lightpath, which is established between any two nodes by allocating the same wavelength on all links along the chosen route. The requirement that the same wavelength must be used on all the links along the chosen route is known as wavelength continuity constraint.

Compared to a ring network, a WDM mesh network can provide a wide variety of protection schemes. The trend in the development of optical networks has recently started moving towards a multiservice platform. In such scenario, considering the requirements of different applications/end users, it is essential to provide services with different Qualities. Consequently, systematic methodology to efficiently select a cost-effective protection scheme for each connection while satisfying its quality-of-service (QoS) requirements is highly desired. Usually, QoS can be measured in many different ways: service availability, service reliability, restoration time, service restorability, etc… Service availability is one of the key concerns of customers and it is usually defined in a Service-Level Agreement (SLA). The SLA is a contract between the network operator and a customer. The violation of SLA may entail penalties to be paid by the network operator. Thus, a cost-effective, QoS-aware, connection-provisioning scheme is very desirable such that, for each customer's service request, a proper protection scheme (dedicated, shared, or unprotected) is designed to guarantee the SLA-defined QoS requirement and to reduce overall network cost.

In this Chapter, We first describe the availability analysis for connections with different protection schemes (i.e., unprotected, dedicated protected, or shared protected). Through this analysis we show how connection availability is affected by resource sharing. Based on the availability analysis, we then develop a distributed provisioning framework in which an appropriate level of protection

is provided to each connection according to its predefined availability requirement. We consider networks without wavelength-conversion capability and consider dynamic lightpath provisioning, where a set of traffic demands is not known in advance. We assume that each connection requires the full capacity of a wavelength channel. The network operator needs to provision each connection with minimal network resources while still meeting the connections availability requirements. Our distributed framework includes approaches to control and manage the network resources and lightpath connections in fully distributed fashion, which improves scalability and reduces control overhead.

The chapter will be organized as follows. Section 2 presents the background of the QoS Requirements in Survivable optical Networks. Section 3 presents the availability analysis for connections with different protection schemes in survivable networks. Section 4 presents a distributed controlled availability-aware provisioning framework in which an appropriate level of protection is provided to each connection according to the customer's predefined availability requirement. Section 5 presents distributed controlled schemes to keep track the availabilities of the existing connections while provisioning new connections. The performance evaluation for the framework will be presented in section 6. Finally, section 7 concludes this chapter and gives some directions for future research.

QUALITY OF SERVICE IN OPTICAL NETWORKS

A WDM mesh network may provide different services for customers. The QoS requirements for these services can be different because of their diverse needs of the customers; banking services, on-line trading, and military applications demand high QoS levels, while IP best-effort packet-delivery service may be satisfied with lower QoS levels. Naturally, how to provide a certain QoS per customer requirement and how to guarantee the service qualities become a critical need for the network operators as well as for the customers (Ho et al.,2007 and Zhang et al.,2007). Service availability, service reliability, and restoration time are important QoS parameters in lightpath provisioning in optical networks. As one of the QoS parameter, Service availability is the main focus of This Chapter.

The following two subsection present an overview about the service Availability and Availability-aware lightpath provisioning literature.

Service Availability

As we know, a customer of an optical network operator may buy some bandwidth with certain service-quality requirements—availability is one of them. Availability is defined as the probability that a system will be found in the operating state at a random time in the future. Connection availability can be computed statistically based on the failure frequency and failure repair rate of the underlying network components that the connection is using, reflecting the percentage of time that a connection is "alive" or "up" during its entire service period. It should be clear that a protection scheme will help improve connection availability since traffic on the failed working path will be quickly switched to the backup path. For example, a path-protected connection will have 100 percent availability in the presence of any single failure. Nevertheless, when the more realistic scenario of multiple failures is considered, connection availability depends greatly on the precise details of the failures (locations, repair times, etc.); how many backup resources are reserved (i.e. single backup route or multiple backup routes); and how the backup resources are allocated (i.e. dedicated or shared). Intuitively, the more backup resources (paths) there are, the higher the connection avail-

ability is, while more backup sharing leads to lower connection availability. What we need now is a methodology to estimate the connection availability; then provision the connection with the proper protection scheme. Such a methodology can essentially help us to understand how well a connection should be protected to guarantee requested service quality. Although protecting the connection may help a network operator avoid any violation in SLA, extra resource consumption will be introduced, which may not be necessary if the connection is provisioned properly. As a result, a cost-effective availability-aware, connection provisioning scheme is most desirable; such that, for each customer's service request, a proper protection scheme (dedicated, shared, or unprotected) is designed so that the SLA-defined availability requirement can be guaranteed. At the same time, overall resource utilization can be achieved.

Availability Aware Connection Provisioning in WDM Optical Networks

Increasing attention has been devoted to service availability and reliability in WDM mesh networks. Availability analysis and the idea of providing differentiated reliability in SONET rings have been studied in the optical networks literature (Grover, 1999, Schupke,2000 and Fumagalli and Tacca, 2001). Grover, 1999 has given an extensive review of availability in ring networks. The concept of differentiated reliability has been proposed and studied by Fumagalli and Tacca (2001) to provide multiple reliability degrees using a common protection mechanism in optical ring networks.

Fumagalli and Tacca (2001) extend the concept of differentiated reliability to shared-path protection in mesh networks with the assumption of single network failure. Their idea is to select some links along the primary path and leave them unprotected. Willems et al.(2003) and Tacca, et al. (2003) resources studied the tradeoff between capacity requirements and service availability provided by reserved protection. Recently, Tornatore, et al. (2006) proposed an availability design scheme for dedicated and shared protection schemes. A detailed comparative study for Dedicated Protection based availability-aware connection provisioning schemes done by Mykkeltveit and E. Helvik (2008).

The Availability Guaranteed Service Differentiated Provisioning *(AGSDP)* algorithm is proposed by Song, et al. (2007) to enhance the performance of the availability-unaware routing protocol proposed by the same researchers (CA-FES). In the AGSDP, if a connection cannot be provisioned unprotected, a backup path must be found. Holding-time-aware AGSDP *(HT*-AGSDP) is proposed as an adaptation of the fundamental holding time-aware routing scheme, PHOTO (Tornatore, et al.,2006), into AGSDP (Cavdar, et al.,2007). PHOTO is based on the assumption that the holding time for each connection request is known at the time of arrival. Upon a connection setup request, the working path is searched by using the same strategy as when searching in CAFES. However, a backup path search considers connection holding times to better utilize the shared backup resources.

Another similar time-aware approach for availability-guarantee has also been proposed in (Wei, et al., 2003). The proposed scheme uses the fact that the connection availability requirement varies with its SLA requirement during the holding time. Zhang et al. (2007) propose a heuristic for SLA-constrained sharing. The proposed heuristic algorithm is tested under several provisioning strategies, such as the most reliable working/backup pair; the working/backup pair that leads to an availability just above a threshold value. The provisioning strategies set up the connections either as unprotected or by dedicated protection path (DPP).

Mykkeltveit and E. Helvik, (2008) proposed a conservative sharing protocol and a preemptive sharing protocol for availability-aware connection provisioning. The schemes are proposed to be centralized. Ho et al. (2007) proposed an availability model for SBPP based on spare capacity availability within the partial protection/restorability concept for Generalized Multi-Protocol Label Switching networks.

Based on the literature survey, it seems that most of the availability-aware connection provisioning schemes consider DPP and SBPP, and use linear connection availability analysis approaches. The majority of the proposed schemes are centralized rather than distributed. HT-AGSDP provides enhancement to the conventional connection provisioning scheme CAFES. Moreover, most of the published work deals with networks that have either a centralized provisioning system or nodes that have global information regarding the resource usage. Furthermore, all of the existing work deals with nodes that have full wavelength conversion capability. As a result, there is a high demand for a distributed protocol to provide scalable and robust availability-aware provisioning.

AVAILABILITY ANALYSIS IN WDM MESH NETWORKS

Many researchers have proposed different analysis to compute connection availability. We present here the analysis proposed by Trivedi (1982) and Arci, et al. (2003). Arci, et al. (2003) analyze the availability of a system (e.g., a component, path, connection) in a mesh networks. The availability of a system is the fraction of time in which the system is "up" during the entire service time. If a connection c is carried by a single path, its availability (denoted by A_c) is equal to the path availability. If c is dedicated or shared protected connection, then A_c will be determined by both the primary and the backup paths.

Table 1. Failure rates and repair times

Metric	Bellcore Statistics
Equipment MTTR	2 hrs
Cable-Cut MTTR	12 hrs
Cable-Cut Rate	4.39/yr/1000 sheath miles
Tx failure rate in *FIT*	10867
Rx failure rate in *FIT*	4311

Network Component Availability

A network component's availability can be estimated based on its failure characteristics. Upon the failure of a component, it is repaired and restored to be "as good as new". This procedure is known as an alternating renewal process. Consequently, the availability of a network component j (denoted as a_j) can be calculated as follows (To and Neusy, 1994 and Arci, et al.,2003):

$$a_j = \frac{MTTF}{MTTF + MTTR} \quad (1)$$

In particular, the *MTTF* of a fiber link is distance-related and can be derived according to measured fiber-cut statistics. Table 1 shows some typical data on the failure rates and failure repair times of network components (transmitters, receivers, fiber links, etc.) (To and Neusy, 1994). In Table 1, *FIT* (failure-in-time) denotes the average number of failures in 10^9 hours. *Tx* denotes optical transmitters while *Rx* denotes optical receivers.

End-to-End Path Availability

Given the route of path *i*, the availability of *i* (denoted as A_i) can be calculated based on the known availabilities of the network components along the route. Path *i* is only available when all the network components along its route are available. Let a_j denote the availability of network component j. Let G_i denote the set of network components used

by path *i*. Then, A_i can be computed as follows (Arci, et al., 2003):

$$A_i = \prod_{j \in G_i} a_j \quad (2)$$

Availability for a Dedicated-Path-Protected Connection

In path protection, connection *c* is carried by one primary path *p* and protected by one backup path *b* that is link disjoint with *p*. By link disjoint, we mean that the backup path for a connection has no links in common with the primary path for that connection. Node failures can also be accommodated by making the primary and the backup paths node disjoint as well. However, node failures are important to protect against in scenarios where an entire node (or a collection of nodes in a part of the network) may be taken down, possibly due to a natural disaster or by a malicious attacker. In this study, we require the primary and backup paths of a connection to be link-disjoint and only consider link failures in the availability analysis. Extensions to include node failures when computing connection availability are open problems for future research.

If the wavelength(s) of the backup path *b* are dedicated to connection *c*, then, when primary path *p* fails, traffic will be switched to *b* if *b* is available; otherwise, the connection becomes unavailable until the failed component is restored. *c* is "down" only when both paths are unavailable, so it can be computed straightforwardly as follows (Arci, et al., 2003):

$$A_c = 1 - (1 - A_p) \times (1 - A_b) = A_p + (1 - A_p) \times A_b \quad (3)$$

where A_p and A_b denote the availabilities of paths *p* and *b*, respectively. Note that a connection may employ multiple backup paths to increase its availability. If all backup paths are disjoint and dedicated to this connection, the connection availability can be derived following the same principle used in the previous equation.

Availability for a Shared-Path-Protected Connection

In shared-path protection, connection c is carried by primary path p and protected by a link-disjoint backup path b; however, the reserved wavelength on each link of b can be shared by other connections as long as SRLG constraints can be satisfied. Let SGc contain all the connections that share some backup wavelength on a link with c. We denote the sharing group of c as SGc.

The availability of connection c will be affected by the size of SGc and the availabilities of the connections in SGc. When one or more primary paths of the connections fail together with c, either c or some of the failing connections in it can acquire the shared backup wavelengths. With the values of SGc connection availabilities, we can now compute the availability of a shared-path-protected connection. A connection is available if: 1) path p is available; or 2) p is unavailable, b is available, and other primary paths of connections in the sharing group are also available. Therefore, A_c can be computed as follows (Arci, et al., 2003):

$$A_c = A_p + (1 - A_p) \times A_b \times \prod_{t_i \in SGc} A_{t_i} \qquad (4)$$

where A_p and A_b denote the availabilities of paths p and b respectively, and A_{t_i} is the availability of the primary path of the connection in SRLG.

DISTRIBUTED AVAILABILITY-AWARE PROVISIONING FRAMEWORK

Based on the availability analysis, we have developed a distributed connection-provisioning framework in which differentiated protection

services can be provided to each connection according to its predefined availability requirement. We first formulate the problem statement. Then we discuss how to compute the paths with the highest availability between a node pair in the network, which is referred to as the K-Most-Reliable Paths (KMRPs). Then, we propose a distributed framework to provision connections cost-effectively while satisfying the connections availability requirements by choosing appropriate protection schemes in a distributed manner.

Problem Statement

We present a distributed availability-aware provisioning framework for WDM networks, including a distributed approach with dedicated-path protection, shared-path protection, and no protection as the candidate protection services. We are given the following inputs to the problem:

1. Let $T = (V, E, A)$, T is the physical network topology where V is the set of nodes, E is the set of unidirectional fiber links, and A is set of link availabilities (it denotes the set of real numbers between 0 and 1). We assume that the nodes do not have wavelength converters.

2. Let $c = (s, d, A_c, h)$ a connection request that needs to be provisioned, where s is the source, d is the destination, and A_c is the availability requirement of request c. We assume that each connection c requires one full wavelength channel capacity. Under a dynamic traffic pattern, a path which has been set up between the members of a node pair to satisfy a connection request is taken down after a period of time h called the connection holding time.

Our goal is to provision differentiated services, i.e., provide either an unprotected, shared-path protected, or dedicated-path protected connection; such that the SLA requirement is met while minimizing the total network cost (wavelength links in particular). To utilize network resource usage, the framework attempts to categorize the connection requests into three categories by comparing the availabilities of MRPs with A_c as described above. The three categories are: *C1*, containing unprotected connections; *C2*, containing shared protected connections; and *C3*, containing dedicated protected connections. Algorithm 1 provides different treatments for different connections, as follows:

- In *C1*, one path is needed to carry each connection. The algorithm tries to find the path that can satisfy the connection availability requirements while minimizing the resources consumption.
- Shared-path protection is considered to protect connections in *C2*. The problem is to provide shared-path protection while satisfying the connections availability requirements.
- Dedicated-path protection is considered to protect connections in *C3*. The problem is to provide dedicated-path protection while satisfying the connections availability requirements.

The algorithm categorizes and provisions the new incoming connection as a *C1* connection as long as the connection availability requirement can be met. If not, the algorithm tries to treat the connection as a *C2* connection as long as the connection availability requirement can be met by the shared-protection scheme. Otherwise, the algorithm tries to categorize the connection as a *C3* connection before deciding to block it if the connection availability requirement cannot be met.

Compute the K Most Reliable Paths

In order to meet the SLA availability requirement, each node in the proposed framework computes the *k* most reliable paths (KMRPs)—the paths with the highest availability—to each other node. The node then saves these paths into the local database in order to use them in the routing process as fixed alternative paths. In the proposed framework, we set *k* equal to three as is recommended in the literature. To compute the most reliable paths, we use the Multiplication-to-Summation conversion technique proposed by Zhang et al. (2007). Suppose that a single path p is used to carry connection c. The availability of c is equal to the multiplication of the availabilities of the components it traverses. Suppose that path p traverses links $l_1, l_2, ..., l_n$:

$$1. A_p = A_{l_1} \times A_{l_2} \times A_{l_3} \times ... \times A_{l_n} \qquad (5)$$

where A_{l_i} is the availability of link i. If we compute the logarithm of both sides of (5), we can convert the multiplication to summation and obtain

$$\log A_p = \log A_{l_1} + \log A_{l_2} + \log A_{l_3} + ... + \log A_{l_n} \qquad (6)$$

Since A_p and A_{l_i} are between 0 and 1, $\log A_p$ and $\log A_{l_i}$ have negative values. Multiplying both sides by -1, we get

$$-\log A_p = -\log A_{l_1} - \log A_{l_2} - \log A_{l_3} - ... - \log A_{l_n} \qquad (7)$$

Now we can observe that, if the cost of a link is defined as a function of its availability (i.e. $-\log A_{l_i}$), the cost is additive and the path with the minimum cost will be the path with maximum availability (the most reliable path). Through this Multiplication-to-Summation conversion technique, a standard modified shortest path algorithm (such as a modified Dijkstra or Bellman-Ford algorithm) can be applied to compute the KMRPs.

Each node saves each of the KMRPs into the local database with its availability by computing the exponential of -1 multiplied by the cost of the path (i.e. $e^{-(-\log A_p)}$).

If the availability of one of the paths is larger than A_c, we know that protection is not needed for connection c. Therefore, we can categorize a connection as either an unprotected connection whose availability requirement can be satisfied without using any backup path, or as a protected connection if otherwise.

Distributed Availability-Aware Routing Protocol

In a wavelength-routed WDM network, a lightpath must be established between a pair of source and destination nodes before the data can be transferred. To establish a lightpath under distributed control, the network must first decide on a route for the connection and then reserve a suitable wavelength on each link along the chosen route. In the proposed distributed framework, each node in the network is required to maintain a routing table that contains an ordered list of KMRPs to each destination node. The routing is fixed-alternate-routing based, i.e. for each node pair, K-link-disjoint candidate routes (KMRPs) are pre-computed, and the availability of each route is calculated. Therefore, for each connection request $c \rightarrow (s, d)$, the source node can select the candidate routes in order to probe them. In addition to the static routing table, each node also maintains another local database for dynamic information. This database reflects the local resource usage at that node (e.g., the status of wavelength usage) as well as sharing knowledge that contains information about the lightpaths whose shared backup paths traverse that node (see Figure 1). This information is required to assist the framework in determining whether a wavelength on the outgoing link is shareable, and to compute the availability in the case of shared protection. For the purpose of

computing the availability of the shared protected connection, we add a new field "*WP_Availability*" to the dynamic part of the local database as shown in Figure 1. This field represents the availability of the working paths which is protected by the outgoing wavelength. Algorithm 1 describes the control and management mechanism.

The basic signaling components to establish a lightpath are built on top of destination routing with backward reservation protocols. However, for availability-aware routing purpose, extra work has been done by each node in the candidate paths. That is in addition to use Parallel Multi-Purpose Probe messages (PMP Probe message). Therefore, the procedure to establish an availability-aware connection can be described as follows:

- When a new connection request arrives in the network, the source node prepares PMP probe messages (PROB), see Figure 2, for each candidate path (i.e. for each MRP). These messages contain information about candidate paths including the connection ID, the requested availability level of the connection, the availability of the path, the link state of outgoing links, and the other two alternative paths each with a vector of values (say, CSfCP2 and CSfCP3) for the purpose of probing the path as a shared protection path for other candidates. Each value of the vector belongs to one of the outgoing wavelengths and represents the multiplication of the availabilities of the working paths of the connections protected by that wavelength. After preparing the PROB messages, the source node sends these k PROB messages toward the destination node through the link-disjoint KMRPs in parallel to collect the recent link-state information.

- Upon receiving the PROB message, the intermediate node examines the local link-state information and updates the PROB message. It updates the shareability infor-

Figure 1. Network architecture and local database

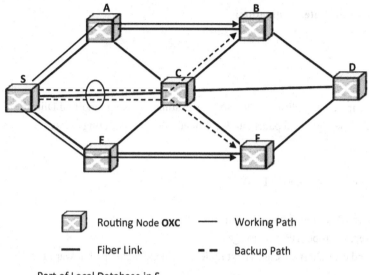

Part of Local Database in S

Usage Information

λ	Status	Connection_ID
0	1	13
1	0	
2	R	
:		

Sharability Information

λ	Connection_ID	Working path	WP_Availablity
2	1	SAB	0.999
2	2	SEF	0.998
:			

mation in the received PROB by examining the shareability of each channel along the probed path with respect to the other candidates. This is done by checking each wavelength in the outgoing probed link; if it is shareable to the candidate working path, then the node updates the value belonging to that wavelength in the received

Figure 2. PMP probe message

Con_ID	S	D	CP1	CP2	CP3	WLS	CSfCP2	CSfCP3

Con_ID: Connection ID   S: Source   D: Destination
CP1 : The First Candidate Path   CP2: The Second Candidate Path
CP3: The Third Candidate Path   WLS: Wavelength Status of the path as working path
CSfCP2: Shareability of the path as Shared Protection for the Second Candidate Path
CSfCP3: Shareability of the path as Shared Protection for the third Candidate Path

Algorithm 1. Distributed availability-aware connection control and management protocol

Status: S = {Source, Intermediate, Destination}
SInitial { Source}
Source_Node
    Spontaneously
        Begin
            • Compute k link disjoint paths to each destination using modified Dijkstra.
            • Save the computed paths into local fixed alternative routing database.
        End

    Receiving (Connection Request (SLA))
        Begin
            • Give ID for the requested connection.
            • Prepare Probe (BRB) massages.
            • Send each PRB messages next node in each candidate path toward the destination.
        End
    Receiving (Primary_Path_Reservation (RESV))
        Begin
            • Primary_Resevation = true;
            • Setup the switch for the outgoing link;
        If (Protection_Reservation) **then**
                • Set Connection_End_Time = Current _Time + connection_Holding_time;
                • Start data Transmission
            Else
                • Wait;
        End
    Receiving (Protection_Path_Reservation (RESV))
        Begin
            • Protection_Resevation = true;
            • Setup the switch for the outgoing link;
        If (Primary_Reservation) **then**
                • Set Connection_End_Time = Current _Time + connection_Holding_time;
                • Start data Transmission
            Else
                • Wait for timeout;
        End
    Transmission_end (ConID)
        Begin
            • Release switch of outgoing link;
            • Send Release (REL) message to the next node in the primary path toward the destination;
        End
Intermediate_Node

continued on following page

Algorithm 1. Continued

    Spontaneously

        Begin

            • Compute three link disjoint paths to each destination using modified dijkastra

            • Save the computed paths into local fixed alternative routing database

        End

    Receiving (PROB)

        Begin

            • Update Probe (BRB) massages based on wavelength availability and the local database;

            • Forward PRB message to the next node toward the destination node;

        End

    Receiving (Primary_Path_Reservation (RESV))

        Begin

            • Setup the switch for the outgoing link;

            • Forward PPRESV message to the next node toward the source node;

        End

    Receiving (Protection_Path_Reservation (RESV))

        Begin

            • Protection_Resevation = true;

            • Setup the switch for the outgoing link (in case of dedicated protection path);

            • Update the sharing database(in case of shared protection path);

        End

    Receiving (REL (conID)

        Begin

            • Release switch of outgoing link;

            • Send Release (REL) message to the next node in the primary path toward the destination;

        End

Destination_Node

    Spontaneously

        Begin

            • Compute three link disjoint paths to each destination using modified Dijkstra

            • Save the computed paths into local fixed alternative routing database

        End

    Receiving (PROB)

        Begin

                 If (Number_of_Receiving_Probe $< k$) **then**

            • Number_of_Receiving_Probe +=1;

            • Wait for timeout;

                 Else

            • Select a route and a wavelength of the primary path;

            • Select a route and a wavelength of the protection path;

            • Create reservation message for the primary path;

continued on following page

Algorithm 1. Continued

        • Send RES message backward to the source node;

      EndIf

  End

Receiving (Primary_Path_Reservation (RESV))

    Begin

        • Setup the switch for the outgoing link;

        • Forward PPRESV message to the next node toward the source node;

    End

Receiving (Protection_Path_Reservation (RESV))

    Begin

        • Protection_Resevation = true;

        • Setup the switch for the outgoing link (in case of dedicated protection path) ;

        • Update the sharing database (in case of shared protection path);

    End

Receiving (REL (conID)

    Begin

        • release process is complete

    End

vector by multiplying it with the availabilities of other primary paths protected by the wavelength. Otherwise, if the wavelength is not shareable, the node sets the value of that wavelength to zero.

Each intermediate node determines the shareability of the wavelength by retrieving the candidate working path of the current connection from the received PROB packet. It then checks with its shareability database to determine whether the candidate working path belongs to the same SRLG of the working paths of the connections protected by this wavelength. If it does not belong to the same SRLG, it is shareable, and therefore the algorithm multiplies the value belonging to that wavelength in the received vector by the availabilities of the working paths of the connections protected by that wavelength, after excluding the redundant working paths. Otherwise, if it is

not shareable, the node sets the value of that wavelength to zero.

• When receiving the *k* connection probes, the destination node first examines the sets of the remaining wavelengths that are free. Then it runs an adaptive routing mechanism to select the optimal path as well as the wavelength for the working path. Furthermore, the destination node also has the ability to select a backup path, either dedicated or shared, at the same time that it selects the primary path. This issue is discussed below in subsection 4.4. After selecting the path(s), the destination node starts backward reservation. Notice that, the reservation process in the case of shared protection path is different since it includes updating the shareability database at intermediate nodes by adding information about the new connection. This information includes the ID of the new connection and the route along which the working path has been selected, as well as the availability of that working path, all of which is added to the RESV message.

Differentiated Protection Service and PMP Probe

A connection can be either unprotected or protected. In order to further reduce the network resource usage without sacrificing service availability, we can protect a connection through either dedicated-path protection or shared-path protection based on the availability requirements of this connection and of all the connections in its sharing group. The destination node decides to assign a route(s) and wavelengths to the connections using the Availability-Aware Routing and Wavelength Assignment (AA-RWA) algorithm being presented. This algorithm selects the working path and decides which type of protection should be provided in order to satisfy the required availability value, say SLA_V, of the connection while minimizing the resource usage. This kind of protection service is called differentiated protection. Algorithm 2 describes path(s) routing and wavelength(s) assignment. This algorithm runs by the destination node of the connection to assign the shortest path that satisfies the SLA_V if such a path exists, and has a free wavelength. Otherwise, it selects the shortest working and shared protection paths to the connection and computes the availability of this candidate combination. If the availability of one of the combinations satisfies the SLA_V and has free wavelengths, the algorithm assigns them to the connection. If no shared protection availability can satisfy the SLA_V, the algorithm tries to select working and dedicated protection paths to satisfy the SLA_V of the connection before deciding to block the connection if none of the above choices are sufficient. AA-RWA needs extensive information about the candidate paths in order to know what each candidate may be (i.e. working or dedicated/shared protection path).

In all previous research, the PROB messages collect information from the visited nodes regarding one purpose: either to collect information about the possibility of using them as a primary, or to collect information about the possibility of using them as a backup path. In this case, numerous probe messages are required to pass several times across each of the candidate paths, which entail massive control overhead.

By using the concept of Parallel Multi-Purpose Probe messages (PMP Probe), the PMP messages probe the outgoing links in the visited node for many purposes:

1. Examining the ability of the outgoing link to be part of the candidate working path;
2. Examining the ability of the outgoing link to be part of the candidate dedicated protection path; and
3. Examining the ability of the outgoing link to be part of the candidate shared protection path for one or more candidate primary paths.

So, with k number of PMP probes, each of them passing one of the k candidate paths, we can generate all the information that the destination node needs to provide to the differentiated protection service.

TRACKING THE AVAILABILITY CONSTRAINTS OF EXISTING CONNECTIONS

Before establishing the new shared protected connection, it is very important to check whether the service availabilities of connections currently participating in the sharing will still be met. In a distributed provisioning framework with no available global information, it is a major challenge to check the availability constraints of the existing connections before allowing the provisioning of the new connection that will share one or more backup links with them. This challenge lies in that if the new connection shares one or more links with the existing connections, then the availability of the existing connections is affected. To deal with this problem, we propose two novel schemes; the Shareability per Spare Channel Controller (SSCC) and the Distributed Availability-Constraints Controller (DACC).

Because tracking the availability of existing connections is very difficult and time-consuming, especially in distributed controlled networks, we propose SSCC to avoid re-computing the availabilities of the existing connections to check whether their availability requirements can still be met. SSCC avoids re-computing the availabilities of the existing connections by controlling the shareability per spare channel, which is the number of connections that share the spare channel. To select the proper shareability per channel, we have searched the literature on availability analysis to learn what other researchers suggested. After combining the effects of backup sharing on the availability and capacity costs in the form of trade-off curves, Doucette, et al. (2003) devised a guideline that suggested limiting SBPP shareability to two or three primary paths per spare channel at most.

In DACC, we do not place any explicit limits on the shareability per spare channel. Instead, the channel shareability is automatically controlled by the availability requirements and the availabilities of connections in the sharing group, which provides more flexibility. Now, let us describe how DACC controls the availability constraints. We start from the availability constraints of the new connection:

$$A_w + (1 - A_w) \times A_b \times \prod_{t_i \in SRLG} A_{t_i} \geq SLA_v \quad (8)$$

If a new connection stratifies the availability constraints, let LP_c be the last product in the availability of the connection c, and EA_c be the extra availability that is assigned to the connection c. LP_c and EA_c can be computed as follows:

$$LP_c = (1 - A_w) \times A_b \times \prod_{t_i \in SRLG} A_{t_i} \quad (9)$$

$$EA_c = SLA_v - A_w \quad (10)$$

DACC appends the value of LP_c and EA_c in the RESV message of the shared protection path. After receiving the RESV message, each node saves the LP_c and EA_c in the shareability database. The node then updates the last product (LP_E) for each existing connection that has been protected by the outgoing channel; this occurs only if the channel can back up the new connection without violating the availability requirement of the connections that is already protecting. As a result, the channel accepts the backup of the new connection if and only if the following condition is satisfied for each existing connection protected by the channel:

$$A_w * LP_E \geq EA_E \quad (11)$$

Where A_w is the availability of the working path of the new connection, and EA_E is the extra availability assigned to the existing connection. With the distributed process mentioned above, the DACC scheme provides availability-guaranteed wavelength provisioning. Notice here that after reserving and releasing a new connection, the value of LP_E changes respectively as follows:

$$LP_E = A_w * LP_E, \text{ and} \quad (12)$$

$$LP_E = LP_E / A_w \quad (13)$$

PERFORMANCE EVALUATION

To evaluate the performance, we carry out a connection setup time analysis and a simulation study to show the performance of the presented

Algorithm 2. AA-RWA

Let*PP1* be the shortest path of the *KMRPs*;

Let*PP2* be the second shortest path of the *KMRPs*;

Let*PP3* be the third shortest path of the *KMRPs*;

Let*SLA$_v$* be the required availability of the connection.

ProtectionType = 0; // No path(s) have been selected for the connection

w =1;

While (w <= 3 and *ProtectionType !=0*)

    If*Avail*(*PPw*) >= *SLA$_v$***and** there is a free wavelength in *PPw***then***ConnectionWorkingPath* = *PPw*;

      ProtectionType = 1; // One of *KMRPs* satisfies the required availability (Unprotected)

    Endif

    w=w+1; // to check next working path candidate

    loop;

If (*ProtectionType* == 0) **then** // nothing assigned yet

    w =1;

    While (w <= 3 and *ProtectionType !=0*)

    s =1;

      While (s <= 3 and *ProtectionType !=0*)

        If (w != s**and** there is a free wavelength in *PPw***and** there is a shareable or free wavelength in *PPs*) **thenLet**

        Avail (*PPw* + *PPs*) be the value of Equation (4) by considering PPw as a working path and *PPs* as a shared

        protection path;

          If*Avail*(*PPw* + PPs) >= *SLA$_v$***then***ConnectionWorkingPath* = *PPw*;

          ConnectionSharedProtectionPath = *PPsProtectionType* = 2; // satisfies the required availability

          (Shared Protection)

        EndIf

        EndIf

      s=s+1; // to check next protection path candidate

    loop;

    w=w+1; // to check next working path candidate

    loop;

If (*ProtectionType* == 0) **then** // nothing assigned yet

    w =1;

    While (w <= 3 and *ProtectionType !=0*)

    d=1;

    While (d <= 3 and *ProtectionType !=0*)

      If (w != d**and** there is a free wavelength in *PPw***and** there is a free wavelength in *PPd*) **thenLet***Avail* (*PPw* +

      PPd) be the value of Equation (3) by considering *PPw* as a working path and *PPd* as a dedicated protection

      path;

      If Avail(*PPw* + *PPd*) >= *SLA$_v$***then***ConnectionWorkingPath* = *PPw*;

        ConnectionDedicatedProtectionPath = *PPd*;*ProtectionType* = 3; // satisfies the required availability (dedi-

        cated Protection)

      EndIf

continued on following page

Algorithm 2. Continued
EndIf

$d=d+1$; // to check next protection path candidate

loop;

$w=w+1$; // to check next working path candidate

loop;

If (*ProtectionType* != 0) **then**

Start the reservation process;

Else

Block the connection;

EndIf

protocol in terms of connections setup time and the connections blocking probability.

Connections Setup Time Analysis

In this section, we describe analytically the Connection Setup Time (CST), the time the framework takes to setup a lightpath. First, some notations and assumptions must be stated:

- Message processing time at each node is *pt*.
- Time to configure, test and setup a switch is *ct*.
- Time to configure, test and reserve as shared resource wavelength is *tr*.
- Average propagation delay on each fiber is *fd*.
- Number of hops along the longer candidate path is h_c.
- Number of hops along a working path is h_w.
- Number of hops along a protection path is h_p.

CST of Unprotected Connections

Let T_{Prob} be the time to probing the candidate paths, let T^w be the time to setup a working path,

and let CST_U be the connection setup time in unprotected scheme.

$$T_{Prob} = h_c \times fd + (h_c + 1) \times pt. \qquad (14)$$

$$T^w = T_{Prob} + h_w \times fd + (h_w + 1) \times (pt + ct). \qquad (15)$$

$$CST_U = T^w. \qquad (16)$$

CST of the Shared Protected Connections

Let T_S^P be the time to setup a shared protection path, and let CST_S be the connection setup time in shared protection scheme.

$$T_{Prob} = h_c \times fd + (h_c + 1) \times pt. \qquad (17)$$

$$T^w = T_{Prob} + h_w \times fd + (h_w + 1) \times (pt + ct). \qquad (18)$$

$$T_S^P = T_{Prob} + h_p \times fd + (h_p + 1) \times pt. \qquad (19)$$

$$CST_S = \text{Max}(T^w, T_S^P) \qquad (20)$$

CST of the Dedicated Protected Connections

Let T_D^P be the time to setup a dedicated protection path, and let CST_D be the connection setup time in dedicated protection scheme.

$$T_{Prob} = h_c \times fd + (h_c + 1) \times pt. \tag{21}$$

$$T^w = T_{Prob} + h_w \times fd + (h_w + 1) \times (pt + ct). \tag{22}$$

$$T_D^P = T_{Prob} + h_p \times fd + (h_p + 1) \times pt + ct). \tag{23}$$

$$CST_D == Max(T^w, T_D^P). \tag{24}$$

Note that the request probing and reservation in both schemes are carried out in parallel because the setup time is the maximum setup time of the working path and protection path. Moreover, unlike the dedicated protection path, the shared protection path does not require a switch in configuration at each node along the path. It does require each node to update its local database.

Simulation Study

The performance of the proposed protocol is also evaluated via extensive simulations of the mesh-based 14-nodes NSFnet (Figure 3), all links in the network are assumed to have one fiber and each fiber has the same number of wavelengths. Here we show the results for the case of a single fiber having 16 wavelengths and no wavelength converters. We also assume the lightpaths to be bidirectional. Connection requests arrive as a Poisson process with mean arrival rate λ. The holding time μ of each connection is exponentially distributed. Thus the load is given by λ/μ. The destination of each request is uniformly distributed. The channel availability model and the corresponding MTTF and MTTR values used by Arci, et al. (2003) are used to obtain the availability. The availability requirements of the requests are uniformly distributed among five classes: 0.999, 0.9993, 0.9995, 0.9998 and 0.9999.

We study the performance of our proposed framework with the multipurpose probing technique and the DACC scheme. Figure 4 shows a comparison between our proposed framework and the source routing sharing protection scheme (SR) in terms of the network blocking probability (BP). The performance of our framework and

Figure 3. 14-node NSFnet backbone topology

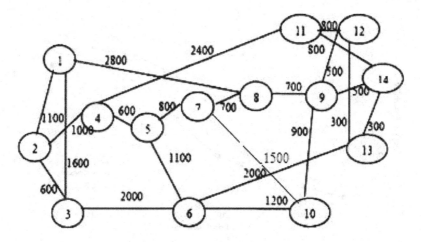

Figure 4. Blocking probability: Our framework vs. source routing

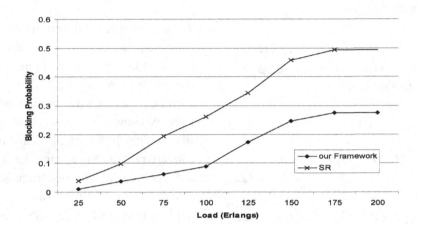

the performance of SR are closed at the beginning. This is due to the light load and the network resources still being sufficient. As can be seen from the figure, when the load becomes heavy the network performance of our framework is remarkably better than that of SC. This is due to the efficient resource utilization performed by the framework, as it is able to provide Differentiated Protection Services as a benefit of using the multipurpose probing. Moreover, as can be seen from Figure 4, when the load is very small, the blocking probabilities and their differences are also very small. As the load increases, the blocking probabilities increase remarkably and the difference becomes significant.

The performance of the proposed DACC scheme is also evaluated by comparing it with SSCC with 2 and 3 connections per spare channel. The performance of DACC and SSCC with both sharing degrees is similar at low traffic (see Figure 5). This is due to the light load and the network resources still being sufficient. However, when the load becomes heavy the network performance under DACC and SSCC with sharing degree 3 is

Figure 5. Blocking probability: DACC vs. SSCC-2 and SSCC-3

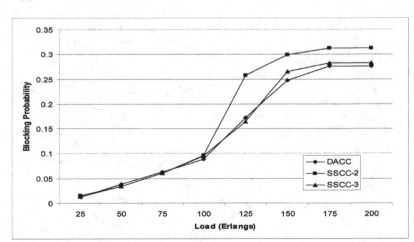

Figure 6. Resource consumption vs. load

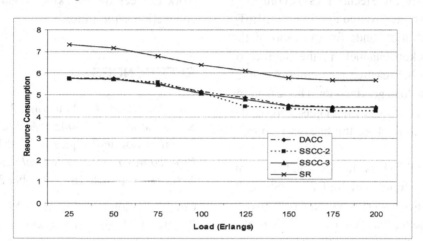

Figure 7. Resource over build vs. load

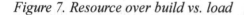

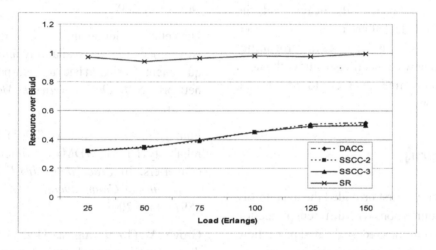

better than the performance under SSCC with sharing degree 2. This is due to the efficient resource utilization preformed by DACC and SSCC with sharing degree 3. Once the load becomes heavier, DACC gives better performance. This is because DACC's dynamicity allows the new connection to share the backup link if the availabilities of the existing connections that are protected by the backup link have not been violated. Moreover, as can be seen from Figure 5, when the load is very small, the blocking probabilities and their difference are also very small. As the load increases, the blocking probabilities increase remarkably and the difference becomes significant.

The average resource consumption (ARC) is the average number of channels needed to support the connection. The proposed framework always requires fewer channels to provision the connection. So, with the proposed framework, we can provision more connections using a smaller number of channels. It can be seen from Figure 6 that the ARC tends to decrease with increasing loads, because there is a higher probability to share at high loads.

In Figure 7, the three techniques are compared in terms of average resource overbuild (ARoB). Resource overbuild stands for the ratio of the number of backup channels to the number of working channels. In each case, resource overbuild decreases as the load gets heavier. The reason for this behavior is that, as the load gets heavier, the connections tend to share the backup resources more.

Figures 6 and 7 show the ARC and ARoB in our provisioning framework for DACC, SSCC-2, SSCC-3, and SR. The performance of both DACC and SSCC is better than that of SR. This is also due to the efficient resource utilization done by the proposed framework, which provides Differentiated Protection Services. The average resource consumption becomes small with the heavy load. This is normal; with the heavy load, the number of channels that are reserved for shared protection is large. This increases the probability of backing up new connections without the need to reserve many free channels to protect the new connections.

CONCLUSION

In this chapter, we first provided the availability analysis for connections with different protection schemes (i.e., unprotected, dedicated protection, or shared protection). Through this analysis, we show how connection availability is affected by resource sharing. Based on the availability analysis, we develop a novel distributed provisioning framework in which an appropriate level of protection is provided to each connection according to its predefined availability requirement. Our distributed framework includes approaches to control and manage connections in a distributed fashion that increase scalability and reduce control overhead. The effectiveness of the proposed provisioning framework is demonstrated through connection setup time analysis and a simulation study. We have shown that the presented frame-

work reduces the blocking probability and the connection setup time.

REFERENCES

Arci, D., Maier, G., Pattavina, A., Petecchi, D., & Tornatore, M. (2003). Availability models for protection techniques in WDM networks. In *Proceedings Design of Reliable Communication Networks*, (pp. 158-166). October 2003.

Cavdar, C., Song, L., Tornatore, M., & Mukherjee, B. (2007). Holding-time-aware and availability-guaranteed connection provisioning in Optical WDM mesh networks. In *Proceedings IEEE International Symposium on High Capacity Optical Networks and Enabling Technologies*, (pp. 1-7). November 2007.

Doucette, J., Clouqueur, M., & Grover, W. (2003, November). On the availability and capacity requirements of shared backup path-protected mesh networks. *SPIE Optical Networks Magazine, 4*(6), 29–44.

Fumagalli, A., & Tacca, M. (2001). Diferentiated reliability (DiR) in WDM ring without wavelength converters. In *Proceedings IEEE International Conference on Communications* (ICC), (pp. 2887-2891). June 2001.

Grover, W. (1999, August). High availability path design in ring-based optical networks. *IEEE/ACM Transactions on Networking, 7*, 558–574. doi:10.1109/90.793028

Ho, P., Mouftah, H., & Haque, A. (2007). Availability constrained shared backup path protection (SBPP) for GMPLS-based spare capacity reconfiguration. In *Proceedings IEEE International Conference on Communications* (ICC), (pp. 2186-2191). June 2007.

Mykkeltveit, A., & Helvik, E. (2008). Comparison of schemes for provision of differentiated availability-guaranteed services using dedicated protection. In *Proceedings IEEE International Conference on Networking* (ICN08), (pp. 78-86).

Schupke, D. (2000). Reliability models of WDM self-healing rings. In *Proceedings Design of Reliable Communication Networks*, Munich, Germany, April 2000.

Song, L., Zhang, J., & Mukherjee, B. (2007). Dynamic provisioning with availability guaranteed for differentiated services in survivable mesh networks. *IEEE Journal on Selected Areas in Communications*, *25*, 35–43. doi:10.1109/TWC.2007.024505

Tacca, M., Fumagalli, A., Paradisi, A., Unghvary, F., Gadhiraju, K., & Lakshmanan, S. (2003). Differentiated reliability in optical networks: Theoretical and practical results. *Journal of Lightwave Technology*, *21*, 2576–2586. doi:10.1109/JLT.2003.819554

To, M., & Neusy, P. (1994). Unavailability analysis of long-haul networks. *IEEE Journal on Selected Areas in Communications*, *12*, 100–109. doi:10.1109/49.265709

Tornatore, M., Maier, C., & Pattavina, A. (2006). Availability design of optical transport networks. *IEEE Journal on Selected Areas in Communications*, *24*, 1520–4532.

Trivedi, S. (1982). *Probability and statistics with reliability, queuing, and computer science applications*. Englewood Cliffs, NJ: Prentice-Hall.

Wei, X., Quo, L., Wang, X., Song, Q., & Li, L. (2008). Availability guarantee in survivable WDM mesh networks: A time perspective. *Elsevier Information Sciences, 178*(11).

Willems, G., Arijs, P., Parys, W., & Demeester, P. (2003). Capacity vs. availability tradeoffs in mesh-restorable WDM networks. In *Proceedings International Workshop on Design of Reliable Communication Networks* (DRCN03), (pp. 158—166). Alberta, Canada, 2003.

Zhang, J., Zhu, K., Zang, H., Matloff, N., & Mukherjee, B. (2007). Availability-aware provisioning strategies for differentiated protection services in wavelength-convertible WDM mesh networks. *IEEE Transaction on Networking, 15*(5), 1177–1190. doi:10.1109/TNET.2007.896232

ADDITIONAL READING

AlSukhni, E., & Mouftah, H. (2008). "Availability-Guaranteed Distributed Provisioning Framework for Differentiated Protection Services in Optical Mesh Networks, *Proceedings IEEE Globecom2008, International Workshop on Optical Networks*, New Orleans, Louisiana, November 2008.

AlSukhni, E., & Mouftah, H. (2008). "'Integrated Routing And Wavelength Assignment And Signaling in Shared Protection Framework For Survivable WDM Optical Mesh Networks", *Proc. IEEE 24th Queen's Biennial Symposium on Communications* (QBSC'2008), Kingston, Canada, June 2008, pp. 103-106.

AlSukhni, E., & Mouftah, H. (2008). "Parallel Fixed-Alternative-Routing Based Provisioning Framework for Distributed Controlled Survivable WDM Mesh Networks", *Proceedings IEEE Conference on Communication Networks and Services Research* (CNSR'2008), Halifax, Nova Scotia, May 2008, pp. 287-294.

AlSukhni, E., & Mouftah, H. (2010). "A Framework for Distributed Provisioning Availability-Guaranteed Least-Cost Lightpaths in WDM Mesh Networks", *In Proceedings IEEE International Symposium on Computers and Communications* (ISCC2010), Riccione, Italy, pp. 1-4, June 2010.

Guo, Q., Ho, P., Haque, A., & Mouftah, H. (2007) "Availability-Constrained Shared Backup Path Protection (SBPP) for GMPLS-Based Spare Capacity Reconfiguration", *In Proceedings IEEE International Conference on Communications* (ICC), 2007.

Ma, H., Fayek, D., & Ho, P. (2007). "Availability-Aware Multiple working-paths Capacity Provisioning in GMPLS Networks," *Springer Lecture Notes in Computer Science*, Vol. 4786/2007, pp.85-94, November 2007.

Mouftah, H., & Ho, P. (2003) "Optical Networks- Architecture and Survivability", Kluwer Academic Publishers, 2003, ISBN 1-4020-7196-5.

Mykkeltveit, A., & Helvik, B. (2008). "On provision of availability guarantees using shared protection," *In Proceedings IEEE International Conference on Optical Network Design and Modeling* (ONDM), pp. 1-6., March 2008.

Naser, H., & Mouftah, H. (2004). "Availability Analysis and Simulation of Shared Mesh Restoration Networks", *Proceedings Ninth IEEE International Symposium on Computers and Communications* (ISCC2004), Alexandria, Egypt, June 2004, pp. 779-785.

Tornatore, M., Lucerna, D., Song, L., Mukherjee, B., & Pattavina, A. (2008). "SLA Redefinition for shared-path-protection Connections with Known Duration," *In Proceedings IEEE Optical Fiber communications/ National Fiber Optic Engineers Conference* (OFC/NFOEC 2008), pp. 1-3, February 2008.

Tornatore, M., Major, C., & Pattavina, A. (2006, November). Capacity versus availability trade-offs for availability-based routing. *GSA Journal of Optical Networking, 5*, 858–869. doi:10.1364/JON.5.000858

KEY TERMS AND DEFINITIONS

Availability: The probability that a system will be found in the operating state at a random time in the future.

The Average Resource Consumption (ARC): Average number of channels needed to support the connection.

Differentiated Reliability: provide multiple reliability degrees using a common protection mechanism in optical networks.

K-**Most-Reliable Paths (KMRPs):** The paths with the highest availability between a node pair in the network.

Chapter 12
Self–Healing on Transparent Optical Packet Switching Mesh Networks:
Overcoming Failures and Attacks

Iván S. Razo-Zapata
ITESM, Mexico

Gerardo Castañón
ITESM, Mexico

Carlos Mex-Perera
ITESM, Mexico

ABSTRACT

This work presents a novel approach for dealing with failures and attacks on Transparent Optical Packet Switching (TOPS) mesh networks. The approach is composed of two phases, whereas the first one dynamically dimensions the resources in the network, the second one applies an incremental learning algorithm that generates an intelligent policy. At each node, such a policy allows a self-healing behavior when there are failures or attacks in the network. Finally, the performance of this approach is presented as well as future research lines.

INTRODUCTION

Transparent optical packet switching (TOPS) networks are becoming more and more attractive due to their ability to reduce power consumption and total cost; this cost reduction is obtained through the use of a lower number of transponders.

TOPS networks can also avoid the bottleneck of optoelectronic conversion and switching at each node (Yoo, 2006). However, transparency raises many security vulnerabilities as well as reliability issues that do not exist in traditional optoelectronic networks (Mas *et al.*, 2005).

Security and reliability issues are of utmost importance in transparent optical networks given

DOI: 10.4018/978-1-61350-426-0.ch012

the extremely large fiber throughput, the lack of optoelectronic conversions and the rapid propagation of attacks among other features. Fast and successful reaction and restoration mechanisms performed by failure management can prevent loss of large amounts of critical data, which can cause severe service disruption (Tomkos *et al.*, 2004).

In order to deal with failures and attacks (Skorin-Kapov *et al.* 2007) have already considered intelligent and self-organized mechanisms. The main argument is that autonomous elements of the network, such as optical switches, can dynamically adapt to changes due to failures and attacks by, for instance, reconfiguring data lightpaths.

In this work we propose a self-organized approach to deal with failures in transparent networks. Our approach relies on two phases. The first one is a dynamic dimensioning where TOPS resources are determined; one of our claims is that by a good dimensioning, the network provides higher flexibility to deal with failures and attacks. The second phase is an incremental learning process in which the TOPS network continuously learns a self-healing strategy to overcome failures and attacks. Multiple Path Routing (MPR) is used on both phases to route packets, *i.e.* nodes have knowledge of different paths for different targets (Castañón *et al.*, 2000).

PROBLEM DESCRIPTION

TOPS mesh networks are a relatively new technology for very high data rate communications, flexible switching and broadband application support. More specifically, they provide transparency features allowing routing and switching of data without interpretation or regression of signals within the network, *i.e.* without opto-electronic-opto conversions. Such networks contain only transparent optical components and therefore differ from the optical networks currently used. In particular, the nature of TOPS components and architectures brings about a new set of problems

for network security, such as the design of resilient mechanism for dealing with failures and attacks.

Before explaining some issues regarding design of resilient mechanisms, it is worthwhile making some comments about failures and attacks. As already established by Rejeb *et al.* (2006), failures occurs due to physical natural *fatigue* and *ageing* of optical devices. They occur once and remain within devices until they are repaired. Contrary, attacks appear and disappear often *sporadically anywhere* in the network, causing additional failures and problems in the network. Based on the previous argument, we assume that failures are subsumed by attacks. Consequently in the rest of the text, we refer to failures and attacks just as attacks.

When designing failure-resilient mechanisms for TOPS mesh networks, it is not only fast-response that becomes an issue but also problems as attack detection & location, and adaptability. In the next paragraphs we provide broader explanation regarding these issues.

Attack Detection and Location

Since in TOPS networks multiple optical signals co-propagate in fiber and optical components, possibly affecting each other directly or indirectly, the quality of a signal is sometimes dependent on or degraded by other signals making difficult the task of determining whether such degradation is caused by an attack or not. In addition, once the attack is detected, the localization task is still not trivial. Mas *et al.* (2005) have already described in which ways optical components can mask attacks in the network. For instance, a regenerator can mask an attack regarding power drop that occurred before the regenerator, therefore monitoring devices located after it are not able to locate the attack on that channel.

Even more, as discussed by Medard *et al.* (1998) signals can be maliciously designed to pass through transparent components, causing undesirable effects at remote components and degrading other signals passing through the components. While there are many reliability studies to defeat physical layer impairment problems in TOPS networks, such as optimal allocation of monitoring devices, adjustment of transmitters to avoid power drop, wavelength alignment to overcome In-band and Out-band Jamming among others, these measures require human intervention to adapt the network to attacks that are most likely to happen (Mas *et al.*, 2005). Apart from the component failures and accidental fiber cuts, networks are vulnerable to attackers capable of disrupting a network from the physical level up to the transport level. Usually, attackers are more interested in attacks that can be repeated and controlled especially if those attacks are rare. Furthermore, attackers may use automated software to repeatedly generate attacks. Attackers who artfully exploit well-known or newly discovered or unexpected vulnerabilities may circumvent the existing countermeasures.

Adaptability

Due to the sporadic and unpredictable nature of attacks, TOPS mesh networks must adapt themselves to the changes caused by the attacks. Such adaptability mainly involves two issues, providing an optimum number of resources to the network and iteratively maximizing the utilization of those resources. Whereas the first issue is a proactive measure that provides the flexibility required for adaptability, the second one is indeed a reactive measure to achieve adaptability.

In order to provide the optimum number of resources, the network topology has to be engineered to determine how many resources must be assigned. This process is commonly known as dimensioning and relies on heuristic algorithms which can compute the optimum number of resources for a given topology (Benjamin & Fouad, 2009).

Briefly, the input of this process is the model of a network topology with the minimum number of resources to allow communication among all the nodes. Nevertheless, this assignation of resources does not guarantee a low packet loss. Consequently, after applying some heuristic process, the output of the dimensioning process is the model of the network topology with the number of resources that guarantees a low packet loss (Eramo *et al.*, 2006). In this way, the lower the packet loss is, the more flexible the network is, and the more adaptability can be reached. Afterwards, such a model can be implemented in a real-world TOPS network.

As already explained, dimensioning the network will provide some flexibility to modify the scheduling of resources. For scheduling we mean assignation of resources for the packets within the network. For instance deciding whether a packet is sent through a fiber or retarded (delayed) in a delay line (DL); here fibers and DLs are examples of resources. In this sense by making use of its flexibility, the network can smartly modify the scheduling of resources to overcome attacks. This smart modification implies learning a strategy to schedule resources in such a way that the utilization of resources is nearly optimal and the packet loss is low. In addition, due to the high amount of traffic and the diversity of resources within the network, this learning strategy cannot be applied in a centralized way; therefore at each node a learning process must be applied.

To sum up, the problem of designing an attack-resilient mechanism for TOPS mesh networks relies on two issues: attack detection & location and adaptability. In this work, for sake of simplicity, we assume that detection & location is already solved by letting know the nodes whether TOPS components are properly working or not. Therefore, we only provide an approach for achieving adaptability. Finally, we also aim to achieve adaptability by means of self-organization at each TOPS node.

RELATED WORK

Network security countermeasures are categorized into three types of practices: prevention, detection and reaction (Rejeb *et al.*, 2006). Since attacks are achieved via physical layer impairments, limiting the physical layer vulnerabilities is of common interest to both reliability and security research. In transparent optical networks, prevention schemes that aim to reduce vulnerabilities include network design, component design, provisioning, and operational regulations, etc. In general, two approaches exist to assure reliable optical channels in the presence of physical layer impairments: routing constrained by estimated physical layer impairments and network architecture designed to guarantee the service quality in every possible case in the given network and traffic demand (Castañón *et al.*, 2009).

The purpose of self-organization is that if a network experiences significant physical layer impairment problems under certain network conditions, it learns them, and then tries to keep away from any of such conditions or unforeseen but similar conditions which are expected to produce similar or worse performance (Skorin-Kapov *et al.*, 2007). For example, instead of blindly using the first-fit route, a multi-path routing (MPR) scheme can be applied to choose the safest path, satisfying the packet or the burst of packets. It is well known that multi-path routing has many benefits, such as decreasing the number of components in an all-optical network, decreasing the use of optical memory (DLs) at the routers and decreasing the use of wavelength converters (Castañón *et al.*, 2000). Furthermore MPR also provides a quick way to solve contention of packets, faults, and attacks using an alternate routing (Castañón *et al.*, 2008).

The key difference from previous work, Yeom *et al.*, (2007) and Pavani et al., (2008), is that such intelligence is obtained without human intervention or without instantaneous detailed knowledge of the network component subsystem, apart from

its ability to self-organize autonomously as the network changes. We aim for an incremental learning approach, making the analogy to the human immunization system's primary defense mechanism, this incremental learning will act as the primary defense mechanism reacting timely to network problems and evolving based on the network information.

Basically, we are proposing the use of a self-healing approach as an instinct immediate network reaction against attacks in transparent networks. In this case routes at each node will be continually updated based on the state of the network. Since intelligence is distributed, every node has to send information to all other nodes. Our goal is to address this problem with autonomous adaptation against new vulnerabilities with possible extensions for attack detection & location (Medard *et al.*, 1998).

SELF-HEALING APPROACH

In this work we propose a self-organized approach to deal with attacks in transparent networks. Figure 1 depicts the general idea of our approach, which relies on two phases. The first one is a dynamic dimensioning where TOPS resources are determined. One of our claims is that good dimensioning provides higher flexibility to deal with attacks. Consequently, we also introduce attacks during this phase; in this way the network is provided with enough resources to overcome attacks happening during normal operation. In real-world TOPS networks, those attacks may come from authentic network elements failures or attacks. In our case we deactivate wavelengths in some fibers to simulate cross-talk attacks, whose aim is to block wavelengths in a fiber by injecting malicious connections with very high power energy (Mas et al., 2005).

In the second phase, attacks are also introduced, however this time a learning process is started. With this process the TOPS network continu-

Figure 1. Self-organized approach to deal with attacks

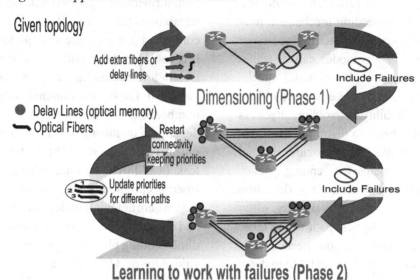

ously learns a self-healing strategy to overcome those attacks. MPR is applied on both phases to route packets, MPR uses a forwarding table with several output fibers ordered by priority. Initially this forwarding table may be created using the *k*-shortest paths based on the minimum hop routing. For example in case of a packet conflict, one of the packets will be forwarded through the output fiber with the best priority and the second packet can be forwarded through the output fiber with second priority in case the node does not have resources for contention resolution such as optical memory and wavelength converters. The final idea is that node's forwarding routing tables are continuously updated based on information delivered by neighbor nodes.

Dynamic Dimensioning

Dimensioning TOPS Networks must deal with issues such as the number and allocation of fibers, number of DLs for buffering, wavelength converters, number of wavelengths, among others that guarantee availability while minimizing the allocated spare capacity (Eramo *et al.*, 2006). Previous work focuses on single-node dimensioning

on the number of fibers and buffers (Aziz *et al.*, 2008). In the same way, (Danielsen *et al.* 1998) address the idea of increasing the number of wavelengths in order to have a bufferless TOPS network; however a bufferless network offers very low fiber utilization, therefore a large amount of bandwidth is wasted. Nevertheless, at this point we are mainly interested in determining the number of extra fibers and DLs. On the one hand, each node can use extra fibers to forward a higher number of packets, but on the other hand, nodes can use extra DLs to retard (delay) packets to provide contention resolution and increase fiber utilization. As can be observed in Figure 1, given a TOPS topology, our approach automatically determines the number of extra fibers and DLs. In addition, since we claim that good dimensioning provides more flexibility to deal with attacks, our approach also considers attacks while determining the number of resources. At each edge of the topology we include attacks that block wavelengths of the nodes connected by such edge.

Resource dimensioning *per se* is not enough for dealing with attacks, each node in the network must have also knowledge of how to forward packets from a current node to a target node. Since forwarding packets also relies on the number of resources, such forwarding mechanism must offer alternative routes for each source-target combination. In this way nodes can distribute traffic among different nodes to avoid attacks. So, the more resources and alternative paths are the greater the probability of avoiding attacks is. In this phase MPR helps to distribute the traffic among nodes and links. MPR provides, at each node, k-alternative routes for sending packets from the current node to a target node. These alternative routes are ordered by priority, where the first option is the shortest path to the target node. In this way whenever a node needs to send a packet, it can look at its routing table to determine through which output fiber the packet can be sent (Castañón *et al.*, 2000). Even more, by distributing traffic we also aim to minimize the amount of resources resulting in an increase (or maximization) of the resource utilization.

The architecture of an optical switch, depicted in Figure 2, is assumed to have N input fibers and M output fibers. Each fiber supports a WDM signal with W wavelengths. In addition,

output fibers are equipped with optical buffering capabilities through B DLs, which are internally connected at each node as depicted in Figure 2. It is considered that nodes are not equipped with DLs before beginning the dimensioning process, *i.e.* $B = 0$. During the dimensioning process, the number of DLs at the output buffers will increase depending on the traffic requirements, where the objective is to decrease the probability of packet loss. In addition, the number of input and output fibers at each node depends on the network topology; where neighbor nodes have at least one input and one output fiber. Consequently the goal of the dimensioning phase is to automatically calculate values for N, M and B at each node.

We have developed a simulator where we can replicate the behavior of TOPS nodes interconnected by a specific topology. Such simulator is based on a Monte-Carlo process, where at each node target nodes for packets are randomly assigned following a normal distribution. In this sense, the simulator takes as input the model of a physical topology where connections among nodes are described as well as the number of

Figure 2. Optical switch architecture

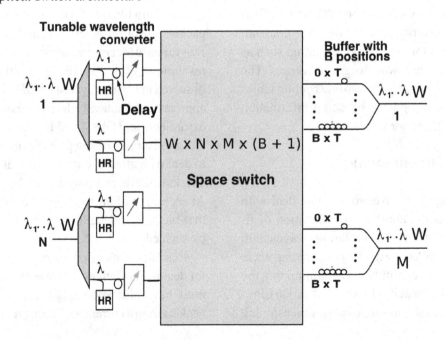

wavelengths. Later on, the simulator can generate traffic among nodes through each one of the wavelengths. Furthermore, such simulator assumes a load of 1.0, so at each time the nodes generate as many packets as wavelengths. For example, if the simulator runs for 10,000 simulation steps and the number of wavelengths is 4, each node generates 40,000 packets in total.

Implemented in the Monte-Carlo network simulator we have a dimensioning heuristic that assigns network resources where are needed in order to decrease the probability of packet loss. The heuristic reasoning to dimension the number of fibers per link and the number of DLs for the corresponding output buffers (Exhibit 1).

As already explained, at this phase attacks are introduced to stimulate the generation of new resources. An attack is considered as a blocking of a set of wavelengths inside an edge; therefore if an edge connecting nodes X and Y is affected by an attack, the node X cannot use the wavelengths blocked by the attack to send packets to Y, the same holds for Y in the opposite direction. Consequently, in such case, the nodes X and Y must find some ways to cope with this attack. DLs and extra fibers are alternatives to overcome attacks. Furthermore, we assume that a single DL is long enough to delay a packet by one unit time. In this way the combination of attacks and the heuristic

dimensioning stimulate the generation of new resources.

The attacks are sequentially introduced into the network, affecting each one of the edges in the TOPS network. Such attacks also have duration intervals, with initial and ending times. These intervals have the same duration for all the attacks. Since we are interested in analyzing the general effect of one single attack at the time, attacks not only affect each edge once but also one edge at time, *i.e.* there are no two or more attacks affecting the network at the same time. Finally, because attacks are bidirectional (blocking wavelengths in both directions), the number of attacks is equal to the number of edges.

Incremental Learning

Once TOPS resources are determined, a learning algorithm is started. Lee and Mukherjee (2004) have already explored the benefits of traffic engineering techniques. Therefore, our algorithm aims to minimize the amount of dropped traffic when fibers are lost by attacks. Our main concern about mesh networks self-healing takes part at this phase; we attempt to achieve a self-organized autonomous behavior at each node by changing routing output priorities from their MPR tables. In order to prove our approach of a self-organized

Exhibit 1.

```
If a packet is lost in a fiber F that goes from node X to node Y
     If the number of DLs at the output buffer servicing the fiber F is less
than the predefined parameter Max_DLs
              Add one (1) DL to the output buffer (i.e. increasing by one
unit the value of B).
     Otherwise
              Remove all the DLs in the output buffers servicing all the fi-
bers from X to Y (eliminate all the output buffers corresponding to the link).
              Insert one extra fiber from X to Y, i.e. increasing M (outgoing
fiber ports) in node X and N (incoming fiber ports) in node Y.
```

autonomous routing we iteratively introduce attacks in the network.

Trends in AON security address to self-organized schemes with a strong utilization of artificial intelligence techniques. In this way our process implements an incremental learning algorithm to dynamically modify priorities based on previous knowledge (Phase 2 in Figure 1). In the end, nodes have different priorities for their paths, which help to distribute the traffic among different paths such as in case of attacks the load can be forwarded to avoid blocked paths.

The process of estimating priorities for each path is similar to well known problems in the field of Reinforcement Learning (RL), in which an agent evolves while analyzing consequences of its *actions* using a reduced information about the system: the actual state s_t of the system, and a simple scalar signal (the reward signal r_t) given out by the environment (Sutton & Barto, 1998). In this sense, the problem of estimating priorities is analogous to the process of determining which actions are more suitable to perform given a state s_t and a reward signal r_t, *i.e.* the output is a set of possible actions ordered according to their suitability.

The reward signal is a very short time evaluation of the goodness of the actions that the agent applies in the environment. The agent uses this signal to determine an operational strategy to reach a long-term objective. In a common learning task, the long-term objective is expressed as the sum of all the possible future rewards signals, called *return*.

The general RL process is schematized in Figure 3. At time step t the environment is in state s_t, known by the agent, and send to the agent a reward signal r_t. The agent chooses a possible action a_t to apply in the environment. At the next time step, the system change to a new state s_{t+1} and send to the agent a new reward signal r_{t+1}. A *policy* is the criterion that the agent uses to determine the action in each state. The agent, then, learns a policy to maximize the possible return of the system by the interaction with the environment.

Q-Learning is a RL method where the learner incrementally learns a policy by building a function $Q(s,a)$ that evaluates the possible total return of an environment if the action a is chosen by the agent at state s. When an action a_t has been chosen and applied, the system is moved to a new state, s_{t+1}, and a reinforcement signal, r_{t+1} is used to update $Q(s_t,a_t)$ by

$$Q(s_t,a_t) = Q(s_t,a_t) + \alpha\delta \qquad (1)$$

Where

$$\delta = r_{t+1} + \gamma \, max_a\{Q(s_{t+1},a)\} - Q(s_t,a_t) \qquad (2)$$

with $0 < \alpha < 1$ as the *learning rate*, which is used to regulate the effect of δ. In this manner, low values of α diminish the impact of δ while high values increase its effect. Furthermore, $0 \leq \gamma \leq$

Figure 3. Agent–environment interaction

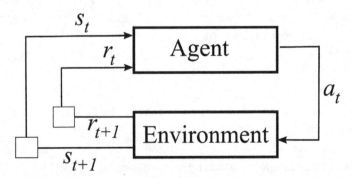

1 is a forgetting factor called *discount*, which determines the present value of future rewards. In this way, a reward received k time steps in the future is worth only γ^{k-1} times what it would be worth if it were received immediately. Whereas high values of γ (near 1) strongly take into account future rewards, a system with $\gamma = 0$ is consider "myopic" and only considers immediate rewards, i.e. only r_{t+1}.

A quasi-optimal policy is then obtained by the agent, choosing the actions to apply in the environment that maximizes the Q function in each state. This policy is called a *greedy* policy. In order to explore the environment in the learning phase, an ε-*greedy* policy is used. In this way, the best actions in the environment are randomly chosen with probability $(1 - \varepsilon)$ while other actions are selected with probability ε.

Following the RL idea we can design each node in the network as an individual agent who must incrementally learn a policy to route packets under undesirable conditions such as attacks. Based on MPR each node has knowledge of which fibers can be used to send packets to specific nodes. Consequently, based on Q-Learning, nodes learn to route packets through a subset A of M output fibers. This subset A is determined by the MPR tables at each node. The idea is that nodes can modify its priorities about the elements in the subset A. For example, assuming a node X with $M = 6$ that wants to send a packet to a node Y, the MPR routing setting can be $A = \{3,2,5,6\}$, where the elements in A represent the numbers of the output fibers through which packets can be forwarded to node Y. Fibers in A connect to different nodes which are not necessarily Y but intermediary nodes. Even more, this A setting is correctly ordered only when there are no attacks in the network, so the first element in A, fiber number 3, represents the shortest path from node X to Y. Nevertheless, when there is an attack the learning task must change the priorities of the subset A in order to reduce the impact of such attack. Consequently, our RL approach will dy-

namically update such priorities to allow a node reacts in case of attacks.

Equation 1, is implemented at each node in the following way, s_t is represented as a configuration given by $s_t = \{x,y\}$ where x is the current node and y is the target node for a packet, which again is not necessarily a neighbor of x but a node at an intermediary hop. The action a_t is then taken from the subset A. Literally $Q(s_t,a_t)$ is the answer to the next question, "what is the expected return if a packet is at the state s_t and the packet is sent through fiber a_t?". Furthermore $Q(s_t,a_t)$ is continuously updated depending on whether the packet is successfully transmitted or not, *i.e.* r_{t+1} is respectively a positive or negative reinforcement signal.

To sum up, our self-organized approach implies two main phases. While the first one deals with dimensioning issues, the second one involves an incremental learning. The first phase relies on a heuristic dimensioning where new resources are added depending on lost packets. At this point, the inclusion of sequential attacks strongly stimulates the generation of new resources. The second phase implements an incremental learning which will allow overcoming link attacks by degrading network's performance gracefully under attacks and gradually healing this performance itself instead of causing abrupt service disruption or service denial. Consequently, the inclusion of attacks at this point allows nodes to learn a policy for routing packets under undesirable conditions.

NUMERICAL RESULTS

Our proposed approach was tested for the TOPS topology shown in Figure 4. This network has 19 optical routers as the one shown in Figure 2. Each node has also connections with neighbor nodes, for each neighbor node there is at least one input fiber and one output fiber. For instance, node number 1, has $N = 3$ input links and $M = 3$ output links since it is connected to nodes 0, 2 and

16. While the number of wavelengths W is fixed to 4, the number of DLs at each fiber is initially established as zero, *i.e. B = 0*, and the maximum number of allowed DLs Max_DLs = 5. In our tests four experiments were performed, one for each wavelength. Each experiment is composed by the two phases previously described. The reasons for performing four experiments rely on the need for analyzing consistent scenarios. The first experiment includes attacks where only one wavelength is affected, both for dimensioning and learning. The second experiment includes attacks affecting two wavelengths, the third one affects three wavelengths and the last one affects four wavelengths.

Once the network topology is given, our two-phase approach is started. Each phase is composed of three interval times: initialization time (IT), core time (CT) and test time (TT). In the IT interval, traffic is generated in all nodes and the network reaches a steady state. In the CT interval the core process of the respective phase is started, *i.e.* dimensioning in the first phase and learning

in the second phase. Finally, in the TT interval the final performance of the network is tested.

In the simulation we have determined the following simulation time units for each time interval, IT = 1×10^4 and CT = TT = 1×10^6. In one million of time units a single node generates 4 million packets, which is a large enough quantity to perform our experiments. Attacks are introduced only at time intervals CT and TT, and depend on the next parameters: node forwarding (NF), node receiving (NR), starting time (ST), duration time (DT), fiber number (FN) and wavelength number (WN). For instance, the following setting: NF = 1, NR = 16, ST = 2000, DT = 4000, FN = 2 and WN = 3, indicates that there is an attack between the nodes 1 and 16, such attack is taking place in the interval [2000, 6000], i.e. with 4000 time units of duration. Furthermore, this attack only affects the third wavelength of the second fiber connecting the affected nodes.

We have introduced equal duration wavelength attacks for all the possible edges in the network, initially taking down only one wavelength until reaching the maximum number of wavelengths,

Figure 4. European topology

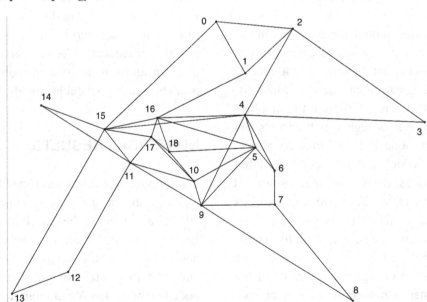

which was previously fixed to 4 wavelengths. In this way we have performed four simulations; one for each blocked wavelength. Also, after an attack is introduced, the network is left working in normal conditions, so without attacks for DT interval and after that the next attack is introduced. Note that the elapsed time between the end of attack and the start of the next attack is equal to the duration of each attack, *i.e.* DT. In our simulations DT was established to 12×10^3 time units for all the attacks, and the first attack is started at time $ST_1 = 2 \times 10^3$. Consequently the second attack is started at simulation time $ST_2 = ST_1 + 2DT = 2 \times 10^3 + 2(12 \times 10^3) = 26 \times 10^3$, and the third one at $ST_3 = ST_2 + 2DT = 50 \times 10^3$, and so on until introducing the last attack at time $ST = 914 \times 10^3$. Although in real systems attacks can occur at anytime, a well-informed attacker can also prioritize some time intervals to introduce the attacks, *e.g.* when traffic is higher. Consequently we want to analyze the global impact of attacks at different intervals.

Dimensioning

Figure 5 shows the number of DLs per node dimensioned during CT interval. As can be observed, to some nodes, whereas the number of affected wavelengths increases the number of DLs does the same. This result seems very reasonable since nodes must either delay or deflect packets, requiring resources that were not initially considered for such situation. Therefore, in the case of DLs the loss of packets in those nodes triggers a generation of more DLs.

As explained before, the first strategy for dealing with attacks is generating new DLs at the outputs connected to output fibers. Nevertheless when the limit of allowed DLs is reached in a fiber connecting node X to Y, the next step is to delete DLs in the fiber under consideration for dimensioning and create an additional fiber to connect X with Y. Figure 6 presents the number of assigned fibers per node. At this point there are two interesting phenomena to analyze:

Figure 5. Number of DLs per node

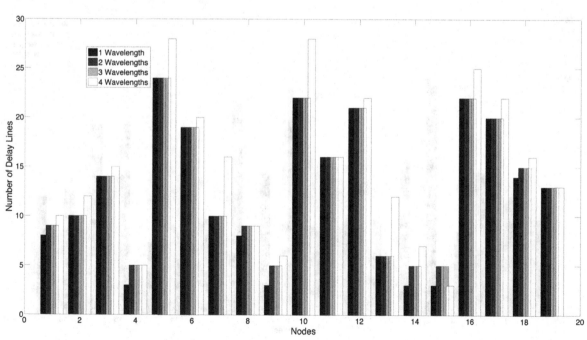

1. the number of fibers remains almost at the same value when 1, 2 and 3 wavelengths fail,
2. the number of fibers increases when all the wavelengths fail.

In the first phenomenon the number of fibers that are distributed among the nodes is 108, and is the same for all the cases but note that the number of DLs is different as depicted in Figure 5. Since at least one wavelength is still working, nodes are solving attacks through DLs, therefore delaying packets helps to alleviate attacks.

In the second phenomenon, since four (4) wavelengths are blocked *i.e.* all the working wavelengths, the packets must be delayed or deflected to alternative fibers. Nevertheless, delaying is no longer useful because almost each packet to be transmitted in the original path must be delayed in the DLs of the alternative fibers. Consequently, in this case the fibers very quickly reach the maximum number of DLs (Max_DLs), which brings about a generation of new fibers,

reaching in total 126 fibers distributed among the nodes. In other words, when blocking all the wavelengths, the dimensioning algorithm assigns more fibers to the nodes.

Learning

In the learning phase we have considered at each node a Q-Learning algorithm with the following parameters: $\alpha = 0.1, \gamma = 0.9, \varepsilon = 0.1$. Furthermore, the value of the reinforcement signal r is determined as follows:

1. 0 if the packet was successfully transmitted.
2. -1 if the packet was delayed at some DLs, and
3. -10 if the packet was dropped.

As can be observed, the above weights allow easy determination of the rewards and punishments for the decisions taken by the nodes. In addition we penalize the most when a packet is dropped.

Figure 6. Number of fibers per node

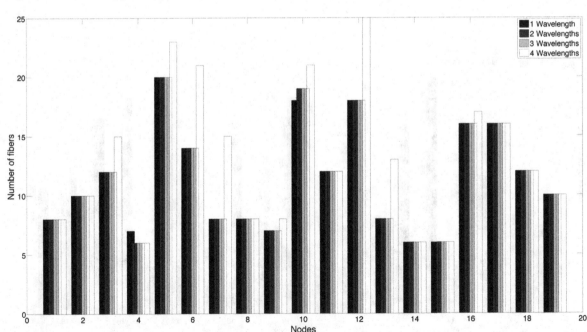

Figure 7. Packet loss

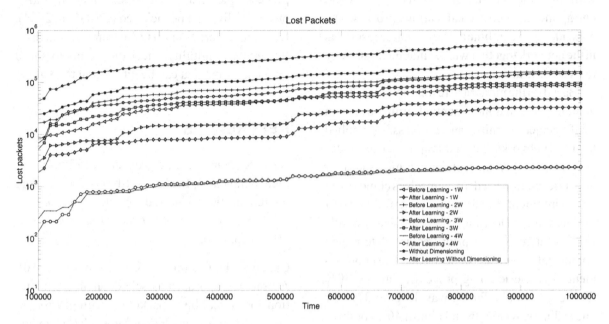

A punishment, in a lower degree, is also applied when a packet is delayed.

Figure 7 shows the performance of the network before and after the learning process. The lines not only represent the number of packets lost in the routing process but also the number of packets that were traveling in the fiber and will not reach the next node, *i.e.* not reaching also the final destination.

In this plot there are three interesting phenomena to be analyzed:

1. The number of lost packets increases as the number of failing wavelengths augments but without blocking all the possible fibers.
2. The number of lost packets is lower when all the wavelengths in a fiber fail.
3. When attacks do not affect all the wavelengths, even without dimensioning process, the learning process can always improve the performance of the network.

The answer for the first two phenomena relies on the number of available resources. In the first

case, while the number of failing wavelengths increases the need for delaying or deflecting also increases, in addition as explained above delaying is also punished in the learning process, therefore nodes learned a policy that prefers deflecting over delaying.

In the second case, since the dimensioning when blocking all the wavelengths yielded more fibers, the availability of fibers to deflect packets is higher, therefore the loss of packets decreases. Nevertheless, at this point there is an open question about the reasons of a similar performance before and after learning.

For the first and third cases it is clear that the learning process always improved the performance of the network. A second open question concerns whether we can improve this performance or not.

CONCLUSION AND FUTURE WORK

We have presented an approach for self-healing on TOPS mesh networks; such an approach is composed of two phases. While the first one deals with the dimensioning of resources, the second one applies an incremental learning process. The main idea is to combine all the available resources in the network together with an intelligent strategy to overcome attacks.

Based on the numerical results obtained, it was demonstrated that incremental learning based on reinforcement learning can be successfully applied to TOPS networks. Under this approach the nodes are considered as independent entities and their only information to adapt themselves comes from its environment. In addition, the performance of this incremental learning was better than a simple MPR strategy where priorities are determined without taking into account attacks. The improvement after the learning process is around 30% for the case of two failing wavelengths, 37% for one failing wavelength and almost 40% for three failing wavelengths. Consequently, an incremental learning shows important improvements to keep exploring.

The final conclusion is that a perfect mixture of resources and intelligent strategies is good enough to deal with attacks. Consequently, the future work must improve the robustness of such mixture.

As next steps in the research, we consider improving the dimensioning process by taking into account not only lost packets but also features such as Hurst, QoS among others (Callegati *et al.* 2006). On the one hand, other switch architectures can also be explored, for instance making use of shared DLs (Gazi and Ghassemlooy, 2007). In addition, different scheduler algorithms can also be contrasted (Callegati *et al.* 2010).

Furthermore, other strategies for introducing attacks can be also explored, such as more than one edge under attack or including attacks in edges that were not dimensioned under attacks. Similarly, attacks can also be introduced by following some random distribution.

Regarding the learning, further improvements can be implemented. For instance, other sources of knowledge can be included in the learning

process, again features like QoS, fiber utilization or time to live can be included (Castañón, 2004). Even more, variations of Q-Learning also include features as eligibility traces or models to speed up the learning that can be interesting to explore.

REFERENCES

Aziz, K., Sarwar, S., & Aleksic, S. (2008). Dimensioning an optical packet-burst switch: More interconnections or more delay lines. *International Conference on Optical Network Design and Modeling,* (pp. 1-6).

Callegati, F., Campi, A., & Cerroni, W. (2010). A practical approach to scheduler implementation for optical burst/packet switching. *Proc. of 14th Conference on Optical Network Design and Modeling*, Kyoto, Japan.

Callegati, F., Cerroni, W., Raffaelli, C., & Savi, M. (2006). QoS differentiation in optical packet-switched networks. *Journal of Computer Communications*, *29*(7), 855–864. doi:10.1016/j.comcom.2005.08.007

Castañón, G. (2004). Performance requirements for all-optical networks. In SPIE Proceedings (Eds.), *Conference on Optical Transmission Systems and Equipment for WDM Networking III: vol. 5596-18.* (pp. 127-134). Philadelphia, USA.

Castañón, G., Razo-Zapata, I., Mex, C., Ramirez-Velarde, R., & Tonguz, O. (2008). Security in all-optical networks: Failure and attack avoidance using self-organization. In IEEE Explorer (Ed.), *International Conference on Transparent Optical Networks, Mediterranean Winter: Vol. 3* (pp. 1-5).

Castañón, G., Razo-Zapata, I., Mozo, J., & Mex, C. (2009). Transparent optical network dimensioning for self-organizing routing. In IEEE Explorer (Ed.), *International Conference on Transparent Optical Networks, Mediterranean Winter,* (pp. 1-5).

Castañón, G., Tancevski, L., & Tamil, L. (2000). Optical packet switching with multiple path routing. *Journal of Computer Networks and ISDN Systems. Special Issue on Optical Networks for New Generation Internet and Data Communication Systems, 32*(5), 653–662.

Chen, B., & Tobagi, F. (2009). Optical network design to minimize switching and transceiver equipment costs. *Optical Switching and Networking, 6*(3), 171–180. doi:10.1016/j.osn.2009.02.002

Danielsen, S., Joergensen, C., Mikkelsen, B., & Stubkjaer, K. (1998). Optical packet switched network layer without optical buffers. *IEEE Photonics Technology Letters, 10*(6), 896–898. doi:10.1109/68.681522

Eramo, V., Listanti, M., & Bovo, L. (2006). Dimensioning models of shared resources for optical packet switching in unbalanced input-output traffic scenarios. *IEICE Trans Commun., E89*(5), 1505–1516. doi:10.1093/ietcom/e89-b.5.1505

Gazi, B., & Ghassemlooy, Z. (2007). Dynamic buffer management using per-queue thresholds. *International Journal of Communication Systems, 20*(5), 571–587. doi:10.1002/dac.834

Lee, Y., & Mukherjee, B. (2004). Traffic engineering in next-generation optical networks. *Communications Surveys & Tutorials IEEE, 6*(3), 16–33. doi:10.1109/COMST.2004.5342291

Mas, C., Tomkos, I., & Tonguz, O. (2005). Failure location algorithm for transparent optical networks. *IEEE Journal on Selected Areas of Communications. Special Series on Optical Communications and Networking, 23*(8), 1508–1519.

Medard, M., Marquis, D., & Chinn, S. (1998). *Attack detection methods for all-optical networks.* Network and Distributed System Security Symposium.

Pavani, G., & Waldman, H. (2008). Addressing self-similarity in optical switching networks by means of ant colony optimization. *Photonic Network Communications, 15*(1), 41–50. doi:10.1007/s11107-007-0086-x

Rejeb, R., Leeson, M. S., & Green, R. J. (2006). Fault and attack management in all-optical networks. *IEEE Communications Magazine, 44*(11), 79–86. doi:10.1109/MCOM.2006.248169

Skorin-Kapov, N., Tonguz, O., & Puech, N. (2007). Self-organization in transparent optical networks: A new approach to security. In IEEE Explorer (Ed.), *International Conference on Telecommunications* (pp. 7-14).

Sutton, R., & Barto, G. (1998). *Reinforcement learning: An introduction.* The MIT Press.

Tomkos, I., Vogiatzis, D., Mas, C., Zacharopoulos, I., Tzanakaki, A., & Varvarigos, E. (2004). Performance engineering of metropolitan area optical networks through impairment constraint routing. *IEEE Communications Magazine, 42*(8), S40–S47. doi:10.1109/MCOM.2004.1321386

Yeom, J., Tonguz, O., & Castañón, G. (2007). Security in all-optical networks: Self-organization and attack avoidance. *IEEE International Conference on Communications,* (pp. 1329-1335).

Yoo, S. (2006). Optical packet and burst switching technologies for the future photonic internet. *Journal of Lightwave Technology, 24*(12), 4468–4492. doi:10.1109/JLT.2006.886060

Chapter 13
Optical Communication:
An Overview

Otto Strobel
Esslingen University of Applied Sciences, Germany

Daniel Seibl
Esslingen University of Applied Sciences, Germany

Jan Lubkoll
Friedrich-Alexander University Erlangen-Nuremberg, Germany

ABSTRACT

The idea of this chapter is to give an overview on optical communication systems. The most important devices for fiber-optic transmission systems are presented, and their properties discussed. In particular, we consider such systems working with those basic components which are necessary to explain the principle of operation. Among them is the optical transmitter, consisting of a light source, typically a low speed LED or a high speed driven laser diode. Furthermore, the optical receiver has to be mentioned; it consists of a photodiode and a low noise, high bit rate, front-end amplifier. Yet, in the focus of the considerations, you will find the optical fiber as the dominant element in optical communication systems. Different fiber types are presented, and their properties explained. The joint action of these three basic components can lead to fiber-optic systems, mainly applied to data communication. The systems can operate as transmission links with bit rates up to 40 Gbit/s. But communication systems are also used for recent application areas in the MBit/s region, e.g. in aviation, automobile, and maritime industry. Therefore—besides pure glass fibers—polymer optical fibers (POF) and polymer-cladded silica (PCS) fibers have to be taken into account. Moreover, even different physical layers like optical wireless and visible light communication can be a solution.

DOI: 10.4018/978-1-61350-426-0.ch013

Figure 1. Basic arrangement of a fiber-optic system

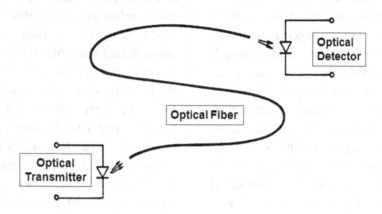

INTRODUCTION

Since the beginning of the sixties, there has been a light source which yields a completely different behavior compared to the sources we had before: This light source is the LASER. The first realized laser was the bulk-optic ruby laser (Maiman 1960). Only a short time after this very important achievement, diode lasers for usage as

optical transmitters had already been developed (see Figure 1) (Quist et al. 1962). Parallel to that accomplishment in the early seventies, researchers and engineers accomplished the first optical glass fiber with sufficient low attenuation to transmit electromagnetic waves in the near infrared region (Kapron, et al. 1970).

The photodiode as detector already worked (Adams & Day 1876), and thus, systems using

Figure 2. Application of fiber-optic systems

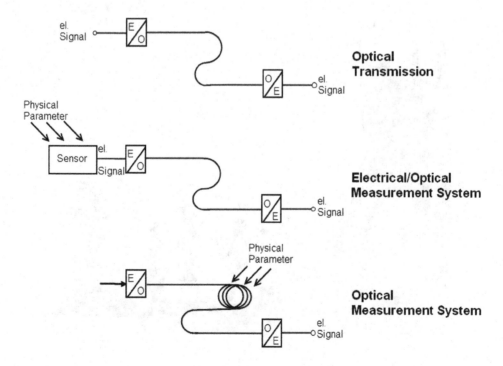

optoelectric (O/E) and electrooptic (E/O) components for transmitters and receivers as well as a fiber in the center of the arrangement could be developed. The main fields of application of such systems are found in the area of fiber-optic transmission and fiber-optic sensors (see Figure 2).

However, the first optical transmission is much older. Indians, for instance, already knew communication by smoke signals a long time ago (see Figure 3, A. Marstaller, personal communication, 1990). Furthermore, it was a very sophisticated and modern system because it already was a digital system, consisting of "binary 1" and "binary 0" (smoke/no smoke).

Charles Kao (Kao & Hockham 1966) and Manfred Börner (Börner 1967) can be regarded as the inventors of fiber-optic transmission systems. Nowadays, their invention would not be very spectacular: Take a light source as transmitter, an optical fiber as transmission medium, and a photodiode as detector (see Figure 1)! Yet, in 1963, it was a revolution because the attenuation of optical glass was in the order of 1000 dB/km, and therefore totally unrealistic for usage in practical systems. Today's fibers achieve attenuation below 0.2 dB/km, which means that after 100 km, there is still more than 1% of light at the end of the fiber. This low value of attenuation is one of the most attractive advantages of fiber-optic systems compared to conventional electrical ones (see Figure 4). In addition, low weight, small size, insensitivity to electromagnetic interference (EMI), electrical insulation, and low crosstalk must be mentioned. Apart from low attenuation, the enormous achievable bandwidth must be pointed out. That leads to a high transmission capacity in terms of the product of fiber bandwidth and length. One of the most important goals is to maximize this product for every kind of data transmission. Figure 4 depicts the attenuation behavior. In particular, we observe independence from modulation frequency of fiber-optic systems in contrast to electrical ones, which suffer from the skin effect.

Figure 3. Digital optical transmission by use of smoke signals

Figure 4. Attenuation of coaxial cables and optical fibers

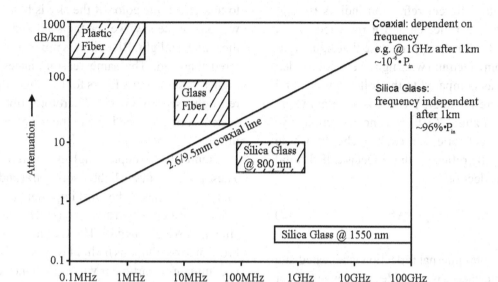

OPTICAL FIBERS

The most important demands on optical fibers are a proper wave guiding, low loss of optical power and low distortion of the transmitted optical signals.

The principle of operation of guiding a light wave can be explained by Snell's law (see Figure 5).

$$\frac{n_2}{n_1} = \frac{\sin \alpha}{\sin \beta} \tag{1}$$

Figure 5. Refraction, reflection, and total internal reflection for light transition between two different media

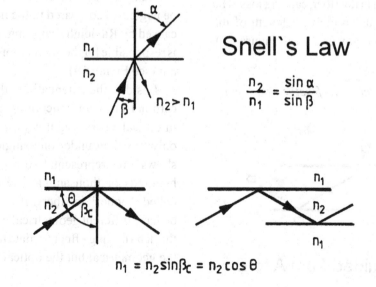

If light is incident on an interface between two media with different refraction indices (n_1 and n_2), there is a reflected and a refracted ray in general. But for the special case that light is incident from a medium with higher refraction index ($n_2 > n_1$) as compared to the following one and furthermore the angle β exceeds a certain value (the cut-off angle β_c), there is no refraction anymore. We get reflection exclusively, the whole light is totally reflected; this effect is called "total internal reflection".

$$n_1 = n_2 \cdot \sin \beta_c = n_2 \cdot \cos \Theta \qquad (2)$$

If this total internal reflection is repeated at a second interface, a waveguide is achieved (Miller et al. 1973). Figure 6 depicts this behavior. In particular, it has to be pointed out that there is no loss due to the multiple reflection because it is a total internal reflection; the coefficient $R = 1$ holds for every repeated reflection. Thus, the attenuation of the fiber is only due to losses inside the fiber.

The most important attenuation mechanisms are Rayleigh scattering and OH⁻ absorption. The scattering effect is due to inhomogeneities in the molecular structure of glass (silicon dioxide: SiO_2). Hence, statistical refraction index changes are caused. This leads to a scattering effect of the traveling light wave in the fiber, causing loss. The loss strongly depends on the wavelength of the light wave (scattered power P_S, see Figure 19).

Figure 6. Total internal refraction and wave guiding

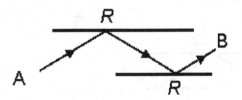

Wave is guided from A to B

Lord Rayleigh discovered and explained that due to this effect, the color of the sky is blue. When we look at the sky, we see the scattered light of the white sunlight. Blue light is much more scattered than red. The same reason causes much higher losses in glass fibers for blue light than for red light (see Figure 19). Therefore, fiber-optic systems operate even beyond the red area, in the infrared (see below).

Figure 19 also depicts high attenuation peaks. These peaks are due to light absorption at undesired molecules in glass. The most important enemy in a fiber is water, which appears as OH⁻ ions in the silicon dioxide structure. The OH⁻ molecules are brought to oscillations by light waves. This effect is dominant in particular when resonance occurs at wavelengths which fit (see peaks). Hence, the energy of a light wave traveling in the fiber is absorbed, which leads to high attenuation. To achieve low fiber attenuation, the demand of purity is very high and the OH⁻ concentration must not exceed a value of 1 ppb. This is one of the reasons why it took a long time from the first idea of fiber transmission, in about 1963, to the first produced fiber in about 1972. Furthermore, it has to be mentioned that the fiber also suffers from SiO_2 self-absorption in the ultraviolet (UV) and infrared (IR) region, which in principle cannot be avoided. Whereas the UV absorption can be neglected compared to the much higher value caused by Rayleigh scattering, the IR absorption is responsible for the attenuation rise beyond 1600 nm (see Figure 19).

Besides the attenuation, there is a second cardinal problem concerning data transmission in optical fibers. Light rays in the fiber are not only traveling under one single angle. Figure 7 shows three representative existing rays (among hundreds or thousands). The existing rays are called "modes" in fibers. It is obvious that they do have different geometrical path lengths L. Yet, the determining effect for data transmission is not the geometrical but the optical path length

Figure 7. Pulse broadening by modal dispersion

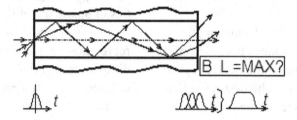

$$g = n \cdot L, \qquad (3)$$

the product of the refraction index and the geometrical path length L. However, this optical path length differs for the three mentioned modes, too, because the refraction index in the fiber core is constant.

Therefore, an optical pulse travels along all the three paths in the fiber. The consequence of this is that they have different transit times and reach the fiber end at different arrival times. The three pulses superimpose and thus, we receive a broader output pulse as compared to the narrow input pulse (see Figure 7). The effect is called "pulse broadening". This behavior causes serious consequences. Since, if we want to transmit a high data rate, we have to place the second input pulse immediately after the first one. As a result, the pulses at the end of the fiber will overlap in such a manner that both pulses cannot be separated any longer. To avoid the overlap, it is necessary to place the second pulse with a greater distance from the first one, which reduces the achievable bandwidth B. The second opportunity is to reduce the fiber length L. Both measures derogate the transmission capacity, the product of fiber bandwidth and length, the most important goal of every data transmission.

To avoid (reduce) this problem, scientists invented the graded-index fiber (see Figure 8). In contrast to the above described fiber, called step-index fiber, the refraction index is not any longer

Figure 8. Fiber types

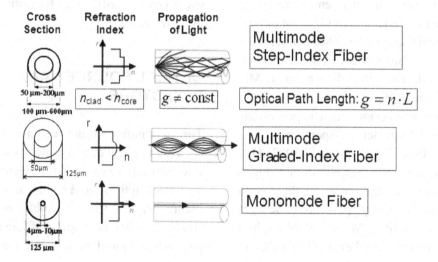

constant across the fiber (Gloge et al. 1973). The latter reveals a gradient behavior in the fiber core, whereas it still remains constant in the envelope, the cladding.

As a consequence, the optical path length $g = n{\cdot}L$ is now constant for each mode because in the fiber center, one can remark the shortest geometrical path length L and the highest refraction index n. In contrast to this result, one can find the longest geometrical path length linked with the lowest refraction index near the cladding. Thus, with a properly chosen index profile, we achieve a constant optical path length for all modes. At this stage, it has to be pointed out that it is not possible to achieve this goal completely. We only obtain a good approximation and thus, there still is a certain modal dispersion left, resulting in a non-negligible remainder of transmission capacity reduction (Marcuse, 1979).

To overcome this problem, another invention was made, the construction of a monomode fiber: If we reduce the fiber core to a diameter below about ten micrometers, there will be only one ray, the ray along the optical axes transmitted, and the modal dispersion problem per se vanishes (Cohen et al. 1982). For very high data rates (about 40 Gbit/s), we have to confess that this disappearance is not completely correct due to polarization effects. Very accurate investigations lead to the result that there is a difference concerning the refraction index between two perpendicularly oriented axes in the fiber. This fact again results in different optical path lengths, and finally as described before, in the same process of pulse broadening; this dispersion is called "polarization mode dispersion" (PMD (Mahlke & Gössing 1987)).

Furthermore, another important dispersion has to be mentioned, the material dispersion (simplified chromatic dispersion) (Cohen et al. 1982). Due to the fact that there is no light source emitting at a single wavelength (see Figure 15), there is no monochromatic but always polychromatic light traveling through a fiber. Moreover, taking into account the dependence of the refraction index on the wavelength, it is obvious that we have always different refraction indices, and therefore, different optical path lengths according to different velocities of pulses traveling along the fiber.

Figure 9 shows three pulses having three representative wavelengths. They suffer from different transit times and reach the fiber end at different times of arrival.

The three pulses superimpose as described above for the modal dispersion process and thus, we again obtain a broader output pulse as compared to the narrow input pulse (see Figure 9). The result is the same as for modal dispersion, just the mechanism is different, and the effect is again pulse broadening and reduction of the transmission capacity.

Figure 10 visualizes the three common fiber types in comparison with a woman's hair. Table 1 shows an overview on fiber types depicting the most important fiber data. Furthermore, two more fiber types must be mentioned. There are also low cost applications concerning fiber-optic transmission, i.e. plastic fibers and PCS fibers (plastic cladding and silica core) are used, too. Their transmission capacity is much lower as compared to pure glass fibers in particular monomode fibers. However, there are applications for such fibers, e.g. if you have low data rates and only some ten meters of spacing. For example, in order to watch a machine tool in an EMI-relevant area, why not use a cheap plastic fiber transmission setup in the kHz-region?

OPTICAL SOURCES AND DETECTORS

The most important demands to optical sources are a high optical output power as well as a small electrical input power. With regard to the fiber, the wavelengths should be in a proper range (see Figure 19). The spectral width has to be small, and for a sufficient coupling efficiency, the beam divergence should be low and the geometrical

Figure 9. Pulse broadening by material dispersion

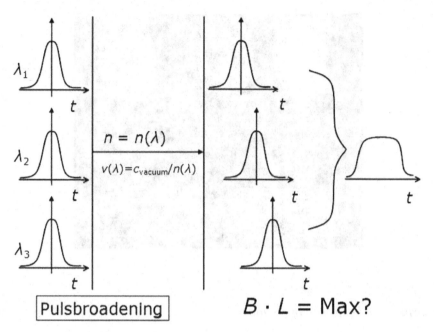

Pulsbroadening

$B \cdot L = $ Max?

(*n*: refractive index; λ: wavelength; c_{vacuum}: free space velocity; *v*: velocity in media)

size should be small. Furthermore, a modulation capability of the injection current at high speed is favorable. To understand the principle of operation concerning optical sources and detectors, fundamental considerations about the interaction between photons and electrons have to be taken into account.

Figure 11 shows the energy band model of the semiconductor material, applied to the optical components. The lower level E_v is the energy level of the valence band, whereas the upper level

E_c denotes the level of the conduction band. The difference ΔE between both levels is the energy gap E_g. There are three dominant effects to be considered:

From the electronic point of view, the sources and detectors are, simply put, p-n diodes following Shockley's well-known current-to-voltage characteristic curve (Burrus & Miller 1971, Panish 1976, Melchior et. al. 1970). LEDs and lasers are driven in forward direction, photodiodes in reverse voltage operation.

Table 1. Fiber types

Type	Profile	Size	Attenuation	Bandwidth Length
Plastic Fiber	Step Index	950/1000 µm	0.2 dB/m	~3 MHz·km
PCS Fiber	Step Index	100-600 µm	6 dB/km	~20 MHz·km
Multimode Glass	Step Index	>100 µm	3-5 dB/km	20 MHz·km
Multimode Glass	Gradient Index	50/125 µm	2 dB/km (0.85 µm) 0.4 dB/km (1.3 µm) 0.2 dB/km (1.55 µm)	500 MHz·km
Monomode Glass		5-10µm		>100 GHz·km

Figure 10. Comparison of fiber types and a woman's hair

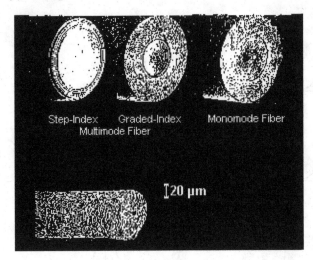

Step-Index   Graded-Index   Monomode Fiber
    Multimode Fiber

$\updownarrow$20 µm

Figure 11. Absorption and emission of photons

Interaction between Electrons and Photons

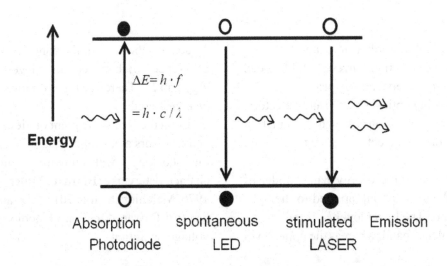

$\Delta E = h \cdot f$

$= h \cdot c / \lambda$

Energy

Absorption   spontaneous   stimulated   Emission
Photodiode   LED   LASER

LASER: Light Amplification by Stimulated Emission of Radiation

Absorption of Light

In single atoms, electrons may occupy only well defined "allowed" energy states. In solids, the electrons do occupy more or less broad allowed bands separated by "forbidden" gaps. A simple

energy-band scheme of a semiconductor is depicted in Figure 11. At temperatures close to 0 K, the highest valence band is filled with electrons. The next higher allowed band, the so-called conduction band, is empty. The two bands are separated by an energy gap of width E_g. This gap is a fundamental quantity of the semiconductor and influences many properties of the material. At room temperature, a considerable number of electrons have been lifted from the valence band into the conduction band, leaving holes in the electron sea of the valence band. The necessary energy is gained from thermal movement.

A transition of electrons from the valence band into the conduction band can also take place if the required energy is supplied by radiation. According to Einstein's theory, radiation can be considered as a flow of optical quantum particles called "photons". Each photon carries a certain amount of energy which is proportional to the frequency ν of the electromagnetic wave and is given by

$$E_{\mathrm{ph}} = h\nu = h\frac{c}{\lambda}; \tag{4}$$

h being Planck's constant, c the velocity of light, and λ the wavelength (Burrus & Miller 1971). It becomes clear that light can only be absorbed in a semiconductor, if the energy of the photons is larger than the band gap, i.e. $E_{ph} \geq E_g$. In other words: The wavelength of the light must be smaller than the wavelength λ_g that corresponds to the energy gap, E_g:

$$\lambda \leq \lambda_{\mathrm{g}} = h\frac{c}{E_{\mathrm{g}}} \tag{5}$$

If this condition for absorption is met, it appears that the optical power of the light wave Φ, is exponentially reduced while traveling through the crystal. If the power which is coupled into the crystal is denoted by Φ_0, the transmitted power that leaves a crystal of thickness d is given by

$$\Phi = \Phi_0 \exp(-\alpha d) \tag{6}$$

α is called "absorption coefficient". From equation (6), it follows that

$$\alpha = -\frac{1}{d}\ln\left(\frac{\Phi}{\Phi_0}\right) \tag{7}$$

Emission of Light

If the conduction band is filled with more than an equilibrium distribution of electrons, then electrons can fall back spontaneously into holes of the valence band. In the course of this recombination process, energy is released either in form of light or heat. In the case of radiative recombination, one photon is emitted for each transition as shown in Figure 11.

The emission of radiation is related to the transitions of electrons from the higher energy level in the conduction band to the lower level in the valence band. The value of the energy gap E_g between these levels is distinctive for every semiconductor, e.g. for GaAs: $Eg \approx 1.43$ eV. The photon energy of the radiative emission is approximately the same as the energy of the forbidden gap:

$$E_{\mathrm{ph}} \approx E_{\mathrm{C}} - E_{\mathrm{V}} = E_{\mathrm{g}} \tag{8}$$

Hence, the center wavelength of the radiation can be calculated using equation (4):

$$\lambda \approx h\frac{c}{E_g}. \tag{9}$$

The wavelength and the color of the emitted light consequently depend on the band gap of the semiconductor, which can be controlled by techniques of crystal growth.

There is no source which emits light at a single wavelength. It is impossible to realize pure monochromatic light, but there is always a polychromatic behavior, which can be understood as follows. The considered electrons and holes possess a thermal energy distribution with an average energy of $\frac{3}{2}kT$ each, where k is Boltzmann's constant and T the absolute temperature. Therefore, the emission bandwidth is expected to be approximately

$$\Delta E \approx 3kT \qquad (10)$$

This energy bandwidth ΔE can be recalculated into a spectral bandwidth $\Delta\lambda$, in terms of wavelengths:

$$\Delta\lambda = \frac{\lambda}{E_g}3kT = \frac{\lambda^2}{hc}3kT \qquad (11)$$

The quantitative description of the bandwidth is determined by the FWHM definition (full width at half maximum). The application of a generally accepted definition by international standards is very important to make results comparable. Otherwise, serious mistakes could occur.

Radiative recombination is the basic underlying principle of light-emitting diodes (LEDs) and semiconductor lasers. An LED consists of a p-n junction, where in the n-doped region, there is an excess of electrons in the conduction band whereas in the p-doped region, an excess of holes in the valence band is registered. By applying a forward current, electrons will flow from the n-type to the neighboring p-type region and holes vice versa. This process allows a huge recombination rate of electrons and holes and therefore leads to the emission of light. The energy difference ΔE is converted into a photon. In particular, the process of a following electron transition causing a second photon has nothing to do with generation of the first photon, i.e. there is a statistical behavior, named "spontaneous emission". This process occurs in an LED (Burrus & Miller 1971, Panish 1976).

In contrast to this process, the stimulated emission is completely different. The electron transition from conduction to valence band is not any longer a statistical process, but an induced or stimulated one. This process is stimulated by an already existing photon, e.g. produced by the spontaneous process. The two photons are not any longer strangers. They know each other, they are coherent, and we receive a coherent radiation caused by stimulated emission. This is the desired behavior in a laser. The two photons, now and again, cause new transitions and multiply themselves. To ensure to have enough electrons in the conduction band, the so-called first laser condition has to be fulfilled—a population inversion. Naturally, only few electrons are in the conduction band due to room temperature's heat. Thus, we need a "pumping" of electrons from the valence band to the conduction band. This can be done by optical means (first part of Figure 11). A diode laser pumps by the injection current like in the LED. The second laser condition is the necessity of an optical resonator. The light is multiply reflected by two partial reflecting mirrors. As we use GaAs or InP as semiconductor materials, the refraction index is about 3.5. Due to the fact that if light transmits between media with different refraction indices, there is not only refraction but also a reflection, the FRESNEL-reflection (see Figure 5):

$$R = \left(\frac{n_{GaAs} - n_{Air}}{n_{GaAs} + n_{Air}}\right)^2 \approx 31\% \qquad (12)$$

Thus, a reflection with a reflection coefficient R is inherently achieved by the front and the rear side of the laser chip without building an extra mirror (see Figure 13).

Due to the multiple reflecting process, an avalanche is produced; we receive a large amplification after exceeding a certain threshold (see Figure 12), we receive a LASER: Light Amplification by Stimulated Emission of Radiation (Panish 1976).

Figure 12 depicts the optical power versus the injection current of a semiconductor laser. Below

Figure 12. Optical power versus injection current

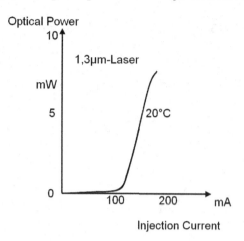

Figure 13. Schematic arrangement of a semiconductor laser

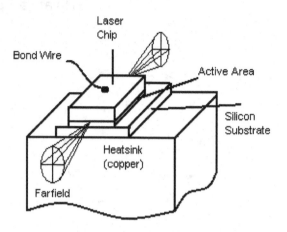

the threshold, the laser operates as an LED. After exceeding a certain injection current, the threshold current, stimulated emission takes place. Unfortunately, there is a strong dependence of the threshold current on temperature. Hence, additional measures such as temperature control or control by a monitor diode are necessary in order to stabilize the optical output power.

Figure 13 shows the far-field distribution of a typical Fabry-Perot semiconductor laser.

Regrettably, semiconductor lasers show beam divergence and astigmatism, whereby the divergence is different in two perpendicular directions. Both effects make the coupling into a fiber - in particular in a monomode fiber—difficult. The laser chip is mounted upside down on a silicon substrate. This mounting enables a close contact of the active area to the heat sink. The chip size is about 300 μm long, 200 μm broad and 100 μm high, the active area is in the order of 1 μm². Figure 14 visualizes a comparison between LED and laser farfield. The LED is a Lambertian light source, and thus it emits the radiation in the half sphere, which makes the fiber coupling even more difficult.

A further interesting comparison between laser and LED is concerned with coherence properties. Figure 15 shows that the spectral width of a laser diode is much smaller than that of an LED, i.e.

the laser coherence length is much larger as compared to the LED.

To depict both spectra in a single diagram, the laser power had to be reduced by a factor of 50. Moreover, it has to be mentioned that the 2 nm of laser width is just a rough idea. It could also be smaller by several decades. This results in a much better behavior concerning the material dispersion of a fiber. Thus, optical power, far-field behavior and spectral size enable the laser much more for highly sophisticated optical transmission systems, and hence exclusively lasers are used for high-speed long distance systems. In contrast to the laser, LEDs are used for low-cost, low-bit-rate and low-distance systems.

The most important demands on optical detectors (Melchior et. al. 1970, Pearsall 1981) are high sensitivity, low noise, linearity (for analog systems only), and small geometrical size. The most famous components are pin photodiodes (see Figure 16 and 17) and avalanche photodiodes (APDs). All photodiodes for transmission systems are used in reverse voltage operation. This operation applies an electric field to the semiconductor material and thus an electron produced by photon absorption is experiencing a force and will be accelerated. This effect is even enlarged by introducing an intrinsic zone into a p-n diode, which results in a

Figure 14. Far-field characteristics of LED and laser

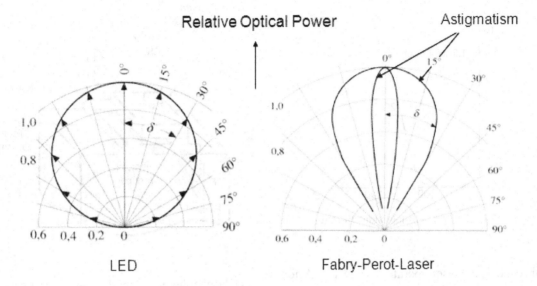

pin diode. This design guarantees a constant and high electric field over the whole absorption zone, whereas a simple p-n diode has only a maximum directly at the p-n junction, which leaves most part of the absorption area in a low electric field. Therefore, the pin diode is able to operate as a high-speed component.

Figure 16 depicts an APD. This component has a highly doped p+ zone, except for the p-n junction and the intrinsic zone. In this case, we receive a very high field at the p-n junction, besides the constant electric field in the i zone (see also voltages for comparison).

This arrangement causes such a high acceleration of charge carriers that collisions produced by the absorption process occur between the electrons and the atoms of the semiconductor material. Thus, further electrons will be freed from the atoms

Figure 15. Spectral width of LED and laser

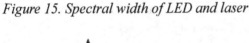

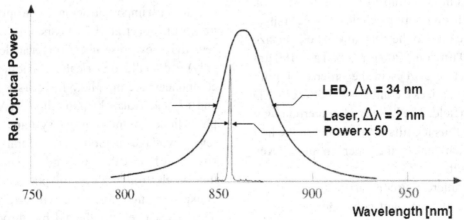

Figure 16. Photodiodes for optical systems

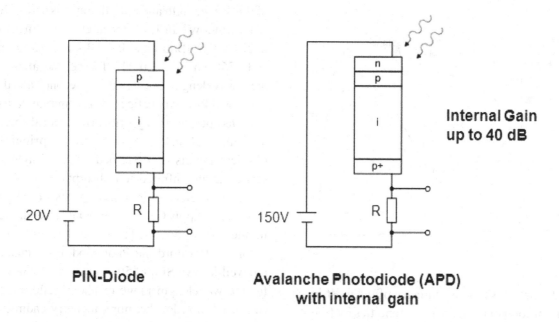

PIN-Diode

Avalanche Photodiode (APD) with internal gain

Internal Gain up to 40 dB

(lifted from the valence band into the conduction band) and we receive secondary electrons. This effect is called "impact ionization". The secondary electrons are now also accelerated and generate tertiary electrons and so on, which leads to an avalanche. As a result, the APD possesses an internal gain and thus, it is a very proper component for optical systems. The small geometrical size enables small junction capacities (see Figure 17)

and therefore, high cut-off frequencies are gained. Typical diameters of high-bit-rate photodiodes are in the order of 50 µm.

The most common materials for application in photodiodes are silicon (Si), germanium (Ge), and gallium indium arsenide phosphate (GaInAsP). Figure 18 shows the proper use according to the relative sensitivity versus wavelength (see also Figure 19).

Figure 17. Schematic structure of a Ge-APD (Ebbinghaus et al. 1985)

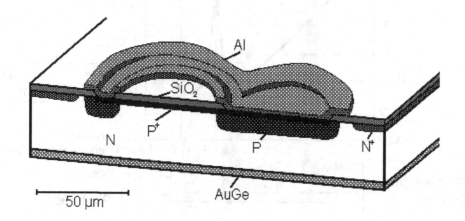

Figure 18. Spectral responsivity of photodiodes

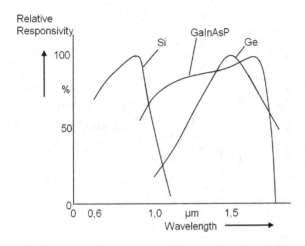

Figure 19 shows a summary of the most important components and their properties for fiber-optic transmission systems: In 1973, there was a fiber featuring a first minimum at a wavelength of about 850 nm in the order of 5 dB/km. There-fore, fiber-optic transmission started at the area of this wavelength and thus, this area is called the "first window". In 1981, the attenuation lowered to about 0.5 dB/km and 0.3 dB/km at 1300 nm and 1550 nm respectively. Hence, the areas at these wavelengths are called the "second" and the "third window" where fiber-optic communication systems operate: Today's fibers reach an absolute minimum of 0.176 dB/km due to principle physical effects as described above: Rayleigh scattering and infrared self-absorption.

As optoelectronic components for light sources, we apply GaAlAs LEDs and laser diodes for the first window, and InGaAsP devices for the second and the third one. Photodiode materials are the well-known Si for 850 nm, Ge and InGaAsP for the wavelength range of about 1200 nm to over 1600 nm. Furthermore, mercury cadmium telluride (HgCdTe) materials are very promising compounds for future optical detectors (Lee 2001).

Figure 19. Spectral attenuation of optical fibers and useful wavelengths of optoelectronic devices

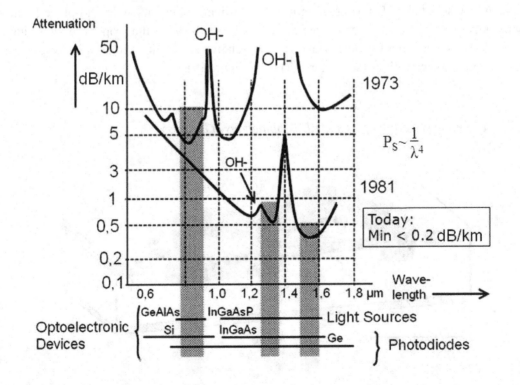

FIBER-OPTIC COMMUNICATION FOR TELECOM APPLICATIONS

Using the devices described above, fiber-optic transmission systems could be developed applying optoelectric and electrooptic components for transmitters and receivers as well as a fiber in the center of the arrangement (see Figure 1).

However, an optical communication system is more than a light source, a fiber, and a photodiode. There is a laser driver circuit necessary to provide a proper high-bit-rate electric signal; this driver, combined with a laser or an LED, builds the optical transmitter. As well, the photodiode (pin or APD), together with the front-end amplifier, forms the optical detector, also called "optical receiver" (see Figure 20).

This front-end amplifier consists of a very highly sophisticated electric circuit. It has to detect a high bandwidth operating with very few photons due to a large fiber length and it is struggling with a variety of noise generators.

However, if the desired link length cannot be realized, a repeater consisting of a front-end amplifier and a pulse regenerator will be inserted (see Figure 21). This pulse regenerator is necessary to restore the data signal before it is fed to a further laser driver followed by another laser.

Figure 22 visualizes the immense capability of the data regeneration: Directly at the front-end amplifier, (1) the eye pattern and the data signal cannot be detected as those. After a first following equalizer circuit, eye pattern and data signal are hardly recognized (2), whereas both are quite well restored after a second equalizer step (3). The non-return to zero signals at 168 Mbit/s can be seen clearly. Finally, a low pass filter is applied to suppress very high frequency noise (4).

At this point, it must be mentioned that an optical communication system is still more than discussed in this chapter: There are further electric

Figure 20. Optical fiber transmission principle

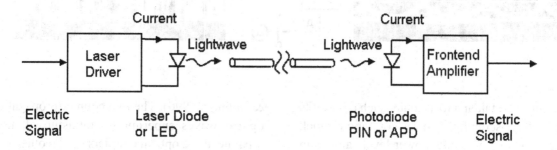

Figure 21. Transmitter, repeater and receiver

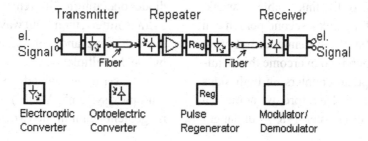

Figure 22. Eye pattern and data signal (N. Kaiser, personal communication, 1984)

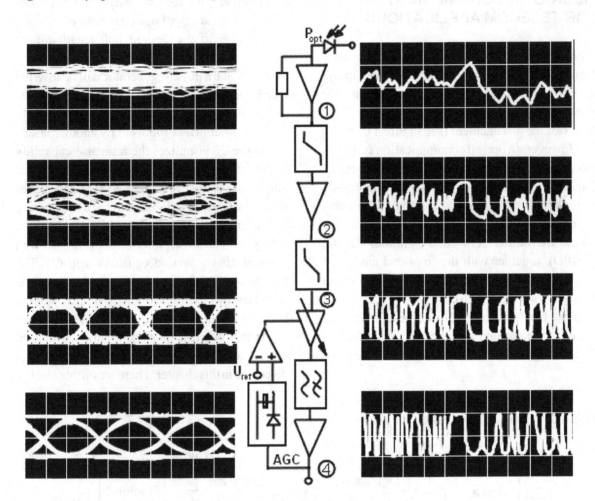

circuits to be taken into account, such as circuits for coding, scrambling, error correction, clock extraction, temperature power-level, and gain controls (Drullmann & Kammerer 1980).

Furthermore, until now, we have described a unidirectional system exclusively (see Figure 23), i.e. we think of a telephone link at which a person at one side of the link is able to speak. At the other side of the link a second person can listen exclusively, but the system does not operate the other way round. To overcome this insufficient situation, optical couplers on both sides of the link are inserted. Therefore, we achieve a bi-directional system (Köster 1983, Fußgänger & Roßberg 1990). The two counter propagating optical waves superimpose undisturbed, they separate at the optical couplers on the other side of the link and reach the according receivers. To improve the transmission capacity drastically, wavelength selective couplers are applied, called "multiplexers" and "demultiplexers". Several laser diodes operating at different wavelengths are used as transmitters; their light waves are combined by the multiplexer and on the other link end separated by the demultiplexer.

The set-up is named "wavelength division multiplex system" (WDM). If we apply this arrangement again in the two counter propagating

Figure 23. Variety of optical transmission system

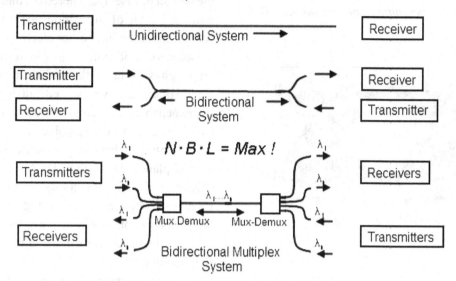

directions, we achieve a bi-directional WDM system (Fußgänger & Roßberg 1990). The transmission capacity is risen by the number N of the channels transmitted over one single fiber.

Figure 24 depicts a scheme to describe the limits of optical transmission systems. For a single channel system, two basic limitations occur. Such systems are called "direct detection systems", consisting of one laser, one fiber, and one detector. The limitations are divided into two groups; the systems suffer from attenuation limitation and dispersion limitation: The attenuation-limited arm is governed by the transmitter power of the applied laser diode, the fiber attenuation, and the receiver sensitivity of the detector. The dispersion-limited arm is governed by the modulation bandwidth of the applied laser diode, the fiber dispersions, and the demodulation bandwidth of the detector.

Thus, for high-bit-rate long-distance transmission systems, exclusively high-speed lasers and photodiodes will be installed as well as a monomode fiber. Figure 25 shows the eye pattern of a 43 Gbit/s data signal transmitted over a single-channel high-bit-rate system. The data rate corresponds to the cut-off frequency of about 30

GHz, which is approximately the highest frequency a single laser diode can be modulated.

COMMUNICATION IN AUTOMOTIVE SYSTEMS

Optical data buses in vehicles are almost exclusively used for infotainment (information and entertainment) applications. The Media Oriented

Figure 24. Limits of optical transmission systems

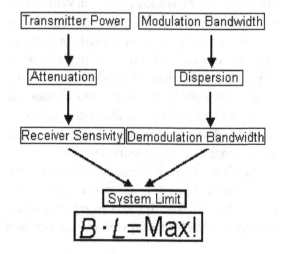

Figure 25. Eye pattern of a high-bit-rate data signal (B. Wedding, personal communication, 2001)

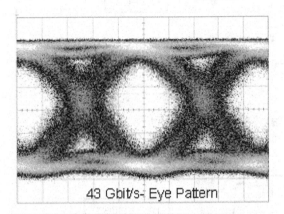

43 Gbit/s- Eye Pattern

Systems Transport (MOST) (MOST Cooperation 2007) is the optical data bus technology currently used in cars with a data rate up to 150 MBit/s. The development of infotainment applications in cars began with a radio and simple loudspeakers. Today's infotainment systems in cars include but are not limited to ingenious sound systems, DVD-changers, amplifiers, navigation, and video functions. Voice input and Bluetooth interfaces complement these packages. Important and basic logical links of these single components are already well-known from a simple car radio. Everybody probably knows the rise of volume in case of road traffic announcements.

However, the integration of more and more multimedia and telematic devices in vehicles led to a large increase in traffic demands. In particular for luxurious classes, a huge need for network capacity and higher complexity by integration of various applications have to be taken into account. Although MOST is the optical data bus technology currently used, alternative solutions for higher data rates that satisfy future automotive applications are highly desirable.

Another serious challenge arises in protecting new generation aircrafts, particularly against lightning strikes (Majkner 2003). This is because new airplanes will be built using carbon fiber to reduce the weight of fuselage. Therefore, these airplanes will lose a lot of protection against lightning, cosmic radiation, and other electrostatic effects. In order to avoid failures in signal transmission on the physical layer, the electrical copper wires should be extremely protected, but this solution is too expensive and increases the weight of cables (Majkner 2003). A reasonable solution is to use glass or plastic fibers as transmission medium in new airplanes. Since the FlexRay bus protocol (FlexRay Consortium 2008) is more adequate for avionic applications, it should be adapted for this transmission medium. Thus, this solution is cost-efficient and offers more safety in the aviation domain.

In this subchapter, we propose an improvement for optical data bus systems that may satisfy the requirements of future automotive applications and safety-relevant operations. First, we give an overview on MOST bus systems. Then, we present the challenges of data transmission that arise in new aircraft generations. After that, we propose two alternative solutions for optical data buses in avionic systems. Subsequently, we discuss the prototyping results and present open directions for future work. Finally, some optical wireless and radar applications are briefly introduced.

MOST Bus Technology

The MOST Corporation was founded in 1998 by automotive manufacturers and several system vendors to establish and refine a common standard for today's and future needs of automotive multimedia networks (MOST Cooperation 2007). MOST is the optical data bus technology currently used in cars with a data rate up to 150 MBit/s. This bus technology offers not only a synchronous transmission for audio and video data, but also makes available the application framework for controlling the system complexity. In particular, MOST specifies interfaces and functions for infotainment applications at a high abstraction level. As shown in Figure 26, different multimedia

Figure 26. MOST network with ring topology (MOST Cooperation 2007)

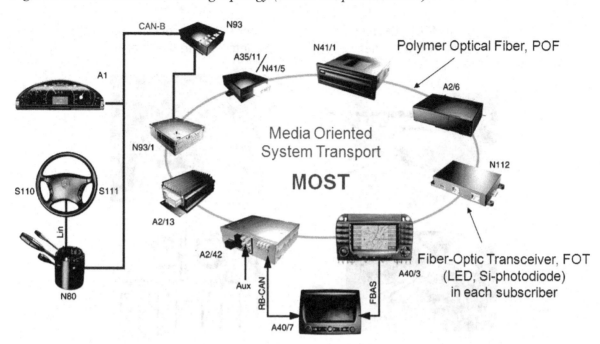

components can be connected in a ring topology. Furthermore, Bluetooth can be used for wireless devices and diagnosis interfaces (Grzemba 2007).

MOST technology has been developed over three generations. The first generation (MOST25) is based on a bit rate of 25 Mbit/s using an optical physical layer. A data rate of 22.58 Mbit/s can be achieved with a frame length of 512 bit. MOST25 uses a 1-mm-Polymethylmethacrylat fiber (PMMA), often called POF, as data transmission medium. An LED is used as transmitter to convert the electrical signal to an optical one, using a driver circuit. The receiver converts the optical signal into an electric current using a Si-photodiode. As shown in Figure 27, transmitter and receiver are combined in a device, called Fiber-Optic Transceiver (FOT).

The second generation, MOST50 is specified with a frame length of 1024 bit and the signal is transmitted by means of an electrical physical layer. The third generation MOST150 was specified to satisfy large data traffic demands. It

is based again on an optical physical layer and a data rate of up to 150 Mbit/s.

As mentioned above, the PMMA fiber is used as transmission medium for MOST bus systems. Figure 28 shows the fiber attenuation curve of PMMA fibers. Although the attenuation is lower at a wavelength of 500 nm MOST components operate at a wavelength of 650 nm because the power of 650 nm LEDs is much higher. Taking into account the required temperature range from -40 °C to 85 °C, the wavelength of the LED varies and therefore the worst case fiber attenuation is approximately 0.4 dB/m (Kibler et al. 2004).

However, using LEDs with 650 nm wavelength, a 1 mm-diameter POF and a large Si-photodiode are sufficient enough for present typical traffic demands in automobile applications. In particular, short transmission distances up to 10 m as well as a bandwidth of 100 MHz are adequate for most available automobile applications. FOT and POF are the most important components of the physical layer for MOST systems. This has been tested using a measurement

Figure 27. MOST physical layer components (MBtech Group 2008)

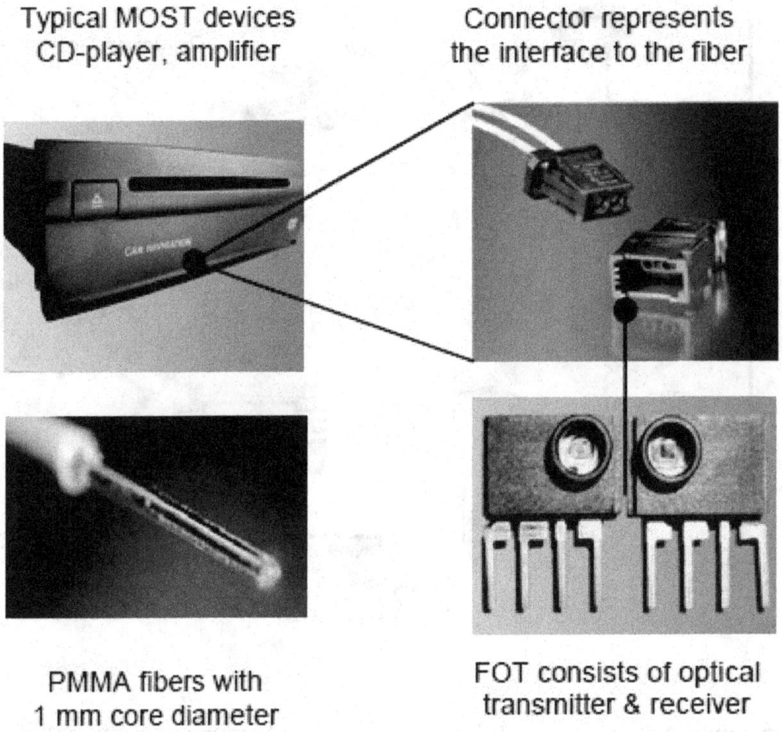

Typical MOST devices
CD-player, amplifier

Connector represents
the interface to the fiber

PMMA fibers with
1 mm core diameter

FOT consists of optical
transmitter & receiver

set-up at MB-technology (Seibl et al.2008). Conventional POF/LED bus systems are capable to achieve 150 Mbit/s and will preliminary remain the solution for cars. In particular, MOST150 is adequate for optical data transmission in cars enabling the cross linking of onboard video cameras, laptops, GPS and cell phones.

For higher data rates, alternative solutions are considered. Several advanced modulation techniques have been proposed recently that

Figure 28. Attenuation and body of PCS-fiber

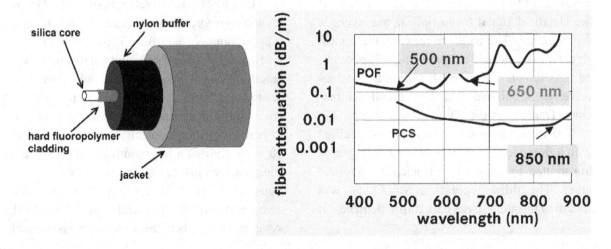

make this step feasible. Especially, by combining multi-carrier modulation with spectrally-efficient quadrature amplitude modulation (QAM), the first demonstration of 1 Gbit/s transmission over 100 m of SI-POF was reported (Lee et al. 2008). An alternative solution is replacing the LED as transmitter by a vertical-cavity surface-emitting laser (VCSEL) and the pure plastic fiber (POF) by a polymer-cladded silica (PCS) fiber (Reiter & Schramm 2008, Strobel et al. 2007). Based on VCSEL and PCS fibers, Section 4 discusses two alternative solutions enabling the use of middle and long transmission distances. Thus, the data rates can be extended into the GBit/s-region. As a result, this optical bus technology can be also used for sensor applications including safety-relevant operations like drive by wire, brake by wire and engine management, and might finally lead to autonomous vehicle driving.

Data Buses in Avionic Systems

In order to reduce the weight of new generation aircrafts, design engineers are going to use more and more carbon-fiber fuselage. Considered over the economic lifetime of an airplane, every saved kilogram affects a fuel economization of several thousand liters of kerosene. On one side this technology may reduce the weight of an aircraft up to 30%, but on the other side it introduces new safety problems and difficulties. A serious problem arises from the fact that many advantages of a closed metal fuselage get lost. An important advantage is the Faraday cage inherent lightning strike and cosmic radiation protection. This could induce transients into wires or equipments that could be possibly disturbed, or totally damaged. However, these complications and failures can be avoided by system redundancy and special protection effects (R. Majkner 2003).

The situation would be fatal in an airplane without a completely closed Faraday cage. Lightning strikes could possibly take different paths through the plane and thus harm or even destroy several

electrical components. These problems can be avoided by complex electrical protections which cause higher costs and increase the weight of the cables. However, a reasonable and cost-effective protection method is the employment of optical wires as transmission medium used in avionic systems based on a mechanism for safety critical systems applying a FlexRay bus protocol.

In order to reduce the total link power budget, the transmitter could be improved to reach a higher fiber input power. This would require a complex electrical circuit and special LEDs. Since, a proper receiver with sufficient sensitivity is not available, the fiber and transmitter have been replaced by more sophisticated components.

Since at a wavelength of 850 nm the PCS fiber attenuation is about 0.008 dB/m, the necessary link budget can be achieved. The LED is replaced by a VCSEL which has a significant smaller output beam divergence as shown in Figure 29. Thus, the VCSEL offers a higher coupling efficiency and in addition a low current consumption. A light source at 850 nm also enables a gain of a higher receiver sensitivity. Figure 30 shows the receiver responsivity at 650 nm (0.47 A/W) and at 850 nm (0.63 A/W), resp. Thus, the total link budget is 30 dB. Therefore, components with an attenuation of up to 27 dB (3 dB margin) can be used.

CONCLUSION AND OUTLOOK

The aim of this chapter was to give an introduction to fiber transmission systems, working with basic components. The reader should be familiarized with the fundamental optical techniques for communication systems.

However, for more comprehensive considerations, there are further components to be dealt with (Strobel, to be published 2012), e.g. the optical amplifier to enhance the link length over the conventional limits described above. In order to do that, Erbium and Raman amplifiers (Payne et al. 1990, Flannery 2001, McCarthy 2001) have

Figure 29. VCSEL output beam

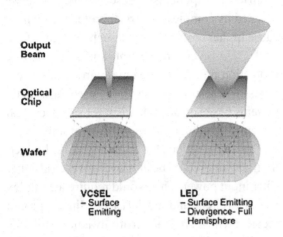

been developed to overcome the problem of attenuation in fibers. Moreover, it is also necessary to avoid signal distortions caused by the dispersion mechanisms in optical fibers. The solution to that problem could be the use of soliton transmission (Kapron et al. 1970, Malyon et. al. 1991). The principle idea has been developed more than 20 years ago but it was just recently introduced as a product to the optical telecom market. For very high data rates such as over 40 Gbit/s, polarization problems in fibers have to be considered. There is a further distortion called "polarization mode dispersion" (PMD), which leads again to pulse broadening and therefore to bandwidth reduction with impact on the transmission capacity, the product of bandwidth, and fiber length (Mahlke & Gössing 1987, Chbat 2000). In this chapter, only point-to-point links have been discussed. Further applications for future optical systems must be taken into account - such as optical networks (Sykes 2001, Weiershausen et al. 2000, Pfeiffer et al. 2001) (LAN and MAN).

Finally, the subject of opening up the last mile for fiber communication is of great interest. Yet, for more than ten years this idea of fiber to the home (FTTH) has been discussed, but it is still too expensive and therefore still waiting to reach the commercial market. Maybe new plastic fibers (Kenward 2001) with sufficiently low attenuation and gradient profile together with high-speed LEDs could solve this problem in the near future.

Concerning automotive application, we presented the state of the art and next-decade technologies for optical data buses in automotive applications. MOST is the optical data bus technology currently used in cars. MOST150 is the current standard with a bit rate of 150 Mbit/s and it is an adequate solution for optical multimedia data transmission in automobiles. However, to

Figure 30. PIN diode responsivity

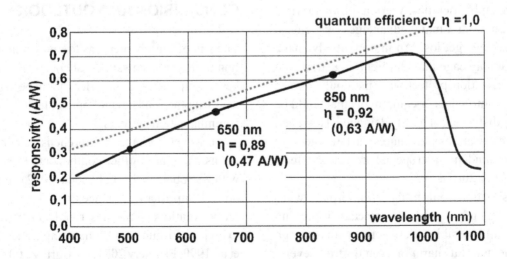

provide the next step to autonomous driving, new bus systems with higher data rates are desirable.

Additional challenges arise in new generation aircrafts. Due to safety problems in data transmission, an optical solution for data transmission is highly needed. We have shown that conventional LED/POF solutions could only perform simple structures, like ring or multimedia point-to-point connections. For more complex or long-ranged structures like a 100 m optical network, LED/POF solutions could reach their limits. Consequently, components to fulfill higher demands have to be taken into account. In particular, low attenuation PCS-fibers, combined with less temperature-critical VCSELs could be a promising solution. Thus, a smaller receiving photodiode diameter can be selected and the transmission data rates can be increased into the range of GBit/s. This combination paves the way for the new generation aircrafts covered by carbon fiber fuselages, having a much better lightning protection and EMC compatibility.

REFERENCES

Adams, W. G., & Day, R. E. (1876). The action of light in Selenium. *Proc. R. Soc.*, *25*, 113. doi:10.1098/rspl.1876.0024

Börner, M. (1967). *Mehrstufiges Übertragungssystem für Pulscodemodulation dargestellte Nachrichten*. Patent DE1254513.

Burrus, C. A., & Miller, B. I. (1971). Small area double heterostructure AlGaAs electroluminescent diode sources for optical-fiber transmission lines. *Optics Communications*, *4*(4), 307. doi:10.1016/0030-4018(71)90157-X

Chbat, M. W. (2000). Managing polarization mode dispersion. *Photonics Spectra*, 100.

Cohen, L. G. (1982). Dispersion and bandwidth spectra in single-mode fibers. *IEEE Journal of Quantum Electronics, QE-18*(1), 49. doi:10.1109/JQE.1982.1071366

Cooperation, M. O. S. T. (2007). *MOST brand book vol. 1.1*, Aug. 2007. Retrieved December 7, 2009, from www.mostcooperation.com

Drullmann, R., & Kammerer, W. (1980). Leitungscodierung und betriebliche Überwachung bei regenerativen Lichtleitkabelübertragungssystemen. *Frequenz*, *34*(2), 45. doi:10.1515/FREQ.1980.34.2.45

Ebbinghaus, G. (1985). Small area ion implanted p+n Germanium avalanche photodiodes for a wavelength of 1.3 μm. *Siemens Research and Development Report*, *14*(6), 284.

Flannery, D. (2001). Raman amplifiers: Powering up for ultra-long-haul. *Fiber Systems*, *5*(7), 48.

FlexRay Consortium. (2008). *Home page*. Retrieved August 7, 2009, from www.FlexRay.com

Fußgänger, K., & Roßberg, R. (1990). Uni and bidirectional 4λ x560 Mbit/s transmission systems using WDM devices and wavelength-selective fused single-mode fiber couplers. *IEEE Journal on Selected Areas in Communications*, 8(6), 1032. doi:10.1109/49.57806

Gloge, D. (1973). Multimode theory of graded core fibers. *The Bell System Technical Journal*, *52*, 1563.

Grzemba, A. (2007). *MOST: Das Multimedia-Bussystem für den Einsatz im Automobil*. Poing, Germany. *Franzis., ISBN-10*, 3772341497.

Kao, C. K., & Hockham, G. A. (1966). Dielectric-fiber surface waveguides for optical frequencies. *Proc. IEE*, *113*(7), 1151.

Kapron, F. P. (1970). Radiation losses in glass optical waveguides. *Applied Physics Letters*, *17*(10), 423. doi:10.1063/1.1653255

Kenward, M. (2001). Plastic fiber homes in/on low-cost networks. *Fiber Systems, 5*(1), 35.

Kibler, T. (2004). Optical data buses for automotive applications. *Journal of Lightwave Technology, 22*, 2184–2199. doi:10.1109/JLT.2004.833784

Köster, W. (1983). Einfluss des Rückstreulichts auf die Nebensprechdämpfung in bidirektionalen Übertragungssystemen. *Frequenz, 37H*(4), 87. doi:10.1515/FREQ.1983.37.4.87

Lee, S. C. J., et al. (2008). *Low-cost and robust 1 Gbit/s plastic optical fiber link based on light-emitting diode technology.* Optical Fiber Conference (OFC), San Diego, CA, USA

Lee, T. P. (2001). *Prospects and challenges of optoelectronic components in optical network systems.* Seminar on Internat. Exchange & Techn. Co-operation, Sept. 22 - 24, Wuhan, China

Lubkoll, J. (2008). *FlexRay with polymer-clad-silica fiber as transmitting medium in aviation electronics. ICTON MW 2008.* Morocco: Marrakech.

Mahlke, G., & Gössing, P. (1987). *Fiber optic cables.* Siemens AG Berlin - Munich: Publicis-MCD-Verlag, Erlangen, 77.

Maiman, T. H. (1960). Optical and microwave-optical experiments in Ruby. *Physical Review Letters, 4*(11), 564. doi:10.1103/PhysRevLett.4.564

Majkner, R. (2003). *Overview - Lightning protection of aircraft and avionics.* Sikorsky Corp. Retrieved December 7, 2009, from http://ewh.ieee.org/r1/ct/aess/aess_events.html

Malyon, D. J. (1991). Demonstration of optical pulse propagation over 10 000 km of fiber using recirculating loop. *Electronics Letters, 27*(2), 120. doi:10.1049/el:19910080

Marcuse, D. (1979). Calculation of bandwidth from index profiles of optical fibers. *Theory Appl. Opt., 18*(12), 2073. doi:10.1364/AO.18.002073

McCarthy, D. C. (2001). Growing by design. *Photonics Spectra,* **88.**

Melchior, H. (1970). Photodetectors for optical communication systems. *Proceedings of the IEEE, 58*(10), 1466. doi:10.1109/PROC.1970.7972

Miller, S. E. (1973). Research toward optical-fiber transmission systems. *Proceedings of the IEEE, 61*(12), 1703. doi:10.1109/PROC.1973.9360

Mollenhauer, L. F., & Stolen, R. H. (1982). Solitons in optical fibers. *Fiberoptic Technol., 193.*

Panish, M. B. (1976). Heterostructure injection lasers. *Proceedings of the IEEE, 64*(10), 1512. doi:10.1109/PROC.1976.10367

Payne, D. N., et al. (1990). Fiber optical amplifiers. *Proc. OFC '90,* Tutorial, paper ThFl, S. 335, San Francisco.

Pearsall, T. P. (1981). Photodetectors for optical communication. *J. Opt. Commun., 2*(2), 42. doi:10.1515/JOC.1981.2.2.42

Pfeiffer, T. (2001). Optical packet transmission system for metropolitan and access networks with more than 400 channels. *J. Lightw. Techn., 18*(12), 1928. doi:10.1109/50.908792

Quist, T. M. (1962). Semiconductor Maser of GaAs. *Applied Physics Letters, 1*(4), 91. doi:10.1063/1.1753710

Reiter, R., & Schramm, A. (2008). *Verfügbarkeitsrisiko senken – Neue Physical-Layer-Spezifikation für MOST150.* WEKA Fachmedien GmbH.

Seibl, D. (2008). *Polymer-optical-fiber data bus technologies for MOST applications in vehicles. ICTON MW 2008.* Morocco: Marrakech.

Strobel, O. (2007). *Optical data bus technologies for automotive applications. ICTON MW 2007, Sousse* (p. 1). Tunesia.

Strobel, O. A. (in press). *Limits and new trends in fiber-optic transmission.*

Sykes, E. (2001). Modelling sheds light on next-generation networks. *Fiber Systems, 5*(3), 58.

Weiershausen, W., et al. (2000). Realization of next generation dynamic WDM networks by advanced OADM design. *Proc. Europ. Conf. on Networks and Optical Comm.* 2000, (p. 199).

Compilation of References

Adams, W. G., & Day, R. E. (1876). The action of light in Selenium. *Proc. R. Soc., 25*, 113. doi:10.1098/rspl.1876.0024

Agrawal, G. P., & Olsson, N. A. (1989). Self-phase modulation and spectral broadening of optical pulses in semiconductor laser amplifiers. *IEEE Journal of Quantum Electronics, 25*, 2297–2306. doi:10.1109/3.42059

Ahlswede, R., Cai, N., Li, S. Y. R., & Yeung, R. W. (2000). Network information flow. *IEEE Transactions on Information Theory, 46*(4), 1204–1216. doi:10.1109/18.850663

Ahuja, S. S., Ramasubramanian, S., & Krunz, M. (2008). *SRLG failure localization in all-optical networks using monitoring cycles and paths* (pp. 700–708). IEEE INFOCOM.

Ali, M. (2001). *Transmission efficient design and management of wavelength-routed optical networks.* Kluwer Academic Publisher 2001.

An der Heiden, M., Sortais, M., Scheutzow, M., Reisslein, M., Seeling, P., & Herzog, M. (2007). Multicast capacity of optical packet ring for hotspot traffic. *Journal of Lightwave Technology, 25*(9), 2638. doi:10.1109/JLT.2007.902092

Andersen, R., Chung, F., Sen, A., & Xue, G. (2004). On disjoint path pairs with wavelength continuity constraint in WDM networks. *In IEEE INFOCOM, 1*, (pp. 524–535).

Androulidakis, S., Doukoglou, T., Patikis, G., & Kagklis, D. (2008). Service Differentiation and traffic engineering in IP over WDM networks. *Communications Magazine, IEEE, 46*(5), 52–59. doi:10.1109/MCOM.2008.4511649

Aneja, Y., Jaekel, A., & Bandyopadhyay, S. (2007). Some studies on path protection in WDM networks. *Photonic Network Communications, 14*, 165–176. doi:10.1007/s11107-007-0075-0

Arci, D., Maier, G., Pattavina, A., Petecchi, D., & Tornatore, M. (2003). Availability models for protection techniques in WDM networks. In *Proceedings Design of Reliable Communication Networks*, (pp. 158-166). October 2003.

Arci, D., Pattavina, A., Petecchi, D., & Tornatore, M. (2003). Availability models for protection techniques in WDM networks. *IEEE Workshop on Design of Reliable Communication Networks (DRCN)*, (pp. 159-166).

Asthana, R., Singh, Y. N., & Grover, W. D. (2010). *p*-Cycles: An overview. *IEEE Communications Surveys & Tutorials, 12*(1), 97–111. doi:10.1109/SURV.2010.020110.00066

ATT. (2009). *AT&T high speed internet business edition service level agreements.* Retrieved May 2009, from http://www.att.com/gen/general?pid=6622

Autenrieth, A., & Kirstädter, A. (2002). Engineering end-to-end IP resilience using resilience- differentiated QoS. *IEEE Communications Magazine, 40*(1). doi:10.1109/35.978049

Aziz, K., Sarwar, S., & Aleksic, S. (2008). Dimensioning an optical packet-burst switch: More interconnections or more delay lines. *International Conference on Optical Network Design and Modeling*, (pp. 1-6).

Babarczi, P., Tapolcai, J., & Ho, P. H. (2009). Availability-constrained dedicated segment protection in circuit switched mesh networks. *In Reliable Networks Design and Modeling. RNDM, 09*, 1–6.

Babbit, J., & Best, R. (2006). *Maintaining availability in an optical backbone network.* Paper presented at the National Fiber Optic Engineers Conference / Optical Fiber Communication Conference (NOEC/OFC), Anaheim, CA, USA.

Balasubramanian, S., He, W., & Somani, A. K. (2005). Light-trail networks: Design and survivability. In *Proceedings of IEEE Conference on Local Computer Networks* (pp. 174-181).

Banerjee, A., Park, Y., Clarke, F., Song, H., Yang, S., & Kramer, G. (2005). Wavelength-division-multiplexed passive optical network (WDM-PON) technologies for broadband access: A review [Invited]. *Journal of Optical Networking, 4*(11), 737–758. doi:10.1364/JON.4.000737

Baroni, S., Bayvel, P., Gibbens, R. J., & Korotky, S. K. (1999). Analysis and design of resilient multifiber wavelength-routed optical transport networks. *Journal of Lightwave Technology, 17*(5), 743. doi:10.1109/50.762888

Batchelor, P., Daino, B., Heinzmann, P., Hjelme, D. R., Inkret, R., & Jäger, H. A. (2000). Study on the implementation of optical transparent transport networks in the European environment-Results of the research project COST 239. *Photonic Network Communications, 2*(1), 15–32. doi:10.1023/A:1010050906938

Bergano, N. S. (2005). Wavelength division multiplexing in long-haul transoceanic transmission systems. *Journal of Lightwave Technology, 23,* 4125–4139. doi:10.1109/JLT.2005.858255

Bernstein, G., Rajagopalan, B., & Saha, D. (2003). *Optical network control: Architecture, protocols, and standards.* Boston, MA: Addison-Wesley Longman Publishing Co., Inc.

Bhandari, R. (1998). *Survivable networks: Algorithms for diverse routing.* Norwell, MA: Kluwer Academic Publishers.

Blumenthal, D. J. (2001). Photonic packet switching and optical label swapping. *Optic. Networks Mag.,* 1-12.

Börner, M. (1967). *Mehrstufiges Übertragungssystem für Pulscodemodulation dargestellte Nachrichten.* Patent DE1254513.

Bouillet, E. (2002). Stochastic approaches to compute shared mesh restored lightpaths in optical network architectures. *Proceedings - IEEE INFOCOM,* (June): 801–807.

Bouillet, E., Labourdette, J.-F., Ramamurthy, R., & Chaudhuri, S. (2002). Enhanced algorithm cost model to control tradeoffs in provisioning shared mesh restored lightpaths. In *Proceedings of OFC.*

Bouloutas, A. T., Calo, S., & Finkel, A. (1994). Alarm correlation and fault identification in communication networks. *IEEE Transactions on Communications, 42,* 523–533. doi:10.1109/TCOMM.1994.577079

Bouloutas, A. T., Hart, G., & Schwartz, M. (1992). Fault identification using a finite state machine model with unreliable partially observed data sequences. *IEEE Transactions on Communications, 42,* 523–533. doi:10.1109/TCOMM.1994.577079

Burrus, C. A., & Miller, B. I. (1971). Small area double heterostructure AlGaAs electroluminescent diode sources for optical-fiber transmission lines. *Optics Communications, 4*(4), 307. doi:10.1016/0030-4018(71)90157-X

Calabretta, N., Contestabile, G., Kim, S. H., Lee, S. B., & Ciaramella, E. (2006). Exploiting time-to-wavelength conversion for all-optical label processing. *IEEE Photonics Technology Letters, 18,* 436–438. doi:10.1109/LPT.2005.863207

Callegati, F., Cerroni, W., Raffaelli, C., & Savi, M. (2006). QoS differentiation in optical packet-switched networks. *Journal of Computer Communications, 29*(7), 855–864. doi:10.1016/j.comcom.2005.08.007

Callegati, F., Campi, A., & Cerroni, W. (2010). A practical approach to scheduler implementation for optical burst/packet switching. *Proc. of 14th Conference on Optical Network Design and Modeling,* Kyoto, Japan.

Castañón, G., Tancevski, L., & Tamil, L. (2000). Optical packet switching with multiple path routing. *Journal of Computer Networks and ISDN Systems. Special Issue on Optical Networks for New Generation Internet and Data Communication Systems, 32*(5), 653–662.

Castañón, G. (2004). Performance requirements for all-optical networks. In SPIE Proceedings (Eds.), *Conference on Optical Transmission Systems and Equipment for WDM Networking III: vol. 5596-18.* (pp. 127-134). Philadelphia, USA.

Castañón, G., Razo-Zapata, I., Mex, C., Ramirez-Velarde, R., & Tonguz, O. (2008). Security in all-optical networks: Failure and attack avoidance using self-organization. In IEEE Explorer (Ed.), *International Conference on Transparent Optical Networks, Mediterranean Winter: Vol. 3* (pp. 1-5).

Castañón, G., Razo-Zapata, I., Mozo, J., & Mex, C. (2009). Transparent optical network dimensioning for self-organizing routing. In IEEE Explorer (Ed.), *International Conference on Transparent Optical Networks, Mediterranean Winter,* (pp. 1-5).

Cavdar, C., Song, L., Tornatore, M., & Mukherjee, B. (2007). *Holding-time-aware and availability-guaranteed connection provisioning in optical WDM mesh networks. High-Capacity Optical Networks and Enabling Technologies (HONET), Dubai. UAE.*

Cavdar, C., Tornatore, M., Buzluca, F., & Mukherjee, B. (2010). Shared-path protection with delay tolerance (SDT) in optical WDM mesh networks. *IEEE/OSA. Journal of Lightwave Technology, 28*(14), 2068–2076. doi:10.1109/JLT.2010.2051414

Cavdar, C., Tornatore, M., & Buzluca, F. (2009). *Availability-guaranteed connection provisioning with delay tolerance in optical WDM mesh networks.* Optical Fiber Communication Conference and Exposition, San Diego, CA, USA.

Chan, T.-K., Chan, C.-K., Chen, L.-K., & Tong, F. (2003, November). A self-protected architecture for wavelength-division-multiplexed passive optical networks. *IEEE Photonics Technology Letters, 15*(11), 1660–1662. doi:10.1109/LPT.2003.818657

Chao, C. S., Yang, D. L., & Liu, A. C. (2001). An automated fault diagnosis system using hierarchical reasoning and alarm correlation. *Journal of Network and Systems Management, 9,* 183–202. doi:10.1023/A:1011315125608

Chao, C. S., Yang, D. L., & Liu, A. C. (1999). Alarm correlation view (ACView). *Proceedings of IASTED International Conference on modelling and Simulation,* Philadelphia, (pp. 291-253).

Chbat, M. W. (2000). Managing polarization mode dispersion. *Photonics Spectra,* 100.

Chen, Y., Qiao, C., & Yu, X. (2004). Optical burst switching (OBS): A new area in optical networking research. *IEEE Network Magazine, 18*(3), 16–23. doi:10.1109/MNET.2004.1301018

Chen, J., & Wosinska, L. (2007). Analysis of protection schemes in PON compatible with smooth migration from TDM-PON to hybrid WDM/TDM-PON. *OSA Journal of Optical Networking, 6*(5), 514–526. doi:10.1364/JON.6.000514

Chen, J., Wosinska, L., & He, S. (2008, March). High Utilization of wavelengths and simple interconnection between users in a protection scheme for passive optical networks. *IEEE Photonics Technology Letters, 20*(6), 389–391. doi:10.1109/LPT.2007.915655

Chen, J., Chen, B., & He, S. (2006). Self-protection scheme against failures of distributed fiber links in an ethernet passive optical network. *OSA Journal of Optical Networks, 5*(9), 662–666. doi:10.1364/JON.5.000662

Chen, J., Mas Machuca, C., Wosinska, L., & Jaeger, M. (2010). Cost vs. reliability performance study of fiber access network architectures. *IEEE Communications Magazine, 48*(2), 56–65. doi:10.1109/MCOM.2010.5402664

Chen, J., & Wosinska, L. (2007). Analysis of protection schemes in PON compatible with smooth migration from TDM-PON to hybrid WDM/TDM PON. *OSA Journal of Optical Networks, 6*(5), 514–526. doi:10.1364/JON.6.000514

Chen, J., Wosinska, L., & He, S. (2008). High utilization of wavelengths and simple interconnection between users in a protection scheme for passive optical networks. *IEEE Photonics Technology Letters, 20*(6), 389–391. doi:10.1109/LPT.2007.915655

Chen, B., & Tobagi, F. (2009). Optical network design to minimize switching and transceiver equipment costs. *Optical Switching and Networking, 6*(3), 171–180. doi:10.1016/j.osn.2009.02.002

Chlamtac, I., Ganz, A., & Karmi, G. (1992). Lightpath communications: An approach to high bandwidth optical WANs. *IEEE Transactions on Communications, 40*(7), 1171–1182. doi:10.1109/26.153361

Chlamtac, I., Faragó, A., & Zhang, T. (1996). Lightpath (wavelength) routing in large WDM networks. In *IEEE J. Select. Areas Commun., 14,* 909–913.

Choi, E., Jang, H., Lee, J., Lee, H., Hwang, S., & Oh, Y.-J. (2005). Modeling and verification of FEC performance for optical transmission systems using a proposed uniformly quantized symbol error probability model. *Journal of Lightwave Technology, 23,* 1100–1104. doi:10.1109/JLT.2005.843452

Choi, J., Choi, M., & Lee, S.-H. (1999). An alarm correlation and fault identification scheme based on OSI managed object classes based on OSI managed object classes. *IEEE International Conference on Communications, 3*, (pp. 1547-1551).

Cholda, P., Mykkeltveit, A., Helvik, B. E., Wittner, O., & Jajszczyk, A. (2007). A survey of resilience differentiation frameworks in communication networks. *IEEE Communications Surveys and Tutorials, 9*(1-4), 32–55. doi:10.1109/COMST.2007.4444749

Chowdhury, A., Huang, M.-F., Chien, H.-C., Ellinas, G., & Chang, G.-K. (2008, February). A self-survivable WDM-PON architecture with centralized wavelength monitoring, protection and restoration for both up- stream and downstream links. In *Conference on Optical Fiber Communication/National Fiber Optic Engineers* (pp. 1-3).

Cinkler, T. (2003). Traffic- and λ-grooming. *IEEE Network, 17*(2), 16–21. doi:10.1109/MNET.2003.1188282

Clemente, R., Bartoli, M., Bossi, M. C., D'Orazio, G., & Cosmo, G. (2005). *Risk management in availability SLA*. Italy: Design of Reliable Communication Networks, Island of Ischia.

Cloqueur, M., & Grover, W. (2005). Availability analysis and enhanced availability design in p-cycle-based networks. S*pringer Science. Photonic Network Communications, 10*(1), 55–71. doi:10.1007/s11107-005-1695-x

Cohen, L. G. (1982). Dispersion and bandwidth spectra in single-mode fibers. *IEEE Journal of Quantum Electronics, QE-18*(1), 49. doi:10.1109/JQE.1982.1071366

Cooperation, M. O. S. T. (2007). *MOST brand book vol. 1.1*, Aug. 2007. Retrieved December 7, 2009, from www.mostcooperation.com

Crawford, D. (1993). *Fiber optic cable dig-ups: causes and failures Network reliability: A report to the Nation: Compendium of Technical Papers*. Chicago: National Engineering Consortium.

Danielsen, S., Joergensen, C., Mikkelsen, B., & Stubkjaer, K. (1998). Optical packet switched network layer without optical buffers. *IEEE Photonics Technology Letters, 10*(6), 896–898. doi:10.1109/68.681522

Datta, P., & Somani, A. K. (2004). Diverse routing for shared risk resource groups (SRRG) failures in WDM optical networks. In *Proceedings of Broadnets* (pp. 120-129).

De Patre, S., Maier, G., Pattavina, A., & Martinelli, M. (2002). Optical network survivability: Protection techniques in the WDM layer. *Photonic Network Communications, 4*(3/4), 251–269. doi:10.1023/A:1016047527226

Dechter, R. (1996). Bucket Elimination: A unifying framework for probabilistic inference" in *Proc. Of the Twelfth Conference on Uncertainty in Artificial Intelligence,* Morgan Kaufmann Publishers.

Dijkstra, E. W. (1959). A note on two problems in connextion with graphs. *Numerische Mathematic, 1*(1), 269–271. doi:10.1007/BF01386390

Dixit, S. S. (2003). *IP over WDM: Building the next generation optical Internet*. Wiley-Interscience.

Doerr, C. R., Chandrasekhar, S., Buhl, L. L., Cappuzzo, M. A., Chen, E. Y., & Wong-Foy, A. (2006). Optical dispersion compensator suitable for use with non-wavelength-locked transmitters. *Journal of Lightwave Technology, 24*, 166–170. doi:10.1109/JLT.2005.860475

Dongvun, Z., & Subramaniam, S. (2000). Survivability in optical networks. *IEEE/OSA. Journal of Lightwave Technology, 14*(6), 16–23.

Dorren, H. J. S., Hill, M. T., Liu, Y., Calabretta, N., Srivatsa, A., & Huijskens, F. M. (2003). Optical packet switching and buffering by using all-optical signal processing methods. *Journal of Lightwave Technology, 12*, 2–12. doi:10.1109/JLT.2002.803062

Doucette, J., Coloqueur, M., & Grover, W. D. (2003). On the availability and capacity requirements of shared backup path-protected mesh networks. *SPIE Optical Networking Magazine, 4*(6), 29–44.

Doucette, J., & Grover, W. D. (2002). Capacity design studies of span-restorable mesh transport networks with shared-risk link group effects. In *Proceedings of OptiComm* (pp. 25-38).

Drullmann, R., & Kammerer, W. (1980). Leitungscodierung und betriebliche Überwachung bei regenerativen Lichtleitkabelübertragungssystemen. *Frequenz, 34*(2), 45. doi:10.1515/FREQ.1980.34.2.45

Ebbinghaus, G. (1985). Small area ion implanted p+n Germanium avalanche photodiodes for a wavelength of 1.3 μm. *Siemens Research and Development Report*, *14*(6), 284.

Elie-Dit-Cosaque, D., Ali, M., & Tancevski, L. (2002). Informed dynamic shared path protection. In *Proceedings of OFC*.

Ellinas, G., Hailemariam, A. G., & Stern, T. E. (2000). Protection cycles in mesh WDM networks. *IEEE Journal on Selected Areas in Communications*, *18*(10).

Ellinas, G., Bouillet, E., Ramamurthy, R., Labourdette, J. F., Chaudhuri, S., & Bala, K. (2003). Routing and restoration architectures in mesh optical networks. *Optical Networks Magazine*, *4*(1), 91–106.

Eramo, V., Listanti, M., & Bovo, L. (2006). Dimensioning models of shared resources for optical packet switching in unbalanced input-output traffic scenarios. *IEICE Trans Commun.*, *E89*(5), 1505–1516. doi:10.1093/ietcom/e89-b.5.1505

Eshoul, A., & Mouftah, H. T. (2009b). Survivability approaches using p-cycles in WDM mesh networks under static traffic. *IEEE/ACM Transactions on Networking*, *17*(2), 671–683. doi:10.1109/TNET.2008.2001467

Eshoul, A., & Mouftah, H. T. (2006). Survivable algorithm for routing and wavelength assignment under dynamic traffic and no wavelength conversion in mesh networks (SAD-RWA). *GESTS International Transactions on Computer Science and Engineering*, *28*(2).

Eshoul, A., & Mouftah, H. T. (2009a). *Performance evaluation of dynamic p-cycle protection methods in WDM optical networks*. Paper presented at International Conference on Transparent Optical Networks (ICTON), Miguel, Azores, Portugal.

FCC. (2001). FCC reportable network outages. *2001 Annual Report*, (pp. 7-17).

Fernandez Vallejo, M., Perez-Herrera, R. A., Elosua, C., Diaz, S., Urquhart, P., & Bariain, C. (2009). Resilient amplified double-ring optical networks to multiplex optical fiber sensors. *Journal of Lightwave Technology*, *27*(10), 1301. doi:10.1109/JLT.2009.2015774

Flannery, D. (2001). Raman amplifiers: Powering up for ultra-long-haul. *Fiber Systems*, *5*(7), 48.

FlexRay Consortium. (2008). *Home page*. Retrieved August 7, 2009, from www.FlexRay.com

Frederick, M. T., Datta, P., & Somani, A. K. (2006). Subgraph routing: A generalized fault-tolerant strategy for link failures in WDM optical networks. *Computer Networks*, *50*(2), 181–199. doi:10.1016/j.comnet.2005.05.025

Fumagalli, A., & Valcarenghi, L. (2000). IP restoration vs. WDM protection: is there an optimal choice? *IEEE Network*, *14*(6), 34–41. doi:10.1109/65.885668

Fumagalli, A., & Tacca, M. (2006). Differentiated reliability (DiR) in wavelength division multiplexing rings. *IEEE/ACM Transactions on Networking*, *14*(1), 159–168. doi:10.1109/TNET.2005.863708

Fumagalli, A., & Tacca, M. (2001). Diferentiated reliability (DiR) in WDM ring without wavelength converters. In *Proceedings IEEE International Conference on Communications* (ICC), (pp. 2887-2891). June 2001.

Fumagalli, A., Tacca, M., Unghvary, F., & Farago, A. (2002). Shared path protection with differentiated reliability. *IEEE International Conference on Communications (ICC)*, (pp. 2157-2161).

Fußgänger, K., & Roßberg, R. (1990). Uni and bidirectional 4λ x560 Mbit/s transmission systems using WDM devices and wavelength-selective fused single-mode fiber couplers. *IEEE Journal on Selected Areas in Communications*, *8*(6), 1032. doi:10.1109/49.57806

Fu-Tai, A., Kyeong Soo, K., Gutierrez, D., Yam, S., Hu, E., & Shrikhande, K. (2004). SUCCESS: A next-generation hybrid WDM/TDM optical access network architecture. *Lightwave Journal of Technology*, *22*(11), 2557–2569. doi:10.1109/JLT.2004.836768

Garey, M. R., & Johnson, D. S. (1979). *Computers and intractability. A guide to the theory of NP-completeness. A Series of Books in the Mathematical Sciences*. San Francisco, CA: WH Freeman and Company.

Gazi, B., & Ghassemlooy, Z. (2007). Dynamic buffer management using per-queue thresholds. *International Journal of Communication Systems*, *20*(5), 571–587. doi:10.1002/dac.834

Gee-Kung, C., Chowdhury, A., Zhensheng, J., Hung-Chang, C., Ming-Fang, H., Jianjun, Y., et al. (2009). Key technologies of WDM-PON for future converged optical broadband access networks [Invited]. *IEEE/OSA Journal of Optical Communications and Networking*, *1*(4), C35-C50.

Georgakilas, K. N., Katrinis, K., Tzanakaki, A., & Madsen, O. B. (2010). *Impact of dual-link failures on impairment-aware routed networks*. In 12th International Conference on Transparent Optical Networks.

Gerstel, O., & Ramaswami, R. (2000). Optical layer survivability: A service perspective. *IEEE Communications Magazine, 38*(3), 104–113. doi:10.1109/35.825647

Gerstel, O., & Sasaki, G. (2001). Quality of protection (QoP): A quantitative unifying paradigm to protection service grades. In *Proceedings of SPIE OptiComm.*

Ghassemlooy, Z., & Ngah, R. (2005). Simulation of 1x2 OTDM router employing symmetric Mach-Zehnder switches. *IEE Proceedings. Circuits, Devices and Systems, 152*, 171–177. doi:10.1049/ip-cds:20041017

Gil Arbues, P., Mas Machuca, C., & Tzanakaki, A. (2007). Comparative study of existing OADM and OXC architectures and technologies from the failure behavior perspective. *Journal of Optical Networking, 6*, 123–133. doi:10.1364/JON.6.000123

Girardin, F., Guekos, G., & Houbavlis, A. (1998). Gain recovery of bulk semiconductor optical amplifiers. *IEEE Photonics Technology Letters, 10*, 784–786. doi:10.1109/68.681483

Gloge, D. (1973). Multimode theory of graded core fibers. *The Bell System Technical Journal, 52*, 1563.

Gradner, L. M., Heydari, M., Shah, J., Sudborough, I. H., Tollis, I. G., & Xia, C. (1994). Techniques for finding ring covers in survivable networks. *Global Telecommunications Conference* (pp. 1862-1866). San Francisco, CA: IEEE.

Green, P. (2001). Progress in optical networking. *IEEE Communications Magazine, 39*, 54–61. doi:10.1109/35.894377

Grover, W. D. (2003). *Mesh-based survivable transport networks: Options and strategies for optical, MPLS, SONET and ATM networking*. Prentice Hall PTR.

Grover, W. D. (2004). *Mesh-based survivable networks: options and strategies for optical. MPLS, SONET and ATM networking*. Prentice Hall PTR.

Grover, W. (1999, August). High availability path design in ring-based optical networks. *IEEE/ACM Transactions on Networking, 7*, 558–574. doi:10.1109/90.793028

Grover, W. D., & Stamatelakis, D. (1998). Cycle-oriented distributed preconfiguration: Ring-like speed with mesh-like capacity for self-planning network restoration. In *Proc. IEEE International Conference on Communications* (pp. 537-543).

Gruber, C. G. (2003). Resilient networks with non-simple p-cycles. *10th International Conference on Telecommunications, 2003* (pp. 1027-1032). IEEE.

Grzemba, A. (2007). *MOST: Das Multimedia-Bussystem für den Einsatz im Automobil*. Poing, Germany. *Franzis.*, *ISBN-10*, 3772341497.

Gumaste, A., & Chlamtac, I. (2004). Light-trails: An optical solution for IP transport. *Journal of Optical Networking, 3*(4).

Guo, L. Q., & Connelly, M. J. (2006). All-optical AND gate with improved extinction ratio using signal induced nonlinearities in a bulk semiconductor optical amplifier. *Optics Express, 14*, 2938–2943. doi:10.1364/OE.14.002938

Guo, Q., Ho, P.-H., Yu, H., Tapolcai, J., & Mouftah, H. T. (2010). Spare capacity re-provisioning for high availability shared backup path protection connections. *Elsevier Computer Communications, 33*(5), 603–611.

Guo, L. (2007). Heuristic survivable routing algorithm for multiple failures in WDM networks. In *Proceedings of 2nd IEEE/IFIP International Workshop on Broadband Convergence Networks.*

Guo, Q., Ho, P.-H., Yu, H., & Mouftah, H. T. (2007). Availability-constrained shared backup path protection (SBPP) for GMPLSBased Spare Capacity Reprovisioning. *IEEE International Conference on Communications (ICC)*, (pp. 2186-2191).

Haider, A., & Harris, R. (2007). Recovery techniques in next generation networks. *IEEE Communications Surveys & Tutorials, 9*(3), 2–17. doi:10.1109/COMST.2007.4317617

Haque, A., Ho, P., Boutaba, R., & Ho, H. J. (2004). Group shared protection (GSP): A scalable solution for spare capacity reconfiguration in mesh WDM networks. *IEEE Global Telecommunications Conference (GLOBECOM), vol. 3*, (pp. 2029-2035).

Harvey, N. J. A., Patrascu, M., Wen, Y., Yekhanin, S., & Chan, V. W. S. (2007). *Non-adaptive fault diagnosis for all-optical networks via combinatorial group testing on graphs* (pp. 697–705). IEEE INFOCOM.

Hauer, M. C., McGeehan, J., Touch, J., Kamath, P., Bannister, J., Lyons, E. R., et al. (2002). Dynamically reconfigurable all-optical correlators to support ultra-fast internet routing. *Proc. OFC 2002, USA*, (pp. 268-270).

He, W., Fang, J., & Somani, A. K. (2004). On survivability design in light-trail optical networks. In *Proceedings of 8th Conference on Optical Networks Design and Modeling (ONDM)*.

He, W., Fang, J., & Somani, A. K. (2005). *A p-cycle based survivable design for dynamic traffic in WDM networks*. Paper presented at IEEE Global Telecommunications Conference (GLOBECOM), St. Louis, Missouri, USA.

Hedrick, C. (1988). *Routing information protocol. STD 34, RFC 1058*. Rutgers University.

Hill, A., Brierley, M., Percival, R., Wyatt, R., Pitcher, D., & Pati, K. (2002). Multiple-star wavelength-router network and its protection strategy. *IEEE Journal on Selected Areas in Communications, 16*(7), 1134–1145. doi:10.1109/49.725184

Hirooka, T., Kumakura, T., Osawa, K., & Nakazawa, M. (2006). Comparison of 40GHz optical demultiplexers using SMZ switch and EA modulator in 160 Gbit/s-500 km OTDM transmission. *IEICE Electronics Express, 3*, 397–403. doi:10.1587/elex.3.397

Ho, P.-H., & Mouftah, H. T. (2002). A framework for service-guaranteed shared protection in WDM mesh networks. *IEEE Communications Magazine, 40*(2), 97–103. doi:10.1109/35.983914

Ho, P.-H., Tapolcai, J., & Cinkler, T. (2004). Segment shared protection in mesh communication networks with bandwidth guaranteed tunnels. *IEEE/ACM Transactions on Networking, 12*(6), 1105–1118. doi:10.1109/TNET.2004.838592

Ho, P.-H., & Mouftah, H. T. (2004). Shared protection in mesh WDM networks. *IEEE Communications Magazine, 42*(1), 70–76. doi:10.1109/MCOM.2004.1262164

Ho, P. H., & Mouftah, H. T. (2002). Allocation of protection domains in dynamic WDM mesh networks. *In 10th IEEE International Conference on Network Protocols*, 2002, (pp. 188–189).

Ho, P., Mouftah, H., & Haque, A. (2007). Availability constrained shared backup path protection (SBPP) for GMPLS-based spare capacity reconfiguration. In *Proceedings IEEE International Conference on Communications (ICC)*, (pp. 2186-2191). June 2007.

Hu, J. Q. (2003). Diverse routing in optical mesh networks. *IEEE Transactions on Communications, 51*(3), 489–494. doi:10.1109/TCOMM.2003.809779

Huang, C., Li, M., & Srinivasan, A. (2007). A scalable path protection mechanism for guaranteed network reliability under multiple failures. *IEEE Transactions on Reliability, 56*(2), 254–267. doi:10.1109/TR.2007.896739

Hunter, D. K., Chia, M. C., & Andnovic, I. (1998). Buffering in optical packet switches. *Journal of Lightwave Technology, 16*, 2081–2094. doi:10.1109/50.736577

IEEE. (2006). Standard for Information Technology-Telecommunications and information exchange between systems- Local and metropolitan area networks-Specific requirements part 3: Carrier sense multiple access with collision detection (CSMA/CD) access method and physical layer specifications.

IEEE 802.3 Ethernet in the First Mile Study Group. (2001). *Ethernet passive optical networks* (EPONs).

ITU-T. (2000). ITU-T recommendation G.983.2: The ONT management and control interface specification for ATM PON.

ITU-T. (2001). ITU-T recommendation G.983.3: A broadband optical access system with increased service capability by wavelength allocation.

ITU-T. (2002). ITU-T recommendation G.983.5: A broadband optical access system with enhanced survivability.

ITU-T. (2003). ITU-T recommendation G.984.1: Gigabit-capable passive optical networks (GPON): General characteristics.

ITU-T. (1998). *ITU-T recommendation G.983.1: Broadband optical access systems based on passive optical networks*. PON.

ITU-T. (1998). *Recommendation G983.1*.

Iwatsuki, K., & Kani, J. (2009, September). Applications and technical issues of wavelength-division multiplexing passive optical networks with colorless optical network units [Invited]. *OSA Journal of Optical Networking*, *1*(4), 17–24.

Jaekel, A. (2006). Lightpath scheduling and allocation under a flexible schedule traffic model. In *Proceedings of IEEE Globecom*.

Jaekel, A., & Chen, Y. (2006). Routing and wavelength assignment for prioritized demand under a scheduled traffic model. In *Proceedings of Broadnets Workshop on Guaranteed Optical Service Provisioning*.

Jaekel, A., & Chen, Y. (2007). Demand allocation without wavelength conversion under a sliding scheduled traffic model. In *Proceedings of IEEE Broadnets*.

Jose, N., & Somani, A. K. (2003). Reconfiguring connections in optical networks. *Proc. of DRCN, October 2003*.

Kamal, A. E. (2010). 1+ N network protection for mesh networks: network coding-based protection using p-cycles. *IEEE/ACM Transactions on Networking*, *18*(1), 67–80. doi:10.1109/TNET.2009.2020503

Kaminow, P., & Koch, T. L. (1997). *Optical fiber telecommunications IIIA*. Academic Press.

Kantarci, B., Mouftah, H. T., & Oktug, S. (2009). Adaptive schemes for differentiated availability-aware connection provisioning in optical transport networks. *IEEE / OSA. Journal of Lightwave Technology*, *27*(20), 4595–4602. doi:10.1109/JLT.2009.2025246

Kantarci, B., & Mouftah, H. T. (2010). SLA-aware protection switching in optical WDM networks. *25th Queen's Biennial Symposium on Communications*, (pp. 230-233).

Kantarci, B., Mouftah, H. T., & Oktug, S. (2008a). Connection provisioning with feasible shareability determination for availability-aware design of optical networks. *International Conference on Transparent Optical Networks (ICTON)*, *vol. 3* (pp. 19-22).

Kantarci, B., Mouftah, H. T., & Oktug, S. (2008b). Arranging shareability dynamically for the availability-constrained design of optical transport networks. *IEEE Symposium on Computers and Communications (ISCC)*, (pp. 68-73).

Kantarci, B., Mouftah, H. T., & Oktug, S. (2008c). *Availability analysis and connection provisioning in overlapping shared segment protection for optical networks*. Paper presented at International Symposium on Computer and Information Sciences (ISCIS), Istanbul, Turkey.

Kao, C. K., & Hockham, G. A. (1966). Dielectric-fiber surface waveguides for optical frequencies. *Proc. IEE*, *113*(7), 1151.

Kapron, F. P. (1970). Radiation losses in glass optical waveguides. *Applied Physics Letters*, *17*(10), 423. doi:10.1063/1.1653255

Katzela, I., & Schwartz, M. (1995). Schemes for fault identification in communication networks. *IEEE/ACM Transactions on Networking*, *3*, 753–764. doi:10.1109/90.477721

Kenward, M. (2001). Plastic fiber homes in/on low-cost networks. *Fiber Systems*, *5*(1), 35.

Kibler, T. (2004). Optical data buses for automotive applications. *Journal of Lightwave Technology*, *22*, 2184–2199. doi:10.1109/JLT.2004.833784

Kiese, M., & Mas Machuca, C. (2007). Optimal placement of different types of monitoring equipment in transparent optical networks. In *Proceedings of the Workshop on Monitoring, Attack Detection and Mitigation (MONAM 2007)* Toulouse, France.

Kilper, D. C., Bach, R., Blumenthal, D. J., Einstein, D., Landolsi, T., & Ostar, L. (2004). Optical performance monitoring. *Journal of Lightwave Technology*, *22*, 294–304. doi:10.1109/JLT.2003.822154

Kim, H. D., Kang, S.-G., & Le, C.-H. (2000, August). A low-cost WDM source with an ASE injected Fabry-Perot semiconductor laser. *IEEE Photonics Technology Letters*, *12*(8), 1067–1069. doi:10.1109/68.868010

Kim, J.-Y., Mun, S.-G., Lee, H.-K., & Lee, C.-H. (2009, November). Self-restorable WDM-PON With a Color-Free Optical Source. *OSA Journal of Optical Networking*, *1*(6), 565–570.

Kim, J., Yang, Y.-M., Park, S., Lee, S., & Chung, B. (2008). Fault localization for heterogeneous networks using alarm correlation on consolidated inventory database. *Lecture Notes in Computer Science, 5297*, 82–91. doi:10.1007/978-3-540-88623-5_9

Kitamura, Y., Lee, Y., Sakiyama, R., & Okamura, K. (2007). Experience with restoration of Asia Pacific network failures from Taiwan earthquake. *IEICE Transactions on Communications. E (Norwalk, Conn.), 90-B*(11), 3095–3103.

Kodialam, M., & Lakshman, T. V. (2000). *Dynamic routing of bandwidth guaranteed tunnels with restoration* (pp. 902–911). Proc. IEEE INFOCOM.

Kodialam, M., & Lakshman, T. V. (2001). *Dynamic routing of locally restorable bandwidth guaranteed tunnels using aggregated link usage information* (pp. 376–385). Proc. IEEE INFOCOM.

Koetter, R., & Medard, M. (2003). An algebraic approach to network coding. *IEEE/ACM Transactions on Networking, 11*(5), 782–795. doi:10.1109/TNET.2003.818197

Köster, W. (1983). Einfluss des Rückstreulichts auf die Nebensprechdämpfung in bidirektionalen Übertragungssystemen. *Frequenz, 37H*(4), 87. doi:10.1515/FREQ.1983.37.4.87

Krishna, G. P., Pradeep, M. J., & Murthy, C. S. R. (2000). A segmented backup scheme for dependable real-time communication in multihop networks. *IEEE International Workshop on Parallel and Distributed Real-Time Systems,* (pp. 678-684).

Kuri, J., Puech, N., Gagnaire, M., Dotaro, E., & Douville, R. (2003). Routing and wavelength assignments of scheduled lightpath demands. *IEEE Journal on Selected Areas in Communications, 21*(8), 1231–1240. doi:10.1109/JSAC.2003.816622

Kuri, J. (2003). *Optimization problems in WDM optical transport networks with scheduled lightpath demands.* Unpublished doctoral dissertation, ENST Paris.

Kuri, J., Puech, N., Gagnaire, M., Dotaro, E., & Douville, R. (2003). Diverse routing of scheduled lightpath demands in an optical transport network. In *Proceedings of Fourth International Workshop on the Design of Reliable Communication Networks.*

Lam, C. (2007). *Passive optical networks: Principles and practice.* Elsevier/Academic Press.

Lastine, D., & Somani, A. K. (2008). Supplementing non-simple P-cycles with preconfigured lines. *IEEE International Conference on Communications* (pp. 5443-5447). Beijing, China: IEEE.

Lee, J. H., Choi, K.-M., Moon, J.-H., Mun, S.-G., Lee, H.-K., & Kim, J.-Y. (2009, October). A seamless evolution method with protection capability for next-generation access networks. *Journal of Lightwave Technology, 27*(19), 4311–4318. doi:10.1109/JLT.2009.2023608

Lee, K., Lee, S., Lee, J., Han, Y., Mun, S., & Lee, S. (2007, April). A self-restorable architecture for bidirectional wavelength-division-multiplexed passive optical network with colorless ONUs. *Optics Express, 15*, 4863–4868. doi:10.1364/OE.15.004863

Lee, K., Mun, S.-G., Lee, C.-H., & Lee, S. B. (2008, May). Reliable wavelength-division-multiplexed passive optical network using novel protection scheme. *IEEE Photonics Technology Letters, 20*(9), 679–681. doi:10.1109/LPT.2008.919445

Lee, K. K., Lim, F., & Ong, B. H. (2005). *Building resilient IP networks.* Cisco Press Networking Technology.

Lee, Y., & Mukherjee, B. (2004). Traffic engineering in next-generation optical networks. *Communications Surveys & Tutorials IEEE, 6*(3), 16–33. doi:10.1109/COMST.2004.5342291

Lee, C., Chan, T., Chan, C., Chen, L., & Lin, C. (2003). A group protection architecture (GPA) for traffic restoration in multi-wavelength passive optical networks. In *Proc. Of European Conference on Optical Communications* (ECOC) (vol. 2).

Lee, H. W., & Modiano, E. (2009). *Diverse routing in networks with probabilistic failures.* IEEE INFOCOM '09.

Lee, S. C. J., et al. (2008). *Low-cost and robust 1 Gbit/s plastic optical fiber link based on light-emitting diode technology.* Optical Fiber Conference (OFC), San Diego, CA, USA

Lee, T. P. (2001). *Prospects and challenges of optoelectronic components in optical network systems.* Seminar on Internat. Exchange & Techn. Co-operation, Sept. 22 - 24, Wuhan, China

Le-Minh, H., Ghassemlooy, Z., & Ng, W. P. (2007). All-optical flip-flop based on SMZ with a feedback-loop and multiple forward set/reset signals. *SPIE Optical Engineering Letters*, *46*, 040501.

Le-Minh, H., Ghassemlooy, Z., & Ng, W. P. (2008). Characterization and performance analysis of a TOAD switch employing a dual control pulse scheme in high-speed OTDM demultiplexer. *IEEE Communications Letters*, *12*, 316–318. doi:10.1109/LCOMM.2008.061299

Le-Minh, H., Ghassemlooy, Z., Ng, W. P., & Chiang, M. F. (2009). All-optical packet routing network based on PPM-HP header processing. *IET Communications Proceeding*, *3*, 465–476. doi:10.1049/iet-com:20070505

Le-Minh, H., Ghassemlooy, Z., & Ng, W. P. (2006a). Multiple-hop routing based on the pulse-position-modulation header processing scheme in all-optical ultrafast packet switching network. *IEEE GLOBECOM 2006*, San Francisco, USA, OPN06-03.

Le-Minh, H., Ghassemlooy, Z., & Ng, W. P. (2006b). Ultrafast all-optical self clock extraction based on two inline symmetric Mach-Zehnder switches. *Proc. of ICTON 2006, Nottingham, UK*, (pp. 64-67).

Li, C.-S., & Ramaswami, R. (1997). Automatic fault detection, isolation, and recovery in transparent all-optical networks. *Journal of Lightwave Technology*, *15*, 1784–1793. doi:10.1109/50.633555

Li, G., Wang, D., Kalmanek, C., & Doverspike, R. (2002). Efficient distributed path selection for shared restoration connections. In *Proceedings of IEEE INFOCOM* (pp. 40–149).

Li, T., & Wang, B. (2005). On optimal survivability design in WDM optical networks under a scheduled traffic model. In *Proceedings of the 5th IEEE International Workshop on Design of Reliable Communication Networks*.

Li, T., & Wang, B. (2006). *Approximating optimal survivable scheduled service provisioning in WDM optical networks with iterative survivable routing*. International Conference on Broadband Communications, Networks, and Systems (Broadnets), San Jose, CA, USA.

Li, T., Wang, B., Xin, C., & Zhang, X. (2005). On survivable service provisioning in WDM optical networks under a scheduled traffic model. In *Proceedings of IEEE Globecom*.

Lin, R., Wang, L., Li, L., & Guo, L. (2005). A new network availability algorithm for WDM optical networks. *International Conference on Computer and Information Technology*, vol. 1 (pp. 480-484).

Liu, Y., Tipper, D., & Siripongwutikorn, P. (2005). Approximating optimal spare capacity allocation by successive survivable routing. *IEEE/ACM Transactions on Networking*, *13*(11), 198–211. doi:10.1109/TNET.2004.842220

Liu, C., & Ruan, L. (2005) p-Cycle design in survivable WDM networks with shared risk link groups (SRLGs). In *Proceedings of 5ᵗʰ International Workshop on the Design of Reliable Communication Networks*.

Liu, S., & Chen, L. (2007). *Deployment of carrier-grade bandwidth-on-demand services over optical transport networks: A Verizon experience*. Optical Fiber Communication Conference and Exposition, San Diego, CA, USA.

Liu, Y., Tipper, D., & Siripongwutikorn, P. (2001). Approximating optimal spare capacity allocation by successive survivable routing. In *Proceedings of IEEE INFOCOM* (pp. 699–708).

Lo, C.-C., Chen, S.-H., & Lin, B.-Y. (2000). Coding-based schemes for fault identification in communication networks. *International Journal of Network Management*, *10*, 157–164. doi:10.1002/(SICI)1099-1190(200005/06)10:3<157::AID-NEM360>3.0.CO;2-G

Lubkoll, J. (2008). *FlexRay with polymer-clad-silica fiber as transmitting medium in aviation electronics. ICTON MW 2008*. Morocco: Marrakech.

Lucerna, D., Tornatore, M., Mukherjee, B., & Pattavina, A. (2009). Availability target redefinition for dynamic connections in WDM networks with shared path protection. *Design of Reliable Communication Networks DRCN 2009*, (pp. 235–242, 25-28).

Luo, X., & Wang, B. (2005). Diverse routing in WDM optical networks with shared risk link group (SRLG) failures. In *Proceedings of IEEE International Workshop on Design of Reliable Communication Networks*.

Mahlke, G., & Gössing, P. (1987). *Fiber optic cables*. Siemens AG Berlin - Munich: Publicis-MCD-Verlag, Erlangen, 77.

Maier, G., Pattavina, A., De Patre, S., & Martinelli, M. (2002). Optical network survivability: protection techniques in the WDM layer. *Photonic Network Communications, 4*(3), 251–269. doi:10.1023/A:1016047527226

Maier, M., Herzog, M., Scheutzow, M., & Reisslein, M. (2005). Protectoration: A fast and efficient multiple-failure recovery technique for resilient packet ring using dark fiber. *Journal of Lightwave Technology, 23*(10), 2816. doi:10.1109/JLT.2005.856165

Maiman, T. H. (1960). Optical and microwave-optical experiments in Ruby. *Physical Review Letters, 4*(11), 564. doi:10.1103/PhysRevLett.4.564

Majkner, R. (2003). *Overview - Lightning protection of aircraft and avionics.* Sikorsky Corp. Retrieved December 7, 2009, from http://ewh.ieee.org/r1/ct/aess/aess_events.html

Malyon, D. J. (1991). Demonstration of optical pulse propagation over 10 000 km of fiber using recirculating loop. *Electronics Letters, 27*(2), 120. doi:10.1049/el:19910080

Marcuse, D. (1979). Calculation of bandwidth from index profiles of optical fibers. *Theory Appl. Opt., 18*(12), 2073. doi:10.1364/AO.18.002073

Mas, C., Tomkos, I., & Tonguz, O. (2005). Failure location algorithm for transparent optical networks. *IEEE Journal on Selected Areas of Communications. Special Series on Optical Communications and Networking, 23*(8), 1508–1519.

Mas Machuca, C., & Thiran, P. (2000). An efficient algorithm for locating soft and hard failures in WDM networks. *IEEE Journal on Selected Areas in Communications, 18,* 1900–1911. doi:10.1109/49.887911

Mas Machuca, C., & Tomkos, I. (2004b). Optimal monitoring equipment placement for fault and attack location in transparent optical networks. *Lecture Notes in Computer Science, 3042,* 1395–1400. doi:10.1007/978-3-540-24693-0_125

Mas Machuca, C., Tomkos, I., & Tonguz, O. (2005). Failure location algorithm for transparent optical networks. *IEEE Journal on Selected Areas in Communications, 23,* 1508–1519. doi:10.1109/JSAC.2005.852182

Mas Machuca, C., Nguyen, H., & Thiran, P. (2004a). Failure location in WDM networks. In Sivalingam, K., & Subramaniam, S. (Eds.), *Optical WDM networks: Past lessons and path ahead.* Kluwer Publishers.

Mas Machuca, C., & Kiese, M. (2007). Optimal placement of monitoring equipment in transparent optical networks. In *Proceedings of the 6th International Workshop on Design and Reliable Communication Networks (DRCN2007),* La Rochelle, France

McCarthy, D. C. (2001). Growing by design. *Photonics Spectra,* 88.

Médard, M., Barry, R. A., Finn, S. G., & He, W. (2002). Generalized loop-back recovery in optical mesh networks. [TON]. *IEEE/ACM Transactions on Networking, 10*(1), 153–164. doi:10.1109/90.986592

Medard, M., Marquis, D., & Chinn, S. (1998). *Attack detection methods for all-optical networks.* Network and Distributed System Security Symposium.

Melchior, H. (1970). Photodetectors for optical communication systems. *Proceedings of the IEEE, 58*(10), 1466. doi:10.1109/PROC.1970.7972

Mello, D. A. A., Schupke, D. A., & Waldman, H. (2005). A matrix-based analytical approach to connection unavailability estimation in shared backup protection. *IEEE Communications Letters, 9*(9), 844–846. doi:10.1109/LCOMM.2005.1506722

Miller, S. E. (1973). Research toward optical-fiber transmission systems. *Proceedings of the IEEE, 61*(12), 1703. doi:10.1109/PROC.1973.9360

Mohan, G., & Murthy, C. S. R. (2000). Light-path restoration in WDM optical networks. *IEEE Network, 14*(6), 24–32. doi:10.1109/65.885667

Mohan, G., Murthy, C. S., & Somani, A. K. (2001). Efficient algorithms for routing dependable connections in WDM optical networks. *IEEE/ACM Transactions on Networking, 9*(5), 553–566. doi:10.1109/90.958325

Mollenhauer, L. F., & Stolen, R. H. (1982). Solitons in optical fibers. *Fiberoptic Technol., 193.*

Monti, P., Tacca, M., & Fumagalli, A. (2004). Resource-efficient path-protection schemes and online selection of routes in reliable WDM Networks. *OSA Journal of Optical Networking, special issue on Next-Generation WDM Network Design and Routing, 3*(4), 188-203.

Mouftah, H. T., & Ho, P. H. (2003). *Optical networks: Architecture and survivability.* Kluwer Academic Publishers.

Mukherjee, B. (2006). *Optical WDM networks.* New York, NY: Springer-Verlag Inc.

Mukherjee, D. S., Assi, C., & Agarwal, A. (2005). An alternative approach for enhanced availability analysis and design methods in p-cycle-based networks. *IEEE Journal on Selected Areas in Communications, 24*(12), 23–34. doi:10.1109/JSAC.2006.258220

Mykkeltveit, A., & Helvik, E. (2008). Comparison of schemes for provision of differentiated availability-guaranteed services using dedicated protection. In *Proceedings IEEE International Conference on Networking* (ICN08), (pp. 78-86).

Nakamura, S., Tajima, K., & Sugimoto, Y. (1994). Experimental investigation on high-speed switching characteristics of a novel symmetric Mach-Zehnder all-optical switch. *Applied Physics Letters, 65,* 283–285. doi:10.1063/1.112347

Ohara, T., Takara, H., Shake, I., Mori, K., Sato, K., & Kawanishi, S. (2004). 160 Gbps OTDM transmission using integrated all-optical MUX/DEMUX with all-channel modulation and demultiplexing. *IEEE Photonics Technology Letters, 16,* 650–652. doi:10.1109/LPT.2003.818953

Olkhovets, A., Phanaphat, P., Nuzman, C., Lichtenwalner, C., Kozhevnikow, M., & Kim, J. (2004). Performance of an optical switch based on 3-D MEMS crossconnect. *IEEE Photonics Technology Letters, 16,* 780–782. doi:10.1109/LPT.2004.823703

Orlowski, S., & Pioro, M. (2009). *On the complexity of column generation in survivable network design with path-based survivability mechanisms.* In International Network Optimization Conference (INOC).

Othonos, A., & Kalli, K. (1999). *Fiber Bragg Gratings: Fundamentals and applications in telecommunications and sensing.* Artech House.

Ou, C., Zang, H., Singhal, N. K., Zhu, K., Sahasrabuddhe, L. H., MacDonald, R. A., & Mukherjee, B. (2004). Subpath protection for scalability and fast recovery in optical WDM mesh networks. *IEEE Journal on Selected Areas in Communications, 22*(9), 1859–1875. doi:10.1109/JSAC.2004.830280

Ou, C., Zhang, J., Zang, H., Sahasrabuddhe, H., & Mukherjee, B. (2004). New and improved approaches for shared-path protection in WDM mesh networks. *Journal of Lightwave Technology, 22*(5), 1223–1232. doi:10.1109/JLT.2004.825346

Ou, C. K. (2003). Traffic grooming for survivable WDM networks--Shared protection. *IEEE Journal on Selected Areas in Communications, 21*(9), 1367–1383. doi:10.1109/JSAC.2003.818233

Ou, C. K. (2004). Traffic grooming for survivable WDM networks-–-Dedicated protection. *OSA Journal of Optical Networking, 3*(1), 50–74. doi:10.1364/JON.3.000050

Ou, C., Rai, S., & Mukherjee, B. (2005). Extension of segment protection for bandwidth efficiency and differentiated quality of protection in optical/MPLS networks. *Elsevier Optical Switching and Networking, 1*(1), 19–33. doi:10.1016/j.osn.2004.10.002

Ou, C., Zhang, J., Zhang, H., Sahasrabuddhe, L. H., & Mukherjee, B. (2004). New and improved approaches for shared-path protection in WDM mesh networks. *IEEE/OSA. Journal of Lightwave Technology, 22*(5), 1223–1232. doi:10.1109/JLT.2004.825346

Page, L. B., & Perry, J. E. (1989). Reliability of directed networks using the factoring theorem. *IEEE Transactions on Reliability, 38*(5), 556–562. doi:10.1109/24.46479

Pal, A., Paul, A., Mukherjee, A., Naskar, M. K., & Nasipuri, M. (2008). Fault detection and localization scheme for multiple failures in optical network. *Lecture Notes in Computer Science, 4904,* 464–470. doi:10.1007/978-3-540-77444-0_48

Pandi, Z., Tacca, M., Fumagalli, A., & Wosinska, L. (2006). Dynamic provisioning of availability-constrained optical circuits in the presence of optical node failures. *IEEE/OSA. Journal of Lightwave Technology, 24*(9), 3268–3279. doi:10.1109/JLT.2006.879505

Panish, M. B. (1976). Heterostructure injection lasers. *Proceedings of the IEEE, 64*(10), 1512. doi:10.1109/PROC.1976.10367

Papadimitriou, D. (2002). *Inference of shared risk link groups.* Internet Draft. Retrieved September 1, 2010, from http://tools.ietf.org/html/draft-many-inferencesrlg-02

Patel, N. S., Rauschenbach, K. A., & Hall, K. L. (1996). 40 Gbps demultiplexing using an ultrafast nonlinear interferometer (UNI). *IEEE Photonics Technology Letters*, *8*, 1695–1697. doi:10.1109/68.544722

Pavani, G., & Waldman, H. (2008). Addressing self-similarity in optical switching networks by means of ant colony optimization. *Photonic Network Communications*, *15*(1), 41–50. doi:10.1007/s11107-007-0086-x

Payne, D. N., et al. (1990). Fiber optical amplifiers. *Proc. OFC '90*, Tutorial, paper ThFl, S. 335, San Francisco.

Pearsall, T. P. (1981). Photodetectors for optical communication. *J. Opt. Commun.*, *2*(2), 42. doi:10.1515/JOC.1981.2.2.42

Pfeiffer, T. (2001). Optical packet transmission system for metropolitan and access networks with more than 400 channels. *J. Lightw. Techn.*, *18*(12), 1928. doi:10.1109/50.908792

Potter, D. (1991). The need for network management. *Computer Communications*, *14*(2), 121–125. doi:10.1016/0140-3664(91)90042-Y

Pramod, S. R., Siddiqui, S., & Mouftah, H. T. (2005). *Novel distributed protocol for dynamic routing and load balancing for optical networks.* Paper presented at OFC/NFOEC Optical Fiber Communication Conference, Anaheim, California.

Provan, J. S. (1986). The complexity of reliability computations in planar and acyclic graphs. *SIAM Journal on Computing*, *15*(3), 694–702. doi:10.1137/0215050

Qiao, C., & Xu, D. (2002). Distributed partial information management (DPIM) schemes for survivable networks— Part I. In *Proceedings of IEEE INFOCOM* (pp. 302-311).

Quist, T. M. (1962). Semiconductor Maser of GaAs. *Applied Physics Letters*, *1*(4), 91. doi:10.1063/1.1753710

Rai, S., Veeraraghavan, M., & Trivedi, K. S. (1995). A survey of efficient reliability computation using disjoint products approach. *Networks*, *25*(3), 147–163. doi:10.1002/net.3230250308

Ramamurthy, S., & Mukherjee, B. (2003). Survivable WDM mesh networks. *Journal of Lightwave Technology*, *21*(4), 870–883. doi:10.1109/JLT.2002.806338

Ramamurthy, S., & Mukherjee, B. (1999). Survivable WDM mesh networks – Part I: Protection. In *Proceedings of IEEE INFOCOM* (pp. 744-751).

Ramaswami, R., & Sivarajan, K. N. (2002). *Optical networks: A practical perspective* (2nd ed.). Morgan Kaufmann Publisher.

Ramos, F., Kehayas, E., Martinez, J. M., Clavero, R., Marti, J., & Stampoulidis, L. (2005). IST-LASAGNE: Towards all-optical label swapping employing optical logic gates and optical flip-flops. *Journal of Lightwave Technology*, *23*, 2993–3011. doi:10.1109/JLT.2005.855714

Rao, N. S. V. (1993). Computational complexity issues in operative diagnosis of graph-based systems. *IEEE Transactions on Computers*, *42*, 447–457. doi:10.1109/12.214691

Ray, M. (2010). *100G DWDM optical networking transport: The telecom industry prepares.* Retrieved December 2010, from http://searchtelecom.techtarget.com/feature/100G-DWDM-optical-networking-transport-The-telecom-industry-prepares

Reiter, R., & Schramm, A. (2008). *Verfügbarkeitsrisiko senken – Neue Physical-Layer-Spezifikation für MOST150.* WEKA Fachmedien GmbH.

Rejeb, R., Leeson, M. S., & Green, R. J. (2006). Fault and attack management in all-optical networks. *IEEE Communications Magazine*, *44*, 79–86. doi:10.1109/MCOM.2006.248169

Report, A. (2005). *Fraunhofer Institut Nachrichtentechnik Heinrich-Hertz-Institut, Berlin, Germany.*

Rubin, I., & Ling, J. (2002). Failure protection methods for optical meshed-ring communications networks. *IEEE Journal on Selected Areas in Communications*, *18*(10), 1950–1960.

Russel, S. J., & Norvig, P. (1995). *Artificial intelligence: A modern approach.* Prentice-Hall Publishing Company.

Saradhi, C. V., & Gurusamy, M. (2007). Scheduling and routing of sliding scheduled lightpath demands in WDM optical networks. In *Proceedings of OFC.*

Saradhi, C. V., & Murthy, C. (2002). Dynamic establishment of segmented protection paths in single and multi-fiber WDM mesh networks. In *Proceedings of OptiComm* (pp. 211-222).

Saradhi, C. V., Wei, L. K., & Gurusamy, M. (2004). Provisioning fault-tolerant scheduled lightpath demands in WDM mesh networks. In *Proceedings of Broadnets* (pp. 150-159).

Sato, K. I., Okamoto, S., & Hadama, H. (1994). Network performance and integrity enhancement with optical path layer technologies. *IEEE Journal on Selected Areas in Communications, 12*(1), 159. doi:10.1109/49.265715

Schupke, D. (2006). Analysis of p-cycle capacity in WDM networks. *Photonic Network Communications, 12*(1), 41–51.

Schupke, D. (2000). Reliability models of WDM self-healing rings. In *Proceedings Design of Reliable Communication Networks*, Munich, Germany, April 2000.

Schupke, D. A., Gruber, C. G., & Autenrieth, A. (2002). *Optimal configuration of p-cycles in WDM networks*. International Conference on Communications, New York, NY, USA.

Sebos, P., Yates, J., Hjalmtysson, G., & Greenberg, A. (2001). Auto-discovery of shared risk link groups. In *Proceedings of Optical Fiber Communication Conference*.

Seibl, D. (2008). *Polymer-optical-fiber data bus technologies for MOST applications in vehicles. ICTON MW 2008.* Morocco: Marrakech.

She, Q., Huang, X., & Jue, J. P. (2006). Maximum survivability under multiple failures. In *Proceedings of IEEE/OSA Optical Fiber Communication Conference*.

Shen, G., & Grover, W. D. (2005). Automatic lightpath service provisioning with an adaptive protected working capacity envelope based on *p*-cycles. In *Proceedings of 5th International Workshop on the Design of Reliable Communication Networks*.

Shen, G., & Grover, W. D. (2004). *A framework for dynamic survivable service provisioning based on p-cycles and the protected working capacity envelope (PWCE) concept*. Edmonton, Canada: TRLabs.

Shepherd, F. B. (2004). Lighting fibres in a dark network. *IEEE Journal on Selected Areas in Communications, 22*(9), 1583–1588. doi:10.1109/JSAC.2004.833850

Shier, D. R. (1991). *Network reliability and algebraic structures*. New York, NY: Clarendon Press.

Sivakumar, M., Fang, J., Sivalingham, K., & Somani, A. K. (2008). Design and analysis of partial protection mechanisms in groomed optical WDM mesh networks. *Journal of Optical Networking, 7*(6), 617–634. doi:10.1364/JON.7.000617

Sivalingam, K. M., & Subramaniam, S. (2004). *Emerging optical network technologies: Architectures, protocols and performance*. Secaucus, NJ: Springer-Verlag, Inc.

Skorin-Kapov, N., Tonguz, O., & Puech, N. (2007). Self-organization in transparent optical networks: A new approach to security. In IEEE Explorer (Ed.), *International Conference on Telecommunications* (pp. 7-14).

SNDlib. (n.d.). *Survivable fixed telecommunications network design library*. Retrieved from http://sndlib.zib.de/home.action

Sokoloff, J. P., Prucnal, P. R., Glesk, I., & Kane, M. (1993). A terahertz optical asymmetric demultiplexer (TOAD). *IEEE Photonics Technology Letters, 5*, 787–790. doi:10.1109/68.229807

Son, E. S., Han, K. H., Lee, J. H., & Chung, Y. C. (2005, March). Survivable network architectures for WDM PON. In *Conference on Optical Fiber Communication/National Fiber Optic Engineers, Technical Digest* (vol. 5).

Song, L., Zhang, J., & Mukherjee, B. (2007). Dynamic provisioning with availability guarantee for differentiated services in survivable mesh networks. *IEEE Journal on Selected Areas in Communications, 25*(3), 35–43. doi:10.1109/TWC.2007.024505

Spragins, J. (2002). Dependent failures in data communication systems. *IEEE Transactions on Communications, 25*(12), 1494–1499. doi:10.1109/TCOM.1977.1093787

Staessens, D., Manousakis, K., Colle, D., Mahlab, U., Pickavet, M., Varvarigos, E., & Demesteer, P. (2010). *Failure localization in transparent optical networks*. 2nd International Workshop on Reliable Networks Design and Modeling, Moscow, Russia.

Stamatelakis, D., & Grover, W. D. (2000). IP layer restoration and network planning based on virtual protection cycles. *IEEE Journal on Selected Areas in Communications, 18*(10), 1938–1949. doi:10.1109/49.887914

Stamatelakis, D., & Grover, W. D. (2000). Theoretical underpinnings for the efficiency of restorable networks using preconfigured cycles (p-cycles). *IEEE Transactions on Communications, 48*(8), 1262–1265. doi:10.1109/26.864163

Stanic, S., Subramaniam, S., Choi, H., Sahin, G., & Choi, H.-A. (2002). On monitoring transparent optical networks. *International Conference on Parallel Processing Workshops* (ICPPW'02), (p. 217).

Steinder, M., & Sethi, A. S. (2002). Increasing robustness of fault localization through analysis of lost, spurious, and positive symptoms. In *INFOCOM 2002. Twenty-First Annual Joint Conference of the IEEE Computer and Communications Societies, 1*, (pp. 322-331).

Sterbenz, J. P. G., et al. (2010). *ResiliNets: Resilient and survivable networks*. Retrieved September 1, 2010, from https://wiki.ittc.ku.edu/resilinets

Stokes, L. F., & Derickson D. (1997). *Lightwave component and system measurements*. Short Course Notes, OFC 97.

Strand, J., Chiu, A. L., & Tkach, R. (2001). Issues for routing in the optical layers. *IEEE Communications Magazine, 39*(2), 81–87. doi:10.1109/35.900635

Strobel, O. (2007). *Optical data bus technologies for automotive applications. ICTON MW 2007, Sousse* (p. 1). Tunesia.

Strobel, O. A. (in press). *Limits and new trends in fiber-optic transmission.*

Su, C., & Su, X. (2001). Protection path routing on WDM networks. In *Proceedings of OFC*.

Sue, C. C. (2006b, July). A novel 1: N protection scheme for WDM passive optical networks. *IEEE Photonics Technology Letters, 18*(13), 1472–1474. doi:10.1109/LPT.2006.877577

Sue, C. C. (2006a, November). 1: N protection scheme for AWG-based WDM PONs. In *IEEE Global Telecommunications Conference* (GLOBECOM) (pp. 1-5).

Sun, X., Chan, C.-K., & Chen, L. K. (2006, February). A survivable WDM-PON architecture with centralized alternate-path protection switching for traffic restoration. *IEEE Photonics Technology Letters, 18*(4), 631–633. doi:10.1109/LPT.2006.870135

Sutton, R., & Barto, G. (1998). *Reinforcement learning: An introduction*. The MIT Press.

Suurballe, J. W. (1974). Disjoint paths in a network. *Networks, 4*(2), 125–145. doi:10.1002/net.3230040204

Suurballe, J. W., & Tarjan, R. E. (1984). A quick method for finding shortest pairs of disjoint paths. *Networks, 14*(2), 325–336. doi:10.1002/net.3230140209

Suurballe, J. W. (1974). Disjoint paths in a network. *Networks, 4*, 125–145. doi:10.1002/net.3230040204

Suzuki, H., Fujiwara, M., & Iwatsuki, K. (2006). Application of super-DWDM technologies to terrestrial terabit transmission systems. *Journal of Lightwave Technology, 24*, 1998–2005. doi:10.1109/JLT.2006.871115

Sykes, E. (2001). Modelling sheds light on next-generation networks. *Fiber Systems, 5*(3), 58.

Tacca, M., Fumagalli, A., & Unghvary, F. (2003). *Double-fault shared path protection scheme with constrained connection downtime*. Banff, Alberta, Canada: Design of Reliable Communication Networks.

Tacca, M., Fumagalli, A., Paradisi, A., Unghvary, F., Gadhiraju, K., & Lakshmanan, S. (2003). Differentiated reliability in optical networks: Theoretical and practical results. *Journal of Lightwave Technology, 21*, 2576–2586. doi:10.1109/JLT.2003.819554

Tacca, M., Monti, P., & Fumagalli, A. (2004). The disjoint path-pair matrix approach for online routing in reliable WDM networks. *IEEE International Conference on Communications (ICC)*, (pp. 1187-1191).

Takahashi, R., Nakahara, T., Takahata, K., Takenouchi, H., Yasui, T., & Kondo, N. (2004). Photonic random access memory for 40-Gb/s 16-b burst optical packets. *IEEE Photonics Technology Letters, 16*, 1185–1187. doi:10.1109/LPT.2004.824987

Takenouchi, H., Takahata, K., Nakahara, T., Takahashi, R., & Suzuki, H. (2004). 40-Gbit/s 32-bit optical packet compressor/decompressor based on a photonic memory. *Proc. Conference on Lasers and Electro Optics, CThQ, San Francisco, California, U.S.A.*

Tanwir, S., Battestilli, L., Perros, H., & Karmous-Edwards, G. (2007). Dynamic scheduling of network resources with advance reservations in optical grids. *International Journal of Network Management, 18*(2), 79–105. doi:10.1002/nem.680

Tapolcai, J., Wu, B., & Ho, P.-H. (2009). *On monitoring and failure localization in mesh all-optical networks* (pp. 1008–1016). IEEE Infocom Proceedings.

Tapolcai, J. (2005). *Routing algorithms in survivable telecommunication networks.* Doctoral dissertation, Budapest University of Technology and Economics, Hungary. LAP Lambert Academic Publishing AG & Co KG. ISBN 978-3-8383-9297-4

Technologies, W. D. M. (2008, January). *WPON white paper*. Retrieved from http://www.ciphotonics.com/PDFs_Jan08/WPON_White_Paper_v10.pdf

Thiagarajan, S., & Somani, A. K. (2001). Traffic grooming for survivable WDM mesh networks. In *Proc. OptiCom'01, 2001,* (pp. 54-65).

To, M., & Neusy, P. (1994). Unavailability analysis of long-haul networks. *IEEE Journal on Selected Areas in Communications, 12,* 100–109. doi:10.1109/49.265709

Tomkos, I., Vogiatzis, D., Mas, C., Zacharopoulos, I., Tzanakaki, A., & Varvarigos, E. (2004). Performance engineering of metropolitan area optical networks through impairment constraint routing. *IEEE Communications Magazine, 42*(8), S40–S47. doi:10.1109/MCOM.2004.1321386

Tornatore, M., Maier, G., & Pattavina, A. (2006a). Capacity versus availability trade-offs for availability-based routing. *Journal of Optical Networking, 5,* 858–869. doi:10.1364/JON.5.000858

Tornatore, M., Maier, G., & Pattavina, A. (2006b). Availability design of optical transport networks. *IEEE Journal on Selected Areas in Communications, 24*(8), 1520–1532.

Tornatore, M., Ou, C. S., Zhang, J., Pattavina, A., & Mukherjee, B. (2005). An efficient shared-path-protection strategy based on connection-holding-time awareness. *IEEE/OSA. Journal of Lightwave Technology, 23*(10), 3138–3146. doi:10.1109/JLT.2005.856174

Tornatore, M., Maier, C., & Pattavina, A. (2006). Availability design of optical transport networks. *IEEE Journal on Selected Areas in Communications, 24,* 1520–4532.

Tran, A. V., Chae, C., & Tucker, R. S. (2005). Ethernet PON or WDM PON: A comparison of cost and reliability. *TENCON*, IEEE Region 10.

Trivedi, S. (1982). *Probability and statistics with reliability, queuing, and computer science applications.* Englewood Cliffs, NJ: Prentice-Hall.

Tucker, R. S., Eisenstein, G., & Korotky, S. K. (1988). Optical time-division multiplexing for very high bit-rate transmission. *Journal of Lightwave Technology, 6,* 1737–1749. doi:10.1109/50.9991

Ueno, Y., Nakamura, S., Hatakeyama, H., Tamanuki, T., Sasaki, T., & Tajima, K. (2002). 168-Gb/s OTDM wavelength conversion using an SMZ-Type all-optical switch. *Proc. ECOC 2002, Munich, Germany,* (pp. 13-14).

Valiant, L. G. (1979). The complexity of enumeration and reliability problems. *SIAM Journal on Computing, 8,* 410. doi:10.1137/0208032

Van Caenegem, B. (1998). Dimensioning of survivable WDM networks. *IEEE Journal on Selected Areas in Communications, 16*(7). doi:10.1109/49.725185

Vasseur, J. P., Pickavet, M., & Demeester, P. (2004). *Network recovery: Protection and restoration of Optical, SONET-SDH, IP, and MPLS.* Morgan Kaufmann Publishers.

Vasseur, J. P. (2004). *Network recovery: Protection and restoration of optical, SONET-SDH, IP, and MPLS.* The Morgan Kaufmann Series in Networking, 2004.

Verbrugge, S., Colle, D., Pickavet, M., Demeester, P., Pasqualini, S., & Iselt, A. (2006, June). Methodology and input availability parameters for calculating OpEx and CapEx costs for realistic network scenarios. *OSA Journal of Optical Networking, 5*(6), 509–520. doi:10.1364/JON.5.000509

Wang, Z., Sun, X., Lin, C., Chan, C.-K., & Chen, L.-K. (2005, March). A novel centrally controlled protection scheme for traffic restoration in WDM passive optical networks. *IEEE Photonics Technology Letters, 17*(3), 717–719. doi:10.1109/LPT.2004.842378

Wang, C., & Schwartz, M. (1992). *Fault detection with multiple observers* (pp. 2187–2196). IEEE Infocom Proceedings.

Wang, C., & Schwartz, M. (1993). Identification of faulty links in dynamic routed networks. *IEEE Journal on Selected Areas in Communications, 11,* 1449–1460. doi:10.1109/49.257936

Wang, B., & Li, T. (2009). Survivable scheduled service provisioning in WDM optical networks with iterative routing. *Optical Switching Network and Networking, 7*(1).

Wang, B., Li, T., Luo, X., Fan, Y., & Xin, C. (2005). Routing and wavelength assignment under a scheduled traffic model in reconfigurable WDM optical networks. In *Proceedings of Broadnets*.

Wang, X., Wang, S., Zhang, A., & Wang, J. (2009, November). A novel highly reliable WDM-PON system. In *Proc. of Communications and Photonics Conference and Exhibition Asia* (ACP) (pp. 1-2).

Wang, Z., Zhang, B., Lin, C., & Chan, C.-K. (2005, September). A broadcast and select WDM-PON and its protection. In *Proc. of European Conference on Optical Communication* (ECOC) (vol. 3, pp. 549-550).

Wei, X., Quo, L., Wang, X., Song, Q., & Li, L. (2008). Availability guarantee in survivable WDM mesh networks: A time perspective. *Elsevier Information Sciences, 178*(11).

Weiershausen, W., et al. (2000). Realization of next generation dynamic WDM networks by advanced OADM design. *Proc. Europ. Conf. on Networks and Optical Comm.* 2000, (p. 199).

Willems, G., Arijs, P., Parys, W., & Demeester, P. (2003). Capacity vs. availability tradeoffs in. mesh-restorable WDM networks. In *Proceedings International Workshop on Design of Reliable Communication Networks* (DRCN03), (pp. 158—166). Alberta, Canada, 2003.

Wosinska, L. (1993). Reliability study of fault-tolerant multiwavelength nonblocking optical cross connect based on InGaAsP/InP laser-amplifier gateswitch arrays. *IEEE Photonics Technology Letters, 5*(10), 1206–1209. doi:10.1109/68.248429

Wosinska, L., Chen, J., & Larsen, P. C. (2009). Fiber access networks: Reliability analysis and Swedish broadband market. *IEICE Transactions on Communications. E (Norwalk, Conn.), 92-B*(10), 3006–3014.

Wosinska, L., Thylen, L., & Holmstrom, R. P. (2001). Large-capacity strictly nonblocking optical cross-connects based on microelectrooptomechanical systems (MEOMS) switch matrices: Reliability performance analysis. *IEEE/OSA. Journal of Lightwave Technology, 19*(8), 1065–1075. doi:10.1109/50.939785

Wosinska, L., & Chen, J. (2008). Reliability performance analysis vs. deployment cost of fiber access networks. 7th International Conference on Optical Internet, Tokyo, Japan.

Xiang, B., Wang, S., & Li, L. M. (2003). A traffic grooming based on shared protection in WDM mesh networks. In *Proc. IEEE PDCAT'03*, Cheng Du, China, (pp. 254-258), August 2003.

Xin, L., Hongxiang, W., & Yuefeng, J. (2008). Resilient burst ring: Extend IEEE 802.17 to WDM networks. *Communications Magazine, IEEE, 46*(11), 74–81. doi:10.1109/MCOM.2008.4689248

Xu, D., Xiong, Y., & Qiao, C. (2003). Novel algorithms for shared segment protection. *IEEE Journal on Selected Areas in Communications, 21*(8), 1320–1331. doi:10.1109/JSAC.2003.816624

Xu, D., Qiao, C., & Xiong, Y. (2002). An ultra-fast shared path protection scheme - Distributed partial information management, part II. *In IEEE ICNP: International Conference on Network Protocols*, (pp. 344–353).

Xue, G., Zhang, W., Wang, T., & Thulasiraman, K. (2007). On the partial path protection scheme for WDM optical networks and polynomial time computability of primary and secondary paths. *Management, 3*(4), 625–643.

Yao, W., & Ramamurthy, B. (2004). Rerouting schemes for dynamic traffic grooming in optical WDM mesh networks. In *IEEE GLOBECOM '04*, vol. 3, (pp. 1793–1797).

Yao, W., & Ramamurthy, B. (2004). Survivable traffic grooming with differentiated end-to-end availability guarantees in WDM mesh networks. In *The 13th IEEE Workshop on Local and Metropolitan Area Networks, LANMAN 2004* (pp. 87–90).

Yeh, C., & Chi, S. (2007). Self-healing ring-based time-sharing passive optical networks. *IEEE Photonics Technology Letters, 19*(15), 1139–1141. doi:10.1109/LPT.2007.900155

Yeom, J., Tonguz, O., & Castañón, G. (2007). Security in all-optical networks: Self-organization and attack avoidance. *IEEE International Conference on Communications*, (pp. 1329-1335).

Yoo, S. J. B. (2003). Optical label switching, MPLS, MPLambdaS, and GMPLS. *Optic. Networks Mag., 4*, 17–31.

Yoo, S. (2006). Optical packet and burst switching technologies for the future photonic internet. *Journal of Lightwave Technology, 24*(12), 4468–4492. doi:10.1109/ JLT.2006.886060

Yoo, M., & Qiao, C. (1999). Supporting multiple classes of services in IP over WDM networks. In *Proceedings of IEEE Globecom* (pp. 1023–1027).

Yuan, S., & Jue, J. P. (2002). Shared protection routing algorithm for optical network. *Optical Networks Magazine, 3*(3), 32–39.

Yuan, S., Varma, S., & Jue, J. P. (2008). Lightpath routing for maximum reliability in optical mesh Networks. [OSA]. *Journal of Optical Networking, 7*(5), 449–466. doi:10.1364/JON.7.000449

Yuan, X. C., Li, V. O. K., Li, C. Y., & Wai, P. K. A. (2003). A novel self-routing address scheme for all-optical packet-switched networks with arbitrary topologies. *Journal of Lightwave Technology, 21*, 329–339. doi:10.1109/ JLT.2003.808755

Yuan, S., & Jue, J. P. (2004). Dynamic path protection in WDM mesh networks under risk disjoint constraint. In *Proceedings of IEEE Globecom* (pp. 1770–1774).

Yuan, S., Varma, S., & Jue, J. P. (2005). Minimum color problem for reliability in mesh networks. In *Proceedings of IEEE INFOCOM* (pp. 2658- 2669).

Yuan, S., Wang, B., & Waller, W., & Delavina, E. (2010). Reliable path routing in mesh networks under multiple link failures. *IEEE Transaction on Reliability*.

Yurong, H., Wushao, W., Heritage, J. P., & Mukherjee, B. (2004). A generalized protection framework using a new link-State availability model for reliable optical networks. *IEEE/OSA. Journal of Lightwave Technology, 22*(11), 2536–2547. doi:10.1109/JLT.2004.836764

Zang, H., & Mukherjee, B. (2001). Connection management for survivable wavelength-routed WDM mesh networks. *SPIE Optical Networks Magazine, 2*(4), 17–28.

Zang, H., Ou, C., & Mukherjee, B. (2003). Path-protection routing and wavelength assignment (RWA) in WDM mesh networks under duct-layer constraints. *IEEE/ACM Transactions on Networking, 11*(2), 248–258. doi:10.1109/ TNET.2003.810313

Zeng, H., Huang, C., & Vukovic, A. (2006). A novel fault detection and localization scheme for mesh all-optical networks based on monitoring cycles. *Photonic Network Communications, 11*, 277–287. doi:10.1007/ s11107-005-7355-3

Zhang, Z., Zhong, W. D., & Bose, S. K. (2005). Dynamically survivable WDM network design with *p*-cycle based PWCE. *IEEE Communications Letters, 9*(8), 756–758. doi:10.1109/LCOMM.2005.1496606

Zhang, J., & Mukherjee, B. (2004). A review of fault management in WDM mesh networks: Basic concepts and research challenges. *IEEE Network, 18*(2), 41–48. doi:10.1109/MNET.2004.1276610

Zhang, J., Wu, J., Feng, C., Xu, K., & Lin, J. (2007). All-optical logic OR gate exploiting nonlinear polarization rotation in an SOA and red-shifted sideband filtering. *IEEE Photonics Technology Letters, 19*, 33–35. doi:10.1109/ LPT.2006.888991

Zhang, J., Zhu, K., Zang, H., Matloff, N. S., & Mukherjee, B. (2007). Availability- aware provisioning strategies for differentiated protection services in wavelength-convertible WDM mesh networks. *IEEE/ACM Transactions on Networking, 15*(5), 1177–1190. doi:10.1109/ TNET.2007.896232

Zhang, Y., Taira, K., Takagi, H., & Das, S. K. (20020). An efficient heuristic for routing and wavelength assignment in optical WDM networks. In *Proceedings of IEEE International Conference Communications* (pp. 2734-2739).

Zhang, Z., Xu, A., & He, Y. (2008). Dynamically survivable WDM network design with shared-cycles-based PWCE. *International Conference on Advanced Infocomm Technology* (pp. 1-5). Shenzhen, China: ACM.

Zhong, W., & Zhang, Z. (2005). P-cycle-based dynamic protection provisioning in optical networks. *IEICE Transactions on Communications, E88*(B(5), 1921-1926.

Zhou, D., & Subramaniam, S. (2000). Survivability in optical networks. *IEEE Network, 14*(6), 16–23. doi:10.1109/65.885666

Zhu, Z., Funabashi, M., Pan, Z., Paraschis, L., & Yoo, S. J. B. (2006). 10000-hop cascaded in-line all-optical 3R regeneration to achieve 1250000-km 10-Gb/s transmission. *IEEE Photonics Technology Letters, 18*, 718–720. doi:10.1109/LPT.2006.871141

Zhu, H. (2003). A novel, generic graph model for traffic grooming in heterogeneous WDM mesh networks. *IEEE/ACM Transactions on Networking, 11*, 285–299. doi:10.1109/TNET.2003.810310

Zhu, K., & Mukherjee, B. (2002). Traffic grooming in an optical WDM mesh network. *IEEE Journal on Selected Areas in Communications, 20*, 122–133. doi:10.1109/49.974667

Zhu, H., et al. (2002). Dynamic traffic grooming in WDM mesh networks using a novel graph model. In *Proc. IEEE GLOBECOM*, (pp. 2681–2685).

Zhu, K., & Mukherjee, B. (2002b). On-line approaches for provisioning connections of different bandwidth granularities in WDM mesh networks. In *Proc. OFC*, Mar. 2002.

About the Contributors

Yousef S. Kavian received the B.Sc. (Hons) degree in Electronic Engineering from the Shahid Beheshti University, Tehran, Iran, in 2001, the M.Sc. degree in Control Engineering from the Amkabir University, Tehran, Iran, in 2003 and the Ph.D. degree in Electronic Engineering from the Iran University of Science and Technology, Tehran, Iran, in 2007. After one year appointment at Shahid Beheshti University, in 2008 he joined the Shahid Chamran University as an Assistant Professor. He worked as a postdoctoral research fellow at Esslingen University and IAER, Germany, in 2010. His research interests include digital circuits and systems design, optical and wireless networking. Dr Kavian has over 50 technical publications including journal and conference papers and book chapters in these fields. He is a senior industrial engineer and trainer with more than 10 years industrial experiences.

Mark Stephen Leeson received the degrees of BSc and BEng with First Class Honors in Electrical and Electronic Engineering from the University of Nottingham, UK, in 1986. He then obtained a PhD in Engineering from the University of Cambridge, UK, in 1990. From 1990 to 1992 he worked as a Network Analyst for the National Westminster Bank in London. After holding academic posts in London and Manchester, in 2000 he joined the School of Engineering at Warwick, where he is now an Associate Professor. His major research interests are coding and modulation, ad hoc networking, optical communication systems, and evolutionary optimization. To date, Dr. Leeson has over 180 publications and has supervised nine successful research students. He is a Senior Member of the IEEE, a Chartered Member of the UK Institute of Physics, and a Fellow of the UK Higher Education Academy.

* * *

Emad M. Al Sukhni received his PhD in Computer Science programme of the University Ottawa in 2011. Dr. Al Sukhni is currently an Assistant Professor at the Computer Engineering department of the Yarmouk University, Jordan.

Péter Babarczi received the M.Sc. degree in Technical Informatics from Budapest University of Technology and Economics (BME), Budapest, Hungary, in 2008 and is currently pursuing the Ph.D. degree at the High-Speed Networks Laboratory, Department of Telecommunications and Media Informatics, BME. His research interests include combinatorial optimization, routing in circuit-switched survivable networks, availability analysis, and routing in delay-tolerant networks.

Gerardo Castañón, Associate Professor, member of the Academy of Science in Mexico. He received the Master of Science degree in Physics (Optics) from the Ensenada Research Centre and Higher Education, México in 1989. He also received the Master and Ph.D. degrees in Electrical and Computer Engineering from the State University of New York (SUNY) at Buffalo in 1995 and 1997, respectively. He has also been part of research centers at Alcatel (1998-2000) and Fujitsu (2000-2002). He is now working in the Department of Electrical and Computer Engineering at ITESM since September of 2002. Dr. Castañón has over 50 publications in journals and conferences and 2 international patents. He frequently acts as a reviewer for IEEE journals. He is a senior member of the IEEE Communications and Photonics societies.

Cicek Cavdar has received her Ph.D. in Computer Engineering Department, Istanbul Technical University, Turkey in 2009. During her Ph.D., she studied as a researcher in the Networks Research Lab. in University of California, Davis from Dec. 2005 to Jul. 2008. She is an Assistant Professor in Computer Engineering Department, Istanbul Technical University, Turkey. Currently, she is a visiting Assistant Professor in Royal Institute of Technology (KTH), Sweden as. Her research interests include design and analysis of wavelength-division-multiplexed networks with focus on survivability, multi-layer resilience, and energy efficiency.

Jiajia Chen received a B.S. degree (2004) in Information Engineering from Zhejiang University, China, and a Ph.D. degree (2009) in Optical Networking from the Royal Institute of Technology (KTH), Sweden. Currently, she is working as a Postdoctoral Researcher in Next Generation Optical Networks (NEGONET) group at KTH. She is the co-author of over 50 publications in international journals and conferences in the area of optical networking. Her research interests include fiber access networks and switched optical networks.

Taisir El-Gorashi received a B.S. degree in Electrical and Electronic Engineering from the University of Khartoum, Sudan, and an M.Sc. degree in photonic and communication systems from University of Wales Swansea, UK in 2004 and 2005, respectively. In 2010 she obtained a Ph.D. degree in optical networking from the University of Leeds, UK. Currently she is a post-doctoral Research Associate in the School of Electronic and Electrical Engineering, University of Leeds. Her research interests include next generation optical network architectures and green ICT.

Jaafar Elmirghani is the Director of the Institute of Integrated Information Systems within the School of Electronic and Electrical Engineering, University of Leeds, UK. He joined Leeds in 2007, and prior to that (2000–2007) as chair in optical communications at the University of Wales Swansea, he founded, developed, and directed the Institute of Advanced Telecommunications and the Technium Digital (TD), a technology incubator/spin-off hub. He has provided outstanding leadership in a number of large research projects at the IAT and TD. He has co-authored *Photonic Switching Technology: Systems and Networks*, (Wiley) and has published over 300 papers. He has research interests in optical systems and networks and signal processing. Dr. Elmirghani is Fellow of the IET and Fellow of the Institute of Physics. He was Chairman of IEEE Comsoc Transmission Access and Optical Systems technical committee and was Chairman of IEEE Comsoc Signal Processing and Communications Electronics technical committee, and an editor of *IEEE Communications Magazine*. He was founding Chair of the Advanced Signal Processing for Communication Symposium which started at IEEE GLOBECOM'99 and has continued since at every ICC and GLOBECOM. Dr. Elmirghani was also founding Chair of the first IEEE ICC/GLOBECOM optical symposium at GLOBECOM'00, the Future Photonic Network Technologies, Architectures and Protocols Symposium. He chaired this Symposium, which continues to date under different names. He received the IEEE Communications Society Hal Sobol award, the IEEE Comsoc Chapter Achievement award for excellence in chapter activities (both in 2005), the University of Wales Swansea Outstanding Research Achievement Award, 2006 and the IEEE Communications Society Signal Processing and Communication Electronics outstanding service award, 2009.

Abdelhamid Eshoul received B.Sc (hons.) degree in electrical and electronic engineering from Aston University, Birmingham, U.K., in 1984 and the M.Sc. degree in electrical and computer engineering from Queen's University, Kingston, Ontario, Canada, in 2002. He completed the Ph.D. degree in electrical and computer engineering at the University of Ottawa, Ottawa, Ontario, Canada, in 2006. Upon completing his PhD degree, he started working as a postdoctoral researcher at the School of Information technology and Engineering of the University of Ottawa. His research interests include development of routing and wavelength assignment algorithms for survivable WDM wavelength-routed mesh networks.

Andrea Fumagalli received his Ph.D. from Politecnico di Torino, Italy, and is a Professor of Electrical Engineering at the University of Texas at Dallas. Prior to joining UTD he was an Assistant Professor of the Electronics Engineering Department at the Politecnico di Torino, Italy. He is a member of the Open Networking Advanced Research (OpNeAR) Lab at UTD (http://opnear.utdallas.edu). His research interests include aspects of wired and wireless networks, and related protocol design and performance evaluation. He has been involved in a number of research projects focusing on packet switched and fault tolerant networks, and has published more than one hundred papers in peer reviewed journals and conferences. He served on the Editorial Board of the *ACM/IEEE Transactions on Networking* and two Elsevier journals. He was on a Distinguished Lecturer Tour for IEEE ComSoc in 2000.

Zabih Ghassemlooy received BSc (Hons) degree in Electrical and Electronics Engineering from the Manchester Metropolitan University in 1981, and MSc and PhD from the University of Manchester Institute of Science and Technology (UMIST), in 1984 and 1987, respectively. He is currently a Professor of Optical Communications and an Associate Dean for Research in the School of Computing, Engineering and Information Sciences at the University of Northumbria at Newcastle, UK. He also heads the Northumbria Communications Research Laboratories within the School. He has supervised a large number of PhD students and has published over 370 papers (123 in journals + 8 book chapters). His research interests are on photonics switching, optical wireless and wired communications, visible light communications, and mobile communications. He is a co-editor of an IET book on "Analogue Optical Fibre Communications," and the founder and chairman of the IEEE, IET International Symposium on Communications Systems, Networks, and DSP. He is a Chartered Engineer of EC UK, a Senior Member of IEEE, and a Fellow of IET. He is currently the Chairman of the IEEE UK&RI Communication Chapter.

Burak Kantarci received his B.Sc., M.Sc, and Ph.D degrees in Computer Engineering at Istanbul Technical University (ITU) in 2002, 2005, and 2009, respectively. During his Ph.D study, he conducted research at University of Ottawa as a visiting scholar under supervision of Prof. H. T. Mouftah. Dr. Kantarci is currently a post-doctoral fellow at the School of Information Technology and Engineering of the University of Ottawa. His research interests are telecommunication networks, WDM networks, optical switching, survivable network design, and green communications.

David Lastine is a graduate student at Iowa State University expecting to earn a Ph. D. in Computer Engineering in 2011. He has completed a Master's of Science majoring in Computer Engineering, as well as Bachelor's that covered Computer Engineering and Physics. Published works are in the areas of fault tolerant routing, traffic scheduling on light trails, and network coding.

Hoa Le Minh is currently a Senior Lecturer in Optical Communications and Computer Networks Groups of School of CEIS. He gained his BEng, MSc, and PhD at Ho Chi Minh University of Technology (HUT, Vietnam) in 1999, Munich University of Technology (Germany) in 2003, and Northumbria University (UK) in 2007. Hoa worked at Siemens AG (Munich, Germany) during 2002-2004 to design the ultra high speed DWDM optical systems. During 2007-2010 he was a research fellow at Oxford University and developed the optical wireless communications system. Hoa's researches focus on Photonics system, optical wired and wireless communications, visible light communications, computer networks, and mobile communications.

Jan Lubkoll received his Bachelor of Engineering degree from Esslingen University of Applied Science, Germany. He has some years of experiences in fiber-optic data buses (FlexRay) in the aviation field. His major research interests include technologies of micro systems, ultrashort pulsed laser and optical communication. He is co-author of several publications in the field of fiber-optic sensors, and optical data buses in cars and aeroplanes. Now he is a student in the elite "Master program of Advanced Optical Technologies" at the Friedrich-Alexander University Erlangen.

Carmen Mas Machuca is a Senior Researcher and Lecturer in the Institute for Communication Networks (LKN) at the Technical University of Munich (TUM), Germany. She holds a Ph.D. degree from the Swiss Federal Institute of Technology, Lausanne (EPFL) and a Telecommunications Engineering degree from the Universitat Politècnica de Catalunya (UPC), Spain. Her research interests include techno-economic studies, network migration, service and network reliability, and control and management of optical networks. She is a member of the IEEE.

Carlos Mex-Perera is with the Department of Electrical and Computer Engineering at the Tecnologico de Monterrey, Mexico. He received a B.Eng. degree from the Instituto Tecnologico de Merida, and a M.Sc. in Electrical Engineering from the Centre for Research and Advanced Studies of the National Polytechnic Institute, Mexico. He holds a Ph.D. in Computer Security from the University of Bradford, UK. His research interests encompass Computer and Telecommunications security. His current focus is on resilient and self-organizing networks: providing resilient mechanisms to telecommunications networks to defense against attacks. He is also interested in artificial intelligence applications to network security.

Paolo Monti received a Ph.D. degree from the University of Texas at Dallas (UTD) in 2005 where he worked as a Research Associate until 2008. He joined the Royal Institute of Technology (KTH) in September 2008 where he is currently an Assistant Professor and a member of the Next Generation Optical Networks (NEGONET) group. He co-authored more than forty papers published in international journals and presented in leading international conferences. Dr. Monti is serving on the editorial board of the Springer *Networking and Electronic Commerce Journal*. His research interests include: network planning, protocol design, performance evaluation, and optimization techniques for both optical and wireless networks.

Hussein T. Mouftah joined the School of Information Technology and Engineering (SITE) of the University of Ottawa in September 2002 as a Canada Research Chair (Tier 1) Professor in Optical Networks. He has been with the ECE Department at Queen's University (1979-2002), where he was, prior to his departure, a Full Professor and the Department Associate Head. He has three years of industrial experience mainly at BNR of Ottawa, now Nortel Networks (1977-79). He served as Editor-in-Chief of the *IEEE Communications Magazine* (1995-97) and IEEE ComSoc Director of Magazines (1998-99), Chair of the Awards Committee (2002-03) and Director of Education (2006-07). He has been a Distinguished Speaker of the IEEE ComSoc (2000-07). He is the author or coauthor of six books, 40 book chapters, and more than 1000 technical papers and 10 patents in this area. Dr. Mouftah is a Fellow of the IEEE (1990), the Canadian Academy of Engineering (2003), the Engineering Institute of Canada (2005), and the Royal Society of Canada RSC: The Academy of Science (2008).

Wai Pang Ng is a Senior Lecturer at Northumbria University, United Kingdom since 2004. He received his BEng (Hons) Communication and Electronic Engineering and IEE Prize from Northumbria University in 1997. Then he pursued his graduate study, in collaboration with BT Labs and completed his PhD in Electronic Engineering at University of Wales, Swansea in 2001. Dr. Ng started his career as a Senior Networking Software Engineer at Intel Corporation. At Northumbria University, he has published over 50 journals and conferences publications in the area of optical switching and optical signal processing. He is a Chartered Engineer of EC UK and Senior Member of IEEE. Dr. Ng currently serves as the vice-chair of IEEE UK&RI Communication Chapter where he was also the secretary from 2007-2008. He was also the co-chair of the Signal Processing for Communications Symposium in IEEE International Conference on Communications (ICC) 2009 (Dresden, Germany).

Gerard Parr holds the Full Chair in Telecommunications Engineering in the School of Computing and Information Engineering at the University of Ulster in Coleraine, UK. He holds a PhD in Self Stabilizing Telecommunications Protocols, aspects of which he developed whilst a Visiting Scientist at the DARPA- funded Information Sciences Institute in Los Angeles, USA and with University College London. Areas of PostDoc and PhD research within the Information and Communications Engineering group include self-stabilizing protocols for security and resilience, interplanetary network protocols, intelligent mobile agents in xDSL and SNMP, real-time data analytics for network management systems, energy aware telecommunications infrastructure, memory and computational resource management protocols, applications performance management in virtualised environments, energy-efficient protocols for the access-customer premises pathway, and dynamic bandwidth allocation strategies for WiMAX-EPON integration. He has extensive experience of managing government and industry-sponsored research contracts and has developed a close working relationship with a number of major ICT companies including Cisco, BT, SAP, Ericsson, InfoSys, Wipro, Sasken, Tejas, IBM, Toshiba, Fujitsum, and Rockwell Inc. He acts as an MSc/PhD External Examiner in CS/EE for University College London, Trinity College Dublin, Xiamen University, P.R. China, Queen's University Belfast, Cardiff University, Queen Mary

University London, Indian Institute of Technology -IIT Madras, and the University of Plymouth. His academic research collaborations include MIT, Georgia Institute of Technology, USC-ISI Los Angeles, UCL, Southampton, Surrey, QMUL, St Andrews, Lancaster, and Cambridge, Beijing University of Posts & Telecommunications (BUPT) and Indian Institutes of Technology in Mumbai, Madras, Kanpur, Hyderabad, Delhi, and IISc Bangalore. He has attracted several £millions of external research and commercial funding and has advised UK and international governments on the allocation of funding to research and infrastructure projects valued in excess of £500 million. He is the UK PI of the EPSRC-DST India-UK Centre of Excellence in Next Generation Networks Systems and Services of which BT is the lead industrial partner and also a PI in the EPSRC funded Project- Sensing, Unmanned, Autonomous Aerial Vehicles- SUAAVE. Recently he was appointed to the UK Engineering and Physical Sciences Research Council Strategic Advisory Team and will take up this position in May 2010.

Ivan S. Razo-Zapata holds B.Sc (2003) and M.Sc. (2006) degrees in computer science both from the Autonomous University of Hidalgo State. In 2007 he joined the Center of Electronics and Telecommunications of the ITESM Campus Monterrey as a Research Assistant. At that time, he worked in projects concerning Intrusion Detection Systems and resilience mechanisms for P2P networks. His current research interests focus mainly on artificial intelligence, optical networks, security, and service science. Since 2008 he is pursuing a PhD in Computer Science.

Ganesh Chennimalai Sankaran received the B.E. degree in Electronics and Communications Engineering from Madurai Kamaraj University in 1999. He joined HCL Cisco Offshore Development Center (part of HCL Technologies Ltd, (India) in 1999. Since then he has been working on Network Management products and services from Cisco. He is currently working towards the M.S degree at Indian Institute of Technology Madras, Chennai, India. His research interests include optical access networks and energy saving mechanisms.

Daniel Seibl is Graduate Engineer and Research Assistant in the laboratory of physics and photonics at the Esslingen University of Applied Sciences. After some years of practical work at Siemens VDO in the fields of quality management, especially in process qualifications and failure mode and effective analysis, he studied Electrical Engineering with the major of Micro-Electro-Mechanical Systems and specialized in Micro Optics and Laser Technology. In October 2008 he received a degree in this study. The main research interests are in the fields of fiber-optic technologies, light-induced fluorescence, and three dimensional measurement techniques. He already has experience in automotive fiber-optic data buses (MOST) and fiber-optic sensors. For that purpose, he is co-author of several papers and publications dealing with MOST and fiber-optic sensors. His actual research work for an aspired Master degree deals with optical 3-D measurement.

Anusha Sivakumar received the B.E. degree in Electronics and Communications Engineering from Prince Shri Venkateshwara Padmavathy Engineering College, affiliated to Anna University, Chennai (India) in 2008. She has been working as a Research Assistant in the Department of Computer Science and Engineering, IIT Madras (India) since 2009. Her research interests include passive optical networks and computer networks.

Krishna M. Sivalingam is a Professor in the Department of CSE, Indian Institute of Technology (IIT) Madras, Chennai, India. Previously, he was a Professor in the Dept. of CSEE at University of Maryland, Baltimore County, Maryland, USA from 2002 until 2007; with the School of EECS at Washington State University, Pullman, USA from 1997 until 2002; and with the University of North Carolina Greensboro, USA from 1994 until 1997. He received his Ph.D. and M.S. degrees in Computer Science from State University of New York at Buffalo in 1994 and 1990, respectively, and his B.E. degree in Computer Science and Engineering in 1988 from Anna University's College of Engineering Guindy, Chennai (Madras), India. His research interests include wireless networks, wireless sensor networks, optical wavelength division multiplexed networks, and performance evaluation. He has published edited books on Next Generation Internet Architectures in 2010, on Wireless Sensor Networks in 2004, and on optical WDM networks in 2000 and 2004. He is serving or has served as a member of the Editorial Board for journals including ACM Wireless Networks Journal, IEEE Transactions on Mobile Computing, and *Elsevier Optical Switching and Networking Journal*. He is presently serving as Editor-in-Chief of *ICST Transactions on Ubiquitous Environments*.

Arun K. Somani is Anson Marston Distinguished Professor of Electrical and Computer Engineering at Iowa State University. Professor Somani's research interests are in the area of fault tolerant computing and networking systems. He has published ~300 technical papers and book chapters, a book, and has supervised more than 100 graduate (35 PhD) students. He has designed and built several systems including anti-submarine warfare system (Indian Navy), Meshkin system and HIMAP analysis tool (Boeing) and Proteus multicomputer system (US Coastal Navy). He is awarded ACM Distinguished Engineer member, IEEE Communication Society Distinguished Lecturer, and IEEE Fellow grades.

Otto Strobel is Head of the Physics Institute and Director of Physics Laboratory, Faculty of Basic Sciences at Esslingen University of Applied Sciences in Germany and Honorary Chair of the IAER group. He received his Dr.-Ing. degree from Technical University of Berlin and his Dr. h.c. degree from Moscow Aviation Institute, Technical University of Moscow in Russia. In 2011 he was awarded as Honorary Professor by the Tecnológico de Monterrey in Mexico. He published about 80 papers in the field of fiber-optic technologies and optoelectronics and performed more than 30 visiting professor stays worldwide. He is author of the textbook (in German language) "Technology of Lightwave-Guides in Transmission and Sensing" (VDE 2002, 2nd edition) and co-author of the text book (in German): "Photonics" (Springer, 1st ed. 2005). Furthermore he is workshop chair at the "International Conference on Transparent Optical Networks ICTON", chair member of the "International Workshop on Telecommunications IWT, Brazil", of "Communication Systems, Networks and Digital Signal Processing, CSNDSP", of the "International Conference Microwave & Telecommunication Technology - CriMiCo" Sevastopol, Ukraine", of the „International Conference on Composites or Nano Engineering – ICCE" Shanghai, China and also member of the Construction Consultative Committee of Wuhan Optics Valley of China. He has more than 10 years experience in companies' R&D, as member and consultant of Daimler, Alcatel-Lucent, HP, Agilent and Siemens.

János Tapolcai (M'09) received the M.Sc. degree in technical informatics and Ph.D. degree in Computer Science from Budapest University of Technology and Economics (BME), Budapest, Hungary, in 2000 and 2005, respectively. He is currently an Associate Professor with the High-Speed Networks

Laboratory, Department of Telecommunications and Media Informatics, BME. He has been involved in several related European and Canadian projects. He is an author of over 40 scientific publications. His research interests include applied mathematics, combinatorial optimization, linear programming, linear algebra, routing in circuit-switched survivable networks, availability analysis, grid networks, and distributed computing. Dr. Tapolcai is the recipient of the Best Paper Award at the IEEE ICC 2006.

Bin Wang earned his Ph.D. from the Ohio State University in 2000. He joined the Wright State University in September 2000, where he is currently Full Professor of Computer Science and Engineering. His research interests include optical networks, real-time computing, mobile and wireless networks, cognitive radio networks, trust and information security, and Semantic Web. He was a recipient of the US Department of Energy Career Award and the AFOSR senior research fellowship. His research has been supported by US Department of Energy, National Science Foundation, Air Force Office of Scientific Research, Air Force Research Laboratories, Ohio Supercomputer Center, and the State of Ohio.

Lena Wosinska received a Ph.D. degree from the Royal Institute of Technology (KTH) in Stockholm, Sweden, where she is currently an Associate Professor heading a research group NEGONET (Next Generation Optical NETworks: http://www.ict.kth.se/MAP/FMI/Negonet/). Her research interests include energy efficient optical networks, photonics in switching, optical network management, reliability/survivability of optical networks, and fiber access networks. She has been involved in a number of professional activities including guest editorship of OSA, Elsevier, and Springer journals, serving in TPC of several conferences. In 2007-2009 she has been an Associate Editor of *OSA Journal of Optical Networking*, and since April 2009, she serves on the Editorial Board of *IEEE/OSA Journal of Optical Communications and Networking*.

Index